Theory and Practice of Ship Handling

By Kinzo Inoue
Professor Emeritus, Kobe University

Sang Hak Dang

Published by
Sang Hak Dang Publishing Co.
Sadang-ro 9ga-gil 6 Dongjak-gu
Seoul, Korea

International Standard Book Number(ISBN):
978-89-6587-084-5

Preface

Ship handling is the operation to shift or stop a ship as intended safely and effectively under any given environmental circumstances. The operations required for this purpose in actual ship handling include course keeping, alteration of course, action to avoid collision, maneuverings for leaving from and entering into port, berthing and un-berthing, mooring and anchoring. A person engaged in ship handling is expected to perform its role safely and effectively as an "expert". For this purpose, any person engaged in ship handling is required to master basic scientific knowledge related to buoyancy, stability, maneuverability, seaworthiness as well as static and dynamic characteristics of the commanding ship so that they can carry out optimum ship handling taking the effects of environmental disturbances such as wind, waves and currents into account. That is to say, the expertise required of the person engaged in ship handling must be a reasonable application of expertise acquired from empirical achievements backed by scientific knowledge.

From the above point of view, this book is prepared not only for persons engaged in ship handling but also for those who are intending to study ship handling from the beginning so that they may study related basic scientific knowledge as theoretically as possible. In explanation, care was exerted as much as possible to keep consistency with the descriptions of preceding textbooks based on static equilibrium in a ship's motion to avoid readers' puzzlement. In addition, I especially tried to give clear answer to the most concerned and probable question of readers "How a ship reacts when a specific external force affects." To enable to forecast a ship's motion in immediate future, dynamical analysis in a certain time span by simulation method is often used. Besides, this book is designed so that you may get a specific ship's motion even if you skip its intricate equations for explanation although there are many of the kind in this book.

Actually the ship type and size you handle may vary. Regardless of the ship type and size you may handle, it is indispensable for you to grasp the characteristics of maneuverability of the ship you will handle beforehand to do ship handling based on its proper data. For this reason, the database of various ship types and sizes is included in Volume 4. I hope this data can serve you to combine "Theory" and "Practice" of Ship Handling.

It can be said that the conventional "Study of Ship Handling" has been a scholarly achievement purposed to improve the individual expertise of person engaged in ship handling. However, it is also important for persons engaged in ship handling today to realize that they are not only expected to fulfill their duties for their own ships, but also expected to give feedback with their own expertise for promotion of safe ship operation to prevent accidents and to protect the marine environment for the global maritime industry. For this purpose I would like to recommend the readers of this book to read "Scientific Marine Safety Management" (Oct., 2008, Seizando Shoten) together with this book.

The ship maneuverability data included in the database of this book were collected through the trials and calculations using a ship handling simulator of my research laboratory, Faculty of Maritime Sciences, Kobe University, and fast time ship maneuvering simulation system produced by Japan Marine Science Inc. in cooperation with the "Working Group for Compilation of Ship Maneuverability Characteristics Database". I would like to express my sincere thanks to the members who cooperated to achieve the Compilation of Ship Maneuverability Characteristics Database by listing their names below.

February, 2011

Dr. Kinzo Inoue
Professor Emeritus
Graduate School of Maritime Sciences
Kobe University

Working Group for Compilation of Ship Maneuverability Characteristics Data Base

Chairman Dr. Kinzo Inoue, Professor Emeritus, Graduate School of Maritime Sciences, Kobe University

Dr. Daichi Hara, Senior Researcher, Japan Marine Science Inc.

Capt. Shinji Ohishi, Member of Osaka Bay Pilot's Association

Dr. Wataru Sera, Associate Professor, Graduate School of Maritime Sciences, Kobe University

Mr. Hironori Ohuchi, Master's Degree Student, Graduate School of Maritime Sciences, Kobe University

Mr. Shuhei Kawamoto, Master's Degree Student, Graduate School of Maritime Sciences, Kobe University

The Publication of the English version of "Theory and Practice of Ship Handling" in Japan

In Japan, it's been our hope to have updated ship handling textbooks to reflect the ever evolving technological innovations in the maritime industry. We'd all been waiting for such definitive textbooks, current enough to pass on to the next generation.

Several decades have passed since the publication of the last ship handling textbooks, and it wasn't until March 2011, when I published and released the Japanese version of "Theory and Practice of Ship Handling" that an updated version for those involved in ship handling was available. It was a wonderful surprise to me that immediately after its publication, people involved in the domestic maritime cluster, here in Japan, told me that the book should be used for maritime education all over the world.

What made me happier was that in November 2011, the book received the Shoichi Sumida Award for the best book from the Japan Shipping Exchange, Inc. The reason for winning the award was its contemporaneity description of recent versatile ship types including larger ships, for which information and data was updated for today's maritime world.

Inside, or even outside Japan, few books exist which systematically connect the "theory" and "actuality" of ship handling necessary for safe navigation. In that sense, it's my greatest pride and joy to offer an English version of the book to maritime clusters and their educational communities all over the world.

In the end, I would like to express my most sincere gratitude to my friend James E Middleton and Misako Middleton for their tireless efforts in preparation of the English version. Their great contributions, including their constant moral support to me made it possible to complete the English version.

With all my heart, I'm hoping this book makes a difference somewhere in the world···

October, 2012

Dr. Kinzo Inoue
Author

The Publication of the English version of "Theory and Practice of Ship Handling" in Korea

Since the Japanese version of "Theory and Practice of Ship Handling" was published in March 2011, and was fortunate enough to receive the Shoichi Sumida Award for the best book in the same year, along with receiving some encouragement from people in domestic maritime cluster, it has become my hope to introduce the book to maritime clusters all over the world. To begin with, the English version of the book was published in Japan in March of 2013.

Immediately after the publication, when he visited Japan on "Hannara", the training ship for the Korean Maritime University, in May 2013, I presented the book to Dr. Park Young-soo who was one of my graduate students in the Doctor's course of the Faculty of Maritime Sciences of Kobe University.

Upon returning to Korea, Dr. Park Young-soo came up with the idea of publishing the book inside his country, which, later on, was accepted and realized by Sang Hak Dang, a Korean publishing company. In August 2013, the final draft of the book was completed. I truly appreciate to Sang Hak Dang for their great effort and creativity in details such as larger fonts, and soft colored figures and charts to make the book even more attractive.

I would like to express my heartfelt thanks to Dr. Park Yong-soo who offered me such a wonderful opportunity for his assistance in realizing my hope for international publication, starting in Korea. Also I would like to express my sincere gratitude to Mr. Il-bok Song, an editor at Sang Hak Dang, for his excellent suggestions pertaining to arrangement and modifications in the evolution of this book.

August 2013

Dr. Kinzo Inoue
Author

Contents

Table of Contents

Volume 1

Introduction

Ⅰ. Ship Handling and Persons Engaged in Ship Handling

Ⅱ. A Ship's Motion, Buoyancy and Stability

Volume 2

Principle of Ship Handling

Ⅰ. Maneuvering by Rudder

Ⅱ. Maneuvering by Main Engine and Propeller

Ⅲ. Maneuvering by Thruster and Tug

Volume 3

Practical Ship Handling

I. Navigation, Berthing and Un-berthing

Ⅱ. Mooring and Anchoring

Volume 4

Ship Handling Characteristics Data Base

Ⅰ. Maneuverability

Ⅱ. Ship Handling Characteristics

Volume 1

Introduction

Ⅰ. Ship Handling and Persons Engaged in Ship Handling

Ⅱ. A Ship's Motion, Buoyancy and Stability

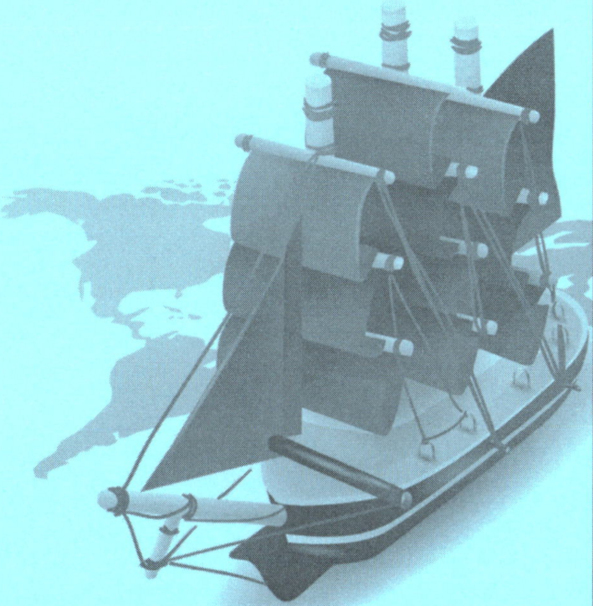

Ⅰ
Ship Handling and Persons Engaged in Ship Handling

Chapter 1 Definition of Ship Handling and Expertise Required of Persons Engaged in Ship Handling

1. The Relation among Ships, Persons and Environment in Ship Handling

It is incumbent on the person engaged in ship handling to maneuver the ship safely under given environmental circumstances. At the same time they must be a subjective executor of expertise for this purpose. Fig. 1.1.1 illustrates the relation among the elements necessary for the process to achieve safe ship handling

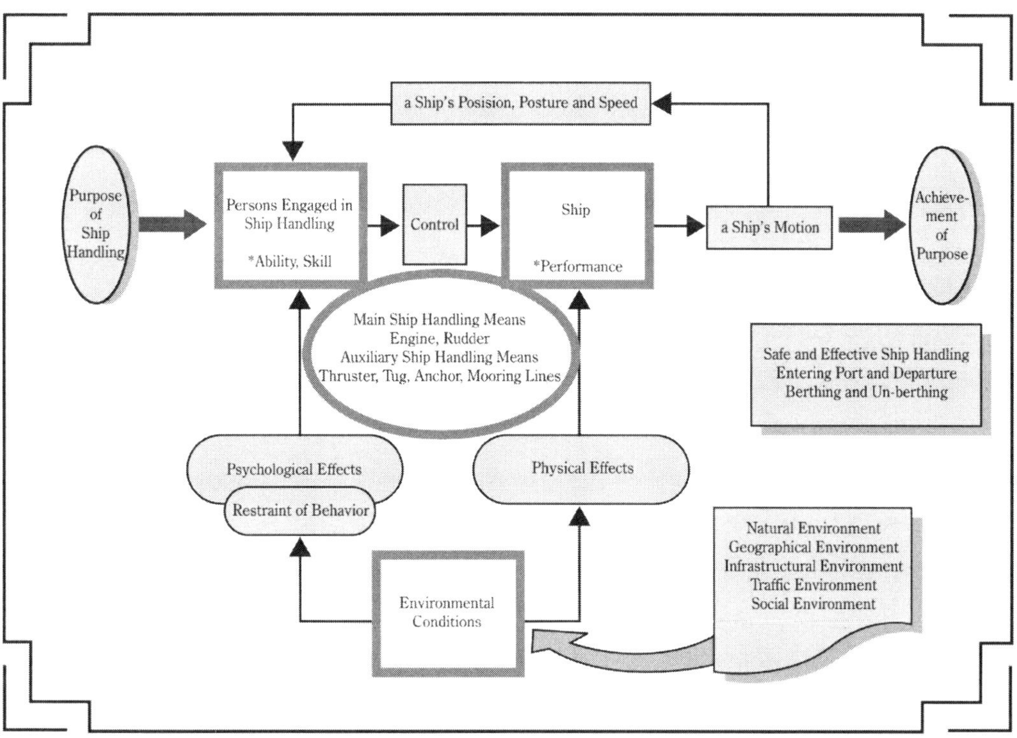

Fig. 1.1.1 [Ship - Person - Environment] Ship Handling co-relation system

centered on 3 main elements, ship, person and environment.

In the co-relation system of ship handling which is comprised of ship, person and environment as shown in Fig. 1.1.1, the person engaged in ship handling is the main technical performer directly participating with that system to achieve safe ship handling. In this context, it can be said that the person engaged in ship handling is the subject, the ship is the object and its environment is an element which affects the deed of the subject.

Table 1.1.1 Environmental elements affecting Ship Handling

Natural environment	Wind, Tide, Current, Waves, Visibility
Geographical environment	Existence of islands, Shoals and Water depth
Infrastructural environment	Maneuvering area, Breakwaters, Wharves, Dredged channels and Roadsteads
Traffic environment	Buoys, Aids to navigation, Traffic density, Operation of fishing and Fisheries
Social & Informational environment	Law, Regulations, Customary practices, Availability of information

The environmental conditions affecting the ship and person engaged in ship handling can be divided into physical element that directly affects the ship's motion and psychological element that indirectly affects the judgment and ensuing behavior of the person engaged in ship handling. These elements are sorted into 5 categories as shown in the above Table 1.1.1.

Wind, tide, current, and waves directly affect a ship's motion just as external disturbances do, while geographical and infrastructural environmental factors affect it indirectly through particular dynamic phenomena such as the shallow water effect or bank effect. For persons engaged in ship handling, geographical and infrastructural environmental factors affect them as the restriction of maneuverability and the traffic environment such as the existence of other ships and objects on the passage gives stress compelling one to take risk avoidance action, along with that the social and informational environment psychologically affects the process of decision.

As these environmental elements affect the ship's motion directly or indirectly as external disturbances, the person engaged in ship handling is expected to acquire knowledge and skill to foresee and control the effects of these external disturbances.

2. Definition of Ship Handling

The role of a person engaged in ship handling is to achieve ship handling as purposed by the best use of their expertise corresponding to the effects of various external disturbances fraught with risk. In the process of ship handling, for the person engaged in ship handling an ability to react quickly, responding to the effects of external disturbances based on the precise assumption of those external forces is required, in addition to the knowledge of their own ship maneuverability from which the behavior of the ship is predictable arithmetically. Moreover, such quick countermeasures must be optimal and effective for the safety of the ship in all aspects of its motion, roll, pitch, and heel without inviting other risks.

From the above, ship handling can be defined as it is an action to shift or stop a ship safely, effectively and economically as intended by controlling its position, posture and speed using its rudder, main engine and other auxiliary means taking environmental conditions into account along with the condition of the ship's buoyancy, stability, maneuverability and seaworthiness.

3. Expertise Required of Persons Engaged in Ship Handling

As discussed in the previous section, the role of a person engaged in ship handling must be an "expert" to achieve the purpose of ship handling safely, effectively and economically. In this context, a person engaged in ship handling is required to master necessary scientific knowledge so that he/she can understand dynamics related to buoyancy, stability, maneuverability and seaworthiness of ship and handle its ship taking external disturbances such as wind, tidal stream, sea current and waves and their cffccts to the ship into account.

A person engaged in ship handling is expected to handle the ship under various conditions from the stormy open sea to the calm sea inside port where maneuvering area is restricted although sheltered from wind and waves. The type and size of ships to be handled will also vary. Thus to perform safe ship handling corresponds to natural changes in environment and various types and sizes of ships, a high level of confidence and expertise are required of the person engaged in ship handling. When a ship is controlled according to the intentions of the person engaged in its handling, using its main engine, rudder and other auxiliary means, its motion changes corresponding to its maneuverability characteristics. And if its subsequent motion differs from that of what was initially expected by the person engaged in its handling, corrective controlling action should be taken. In this instance, the person engaged in ship handling is required to have the ability to foresee the ship's resultant behavior from its performance characteristics and to control the ship as intended. In other words, skills of the person engaged in ship handling must include reasonable application of empirical expertise based on deep scientific understanding of dynamics of ship motion and the effects and performance of its controlling devices and auxiliary means such as main engine, rudder, propeller, mooring ropes, anchors, tugs and thruster.

Chapter 2 Practical Ship Handling and Ability Required of Persons Engaged in Ship Handling

1. Subtle Control Required for Ship Handling

Ship handling is a series of actions to make a ship move or stop safely as intended, as effectively and economically as possible. These actions include course keeping, alteration of course, collision avoidance actions, entering into or leaving from port, berthing or un-berthing, mooring or un-mooring, dropping or weighing anchor, shifting anchorage and anchor watch. Above all, persons engaged in ship handling are required to use subtle control and exquisite judgments in a restricted water area close to breakwaters, wharfs, buoys, dredged channels, or the existence of other ships, and they are obliged to work under formidable psychological pressure where a slight mistake in judgment may not be acceptable.

Narrow channels where many obstacles such as strong and complicated tidal streams, sharp bends, shoals, reefs, traffic congestion and fishing operations exist are well known as difficult places for ships to pass through. Likewise port and harbor areas are also difficult places for ships to maneuver due to their confined maneuvering area and restricted depth. Confined maneuvering area restricts the operational choices of the person engaged in ship handling and shallow depths reduce the responsiveness of the ship due to the shallow water effect. Another major peculiarity of ship maneuvering in harbor is that ships are obliged to proceed at quite a slow speed. At slow speeds, the effects of wind and waves are larger and the rudder effect deteriorates inversely.

Particularly for ships like PCCs which have a larger surface area, the effects of wind are critical in restricted water areas. In addition, today, maneuvering inside harbor has become increasingly harder than ever before because of hypertrophied ship size and deep draft which makes available maneuvering area and depth tighter and tighter. In addition, the maneuvering of container or LNG ships in which the ship's forward views from the navigation bridge are obscured, gives an oppressive feeling for those persons engaged in ship handling and the high raised bridges make them underestimate the ship's real speed by eyesight.

2. Visual Observation of a Ship's Motion

For a person engaged in ship handling in a restricted water area, a different sense from that required for a normal voyage on the open sea is necessary. This is because, for ship handling in a restricted water area, the person engaged in ship handling is required to estimate leeway, swing of bow and residual headway speed together with lateral drift speed of the ship correctly under quite a slow moving state and issue orders to take appropriate counter actions based on instant decisions.

Persons engaged in ship handling in restricted waters running at such slow speeds, are required to carry out delicate and responsive ship handling corresponding to environmental situations. They are obliged to rethink their basis of judgment on the ship's dynamic motion solely on their own recognition by visual observation.

It seems that this may be partly attributed to the fact that they are too busy to spare the time to read the indications of navigational equipment and become obliged to rely on their own empirical skill. On the contrary, this means that the less time the person engaged in ship handling has to refer to mechanical measurements, the more they will be obliged to revert to personal skills based on the spirit of a traditional craftsman.

3. Correction of Information Acquired by Eyesight and Intuition

The reason why maneuvering in a restricted water area is said to be by far more difficult than that in a voyage under normal sea speeds is because the effect of external disturbances are much larger on the poor maneuverability of a ship running at slow speed and persons engaged in ship handling are obliged to make so many observations, decisions and operations successively in quite a limited time. In addition, in such areas close to port facilities and other obstacles, slight errors in visual observations and instant judgments could lead to a fatal accident.

In such cases, collection of information on a ship's motion has been dependent on an individual's intuition by visual observation. However, it is certain that hypertrophying ship size and diversifying ship types make ship handling more difficult than ever before for persons engaged in ship handling to grasp realistic ship motion through eyesight and intuition. Recently various instruments to supplement this shortcoming of human senses and support subjective observation have been introduced and are used effectively.

Such instruments introduced to correct the information recognized by human intuition through visual observation on the delicate motion of a ship moving at quite a slow speed must have highly precise measurement abilities. For instance, Doppler Sonar to measure a ship's head way, lee way and drift angle, ECDIS (Electronic Chart Display and Information System) to indicate integrated position information of the GPS (Global Positioning System) and AIS (Automatic Identification System), Berthing Velocity Indication System to indicate lateral approaching speed and distance between ship and berth, and leading marks and direction lights to detect the deflection of the ship's course and position are the good examples of such supporting systems and devices.

Chapter 3 Societal Roles of Persons Engaged in Ship Handling

1. Safety at Sea

Safety at sea can be achieved when marine accidents are eradicated through the improvements of the infrastructural environment of passages and harbors, and maneuvering skills of persons engaged in ship handling based on the ship's ensured seaworthiness including buoyancy, stability and maneuverability.

Persons engaged in ship handling are the main players in realizing safety and security at sea through their expertise and are expected to exert unrelenting efforts to brush up their professional knowledge and skills. In addition, efforts to improve a ship's performance and the infrastructural environment where ships are operated have to be made harmoniously. That is to say, we should realize that safety at sea could not be achieved by isolated efforts for improvement by any one ship, persons engaged in ship handling, or environmental conditions, but can be realized when harmonious improvements of the co-relational system of "a person engaged in ship handling", "a ship", and "environment" as shown in Fig. 1.1.1 are attained.

2. Elements to Achieve Safety

Safety at sea cannot be achieved when a person engaged in ship handling is lacking in ability even though the ship has good maneuverability. In keeping with that, safety cannot be achieved when a ship is poor in maneuverability even though the person engaged in ship handling has enough experience, expertise, and ability.

In addition, there are other things to take into consideration; when the infrastructural environment is unfavorable, despite the level of expertise of the person engaged in ship handling, and/or the maneuverability the ship is at a high level, there can be serious issues. It is ideal that these three elements must be offered in coordination, i.e., a ship is provided with good performance, environmental conditions for ship handling are designed optimally, and the person engaged in ship handling has an excellent ability to react correctly and in a timely fashion corresponding to changing operational situation.

3. Stance for Safety

The theory of ship handling is a product of systematic research for ship maneuvering techniques purposed to improve individual knowledge and skill of persons engaged in ship handling to prevent marine accidents and to promote safety at sea.

However in the past, skills and knowledge expected of each person engaged in ship handling were limited to the safe handling of the ship in their command; the persons in new era engaged in ship handling are expected to take part in the improvement of their ship's performance and environmental conditions for safe maneuvering of ships as a whole. In addition to their own ships' safe handling, they are also expected to make necessary propositions to more aggressively promote safety at sea and to contribute to society as marine professionals.

As long as persons engaged in ship handling confine themselves in a narrow professional and self-complacent niche dependent on individual techniques nurtured by sea experience, as ever before, participation in a broader field for improvement and innovating ship performance and infrastructural environment of ports and harbors cannot be expected.

4. Self Improvement Expected of Persons Engaged in Ship Handling

Self improvement is also expected of persons engaged in ship handling. It has been long pointed out an evil of a strict and unilateral work system on board a ship and its ill effects in which persons engaged in ship handling are apt to confine themselves to their own duties being indifferent to other's duties. Persons engaged in ship handling in the new era are expected to improve their conscious from a self-complacent and individual working system to group working systems from the view point of a team management.

Although the attention of the person engaged in ship handling is often concentrated on the horizontal area which they can see from the ship's bridge, it would be worth it to try to convert its view to a bird's-eye-view looking around on a much broader area from the sky. Henceforth, it is incumbent on persons engaged in ship handling to make necessary proposals and suggestions to any new construction, design and development projects for ports and harbors, roadsteads, cross-over bridges, traffic lanes including control systems, and rules and regulations to improve environmental infrastructures in addition to self improvement of expertise of themselves and their ship's performance.

5. Challenges and Solutions Expected of Persons Engaged in Ship Handling

5.1 Ideal Training and Discipline for Persons Engaged in Ship Handling

Persons engaged in ship handling are the main players in realizing safety and security at sea through their expertise and are expected to exert unremitting effort to brush up their professional knowledge and skills.

Maneuverings for course keeping, alteration of course, collision avoidance, entering into or leaving from port, berthing or un-berthing, mooring or un-mooring, dropping or weighing anchor, shifting anchorage and anchor watch are fundamental operations necessary at every phase of ship handling regardless of the scale of maneuvering area. It is important for persons engaged in ship handling not only to master ship handling for every maneuvering phase theoretically, but also to endeavor in learning the expertise that their predecessors have built up through long experience at sea.

However, it cannot be hoped that everyone engaged in ship handling has enough sea experience to master practical ship maneuvering before they are commissioned to duties handling a ship. Today, ship handling simulators have been effectively used to train such novices. Training by ship handling simulators is quite useful for novices and updates of expertise of the inexperienced persons engaged in ship handling.

In training with a ship handling simulator, scenarios to familiarize one with the processes for course keeping, alteration of course, collision avoidance, entering into and leaving from port, berthing and un-berthing, mooring and anchoring are generally prepared. But, training to respond to an emergency such as collision, grounding and engine and rudder trouble, and how to avert such situation has not been included. Individual persons engaged in ship handling have been obliged to rely on image training for these countermeasures.

Training on how to overcome such risky situations or how to brush up on, or improve the skills of persons engaged in ship handling is necessary. For such training, scenarios on how to avert collisions with ships, bridge piers, floating objects, ice, groups of fishing boats, or grounding, and how to escape from heavy weather at sea and how to react to engine and rudder trouble as well as how to behave when rescuing a distressed ship or a man overboard should be included to enable improvement of individual skills through trial and error.

In addition, so that these courses may be applied to a broader base of trainees enabling them to respond to possible risks correctly, it is hoped that as many persons engaged in ship handling as possible will take part to develop and improve these scenarios and research programs by providing their know-how.

The above effort will also be useful to enhance the resilience (readiness and responsiveness) of persons engaged in ship handling to emergencies.

5.2 Ideal Information Supply and Support System for Safe Ship Handling

Among the transforming technologies surrounding ship handling, the advancement of information technology may be most prominent in the near future. Persons engaged in ship handling will be able to do their jobs by continually enhancing their own intuition through visual observations and the use of precise information from ARPA Radar (Radar with Automatic Radar Plotting Aids), ECDIS, GPS and AIS in the process of collecting information on which they base their decision making.

On the other hand, an overflow of various information may overwhelm them. Therefore, such information must be evaluated and screened before it is offered to users depending on their needs. Persons engaged in ship handling are well aware of what kind of information should be supplied along with its timing.

In this sense, persons engaged in ship handling should not remain a reticent receiver of information but make clear their requests to suppliers of information about their needs.

5.3 Expected Ship Maneuverability and Infrastructural Environment in View of Persons Engaged in Ship Handling

Most persons engaged in ship handling savor the thrill of achievement after, with all their might, they have overcome a trying task. It is certain that the harder the task, the greater the delight. It is undeniable that such moments are meaningful for the person engaged, and that it concerns personal pride in performing one's duties safely, regardless of a ship's performance under any given circumstance. However, it is also their duty to draw attention from those concerned in the maritime cluster and have them take the proper actions to improve their working environment.

Because persons engaged in ship handling know better than any other persons about how the minimum safety standard of a ship's performance should be and to what extent environmental condition should be improved as a maritime expert. In addition, persons engaged in ship handling should take part more actively in designing maneuvering environments as a marine technical information source.

5.4 Standards of Operational Limits to Warrant Safe Ship Handling

When performing ship handling, the first priority should be placed on achievement of safe maneuvering. Accordingly, a person engaged in ship handling has the right to request safe maneuverability of their ship and a safe maneuvering environment where the ship will be maneuvered. In return, it is the responsibility of a person engaged in ship handling not only to maintain but also to update their expertise for safe ship handling under any given environmental condition. However, it should be noted that human capability of ship handling has its own limits by itself in view of safety and the degree of difficulty.

This is to say, a person engaged in ship handling should endeavor to ascertain potential risks to the maneuvering operation of their commissioned ship and proactively evaluate their physical and psychological load rationally if instructed or proposed maneuvering is practical or not, or otherwise request the minimum safety standards to secure safe ship handling.

The operational safety standards and guidelines for entering and leaving port, berthing and un-berthing including the required number of tugs or horsepower, cargo operations and the strength of berth and mooring facilities have been settled locally based on the study of respective natural environmental conditions. It is important for persons engaged in ship handling to endeavor making necessary proposals and suggestions as appropriate to secure safe ship maneuvering reflected by their experience for these safety standards.

Ⅱ A Ship's Motion, Buoyancy and Stability

Chapter 1 Characteristics of the Ship's Motions

1. A Ship's Motion in Six Degrees of Freedom

A ship's motion has six degrees of freedom, three parallel motions along each axis and three turning motions around each axis, in a three dimensional solid perpendicular with coordinates centered on its center of gravity as the origin. The illustration of these six degrees of freedom of ship motions are given in Fig. 1.2.1. A parallel motion along X-axis is called ***surging***, that along Y-axis is called ***swaying*** and that along vertical axis Z is called ***heaving***. While a turning motion around X-axis is called ***rolling***, that around Y-axis is called ***pitching*** and that around Z-axis is called ***yawing***.

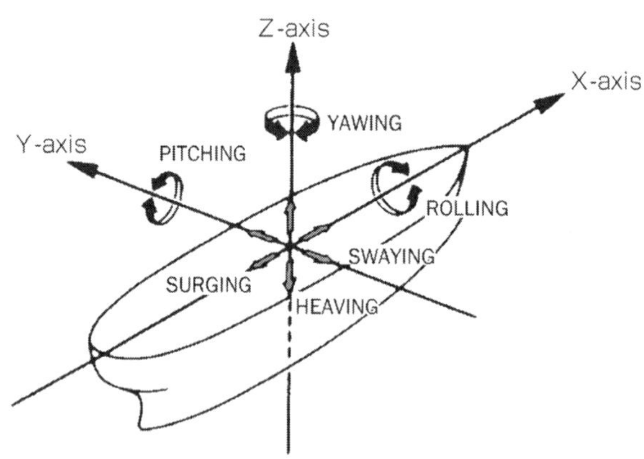

Fig. 1.2.1 A ship's motion in six degrees of freedom

2. A Ship's Motion in Six Degrees of Freedom and Restoring Forces

Among a ship's motions in the 6 degrees of freedom, some are restitutive and others are not. Rolling, pitching and heaving are restitutive and each of them has a respective natural period. The natural rolling period can be found from the equation as shown in (1.2.1)

$$T_R = 2\pi\sqrt{\frac{k_x^2}{g \cdot GM}} \qquad \cdots\cdots (1.2.1)$$

where

T_R : natural rolling period
k_x : radius of gyration of rolling
g : gravitational acceleration (9.8m/sec^2)
GM : transverse metacentric height

Natural pitching period can be obtained by the following equation (1.2.2)

$$T_P = 2\pi\sqrt{\frac{k_y^2}{g \cdot GM_L}} \qquad \cdots\cdots (1.2.2)$$

where

T_P : natural pitching period
k_y : radius of gyration of pitching
GM_L : longitudinal metacentric height

Natural pitching periods range roughly 1/2 to 2/3 of natural rolling periods. And natural heaving periods can be estimated almost the same as its natural pitching period.

By contrast, surging, swaying and yawing motions are not innately restitutive. Therefore, once any of these actions starts to working, subsequent motion will continue in the same direction as long as any reversing force is not given. When a ship is moored to berth, restoring force is generated as mooring ropes and fenders together with the ship itself form a ***spring-mass system***. Accordingly, natural periods of surging, swaying and yawing will also be created on the hull moored to berth. Natural periods while a ship is moored to a berth can be deduced by the following equation (1.2.3).

$$\tau = 2\pi\sqrt{\frac{m}{k}} \quad \cdots\cdots (1.2.3)$$

where

τ : natural period of motion required to find
m : mass of hull
k : for surging, total dynamic spring constant of integrated longitudinal mooring forces
for swaying, total dynamic spring constant of integrated transverse mooring forces
for yawing, substitute moment of inertia for m, and torque required to turn hull 1 radian for k

3. Coupled Motions

In a ship's motion in six degrees of freedom, one motion induces other motions. For instance, when a ship has started to move right to left by swaying, rolling and yawing may be induced. When a ship starts to move up and down by heaving, pitching may be induced. These subsequent motions which are caused due to longitudinal asymmetry of the hull shape, and the deflection between the center of gravity and center of buoyancy are called coupled motions.

However, it should be noted that swaying, rolling and yawing can induce coupled motion by themselves, but be immune from the other motions. In

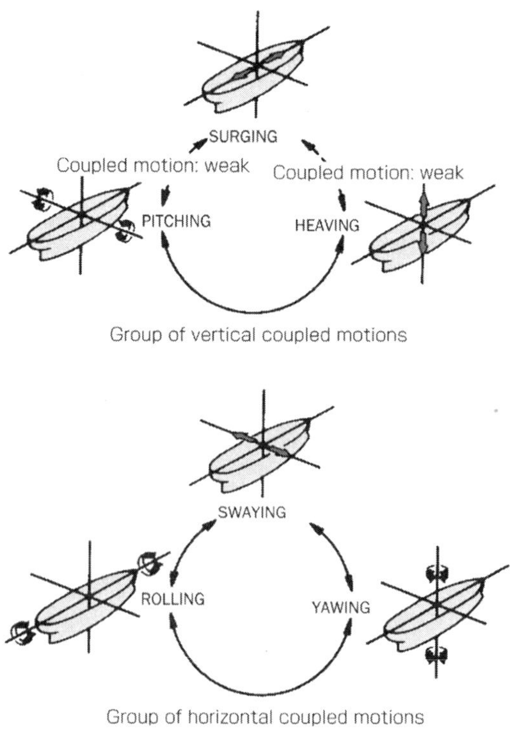

Fig. 1.2.2 Coupled motions

addition, surging, pitching and heaving can induce coupled motion by themselves, but be immune from other motions. The above two groups of motions are sometimes handled and sorted into two groups, the former as "horizontal motion group" and the latter as "vertical motion group".

Chapter 2 Buoyancy and Stability of Ship

1. Buoyancy

To a floating object partly immersed in water, upward force equivalent to the weight of displaced water acts according to ***Archimedes' Principle***. This upward force is called buoyancy and acts on the center of the volume of the floating object. The point where buoyancy acts is called the center of buoyancy.

When an immersed volume is increased, buoyancy will increase corresponding to the additionally immersed volume. If the weight of a ship and buoyancy is balanced before the total volume of the ship is immersed, it means that the ship can float in water. This condition in equilibrium can be substituted by the following equation (1.2.4)

$$W = \gamma \cdot V \qquad (1.2.4)$$

where

W : weight of ship
V : immersed volume of ship
γ : specific gravity of water (weight to volume ratio of water)

The vector of weight of the left side of equation (1.2.4) acts vertically downward while the vector of buoyancy of the right side acts vertically upward.

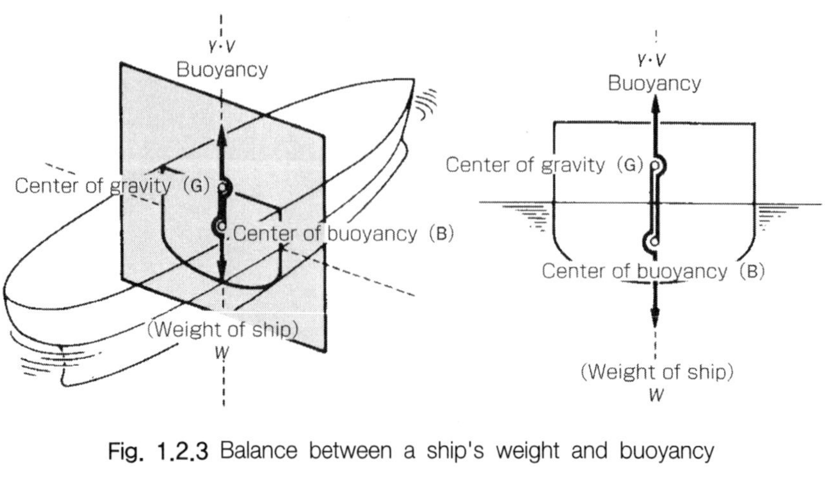

Fig. 1.2.3 Balance between a ship's weight and buoyancy

The scales of two vectors are the same but the directions are opposite. In this condition the ship does not sink nor float any more.

The specific gravity of seawater is 1.025 tonf/m^3 and slightly heavier than that of fresh water, 1.000 tonf/m^3. When a ship moves from the sea where the specific gravity "γ" is heavier than that of fresh water into the fresh water where the specific gravity "γ" is lower than that of sea water, its immersed volume will increase as $W=\gamma \cdot V$(W: constant) demonstrates, and ship draft increases. On the contrary, when a ship moves from a fresh water area to a sea water area, the ship's hull floats proportionally and its draft decreases.

2. Conditions to Maintain Stable Equilibrium

When a ship is afloat, statically in water, its dynamic condition can be summarized as follows.

(1) The center of gravity and buoyancy of a ship are on the same vertical line
(2) The vectors of weight and buoyancy of a ship are equal in scale but their directions are opposite.

Under the above (1) and (2) conditions, a ship can float in water statically. However, these conditions can not solely warrant a ship to float stably in water. To keep a ship afloat with stable equilibrium the following conditions are indispensable in addition to the foregoing (1) and (2).

(3) When a ship afloat in an equilibrium state is listed by any external force, the ship has a potential moment to restitute its posture.

Under the complete set of the above three conditions, the ship can float in water in stable equilibrium. The moment referred to in (3) is called ***righting moment*** and its potential is called ***stability***.

3. "GM" for Stability Index

When a ship is listed at $\theta°$ from its upright state, its draft of the listed side deepens and the immersed shape becomes asymmetrical. And its center of buoyancy as the center of the immersed volume moves to B', toward the listed board from the center line of the ship as shown in Fig. 1.2.4. In this state, the vector of "W" (weight of ship) acts vertically downward on G (ship's center of gravity) while the vector of $\gamma \cdot V$ (buoyancy) acts vertically upward on B' (center of buoyancy when a ship is listed) as the position of the center of gravity of the ship remains unchanged even when it is listed. In this instance the above two vectors which forces are equal and directions are opposite create a ***couple of forces***. The moment created by this couple of forces works to restitute ship's position upright as shown in Fig. 1.2.4. This is the foresaid righting moment. The scale of the righting moment which is a potential to restitute the ship's posture can be calculated by equation (1.2.5).

$$\text{Righting Moment} = W \cdot GZ \qquad (1.2.5)$$

"GZ" substitutes the length of the foot of perpendicular from G to the line of action of buoyancy, which corresponds to the lever of righting moment. From the equation $GZ = GM \cdot \sin\theta$ in a triangle GMZ, righting moment can be expressed as equation (1.2.6) below.

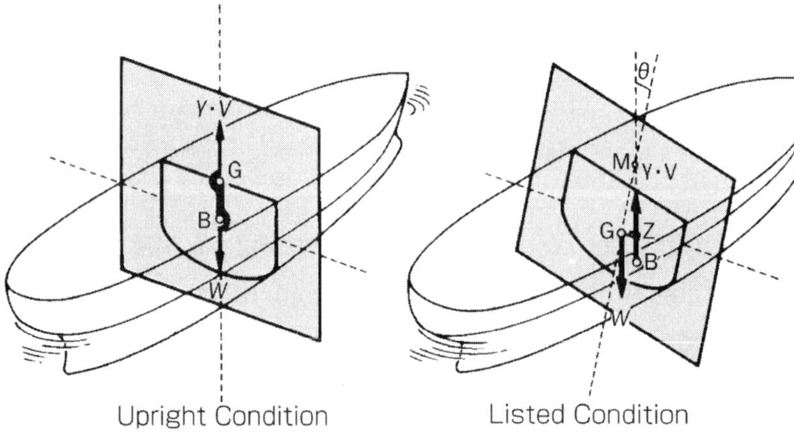

Fig. 1.2.4 Righting moment

$$\text{Righting Moment} = W \cdot GM \cdot \sin\theta \quad \cdots\cdots (1.2.6)$$

In the above, "GM" substitutes the distance between the center of gravity of the ship and its metacenter. The metacenter M is defined as the point of intersection between the line of action of buoyancy passing the center of buoyancy B' when the ship is listed and the vertical center line of the ship. The position of the metacenter can be deemed as unchanged while its list ranges 10° to 15°.

The righting moment for which a ship's metacenter remains unchanged within a small list is called ***initial stability***. Initial stability becomes larger proportionally to the scale of "GM". And the scale of "GM" is decided by the relative vertical distance between metacenter M and the center of gravity of ship G.

In other words, it can be said that the center of gravity of a ship will be decided by its vertical mass distribution onboard. This means that when heavier cargoes are stowed in a lower place, then G moves downward and stability is enhanced by increased "GM" and righting moment to restitute a ship's list. On the contrary, when heavier cargoes are stowed in higher place, then G moves upward and stability is undermined by decreased "GM" and righting moment. From the above, the quality of stability can be estimated by the scale of "GM". In this sense, "GM" is an important index to evaluate a ship's stability.

4. Signs of "GM" and Stability

To keep a ship afloat and in stable equilibrium in water, it is imperative that the center of gravity of the ship is located lower than its metacenter. Therefore, proper caution should be exerted as to vertical distribution of weight in stowage so that the center of gravity of the ship may be placed lower than metacenter for safe voyage. Persons engaged in ship handling need to realize what happens when G is obliged to be placed higher than M. Then, we will see the following three cases, that is, in cases where G is placed lower than M, G is placed higher than M and G coincides with M as shown in Fig. 1.2.5

4.1 When a Ship's Center of Gravity is Located below the Metacenter

The moment created by combined forces of weight and buoyancy acts to restitute a ship's posture from listed condition to upright. When a ship is placed under such conditions, it can be said that the ship is in a stable condition. As such, a "*GM*" positive sign is given. And the larger the absolute value of "*GM*", the higher stability becomes.

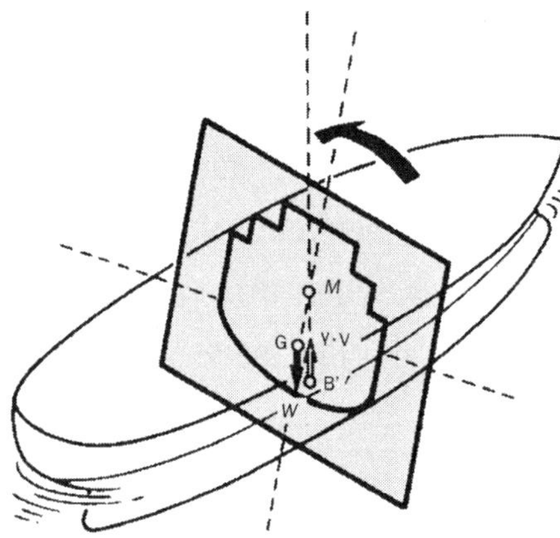

$GM > 0$: Stable Equilibrium

4.2 When a Ship's Center of Gravity is Located above the Metacenter

In this case, the moment created by combined forces of weight and buoyancy acts to increase the ship's list. When the moment created by the combined forces of weight and buoyancy act as a capsizing moment, the ship is placed under unstable conditions and a negative sign is given to "*GM*". The larger the absolute value of "*GM*", the lower the stability becomes.

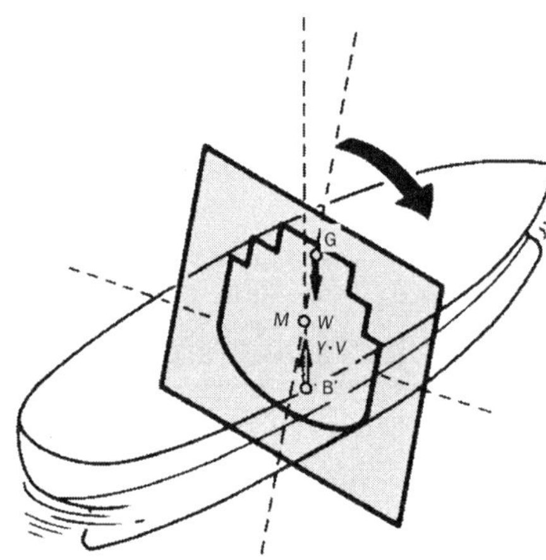

$GM < 0$: Unstable Equilibrium

4.3 When a Ship's Center of Gravity and the Metacenter are Identical

In this case the value of "*GM*" is equal to zero and neither a righting moment nor a capsizing moment will be

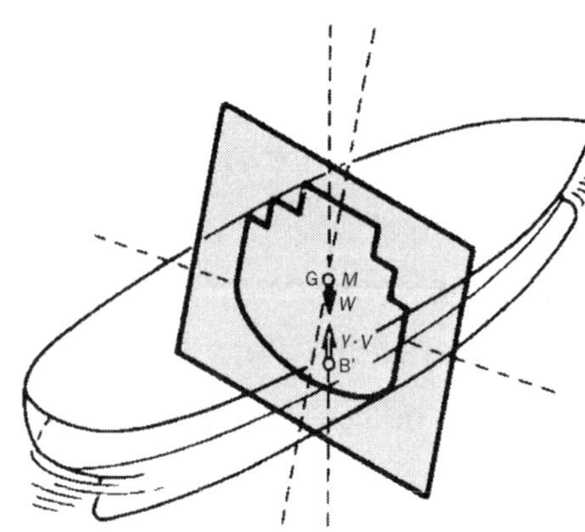

$GM = 0$: Neutral Equilibrium

Fig. 1.2.5 Signs of "*GM*" and stability

created. This condition is called neutral equilibrium.

5. Securing Stability

Most disastrous accidents at sea may be capsizing and subsequent sinking. It may be possible to avert the worst case scenario of total loss by sinking, if at least buoyancy is maintained even after capsizing. To prevent capsizing, it is of primary importance that the ship should not be listed; by doing so the ship is capable of maintaining stability. For this purpose indication of "*GM*" must be positive and its absolute value must be large enough as aforementioned.

To maintain safe "*GM*", consideration to stow heavier cargoes lower in relation to the ship's center of gravity must be exerted. Minimum safe requirements for "*GM*" are prescribed in the National Ship Stability Regulation in accordance with the recommendation of the resolution adopted in 1968 in IMO (International Maritime Organization) in which more than 15cm of "*GM*" is recommended.

In practice, persons engaged in ship handling of ocean going ships seem to be adopting much larger "*GM*" more than 30cm compared with the above recommendation. From past statistical data it appears around 4% of a ship's beam. The followings are the statistical data of "*GM*" by ship type.

General cargo ship : 0.5m to 1m
Container Ship : 1.5m to 2m
PCC : 1m to 1.5m
LNG : 4m to 5m
VLCC : 5m to 10m

Stability may decrease during a voyage not only due to the addition and subtraction of materials onboard such as the consumption of fuel, fresh water, scooped seawater on deck, absorbency of cargo, and accretion of ice to the hull structure, but also due to free water effects on free surfaces of a ship's tanks and scooped seawater, including the bilge. When a ship is obliged to navigate with marginal "*GM*", utmost caution should be paid for these effects because sometimes these integrated effects may suddenly create negative "*GM*".

"GM" can be deduced from ship's rolling period using the equation (1.2.7) derived from the equation (1.2.1)

$$T_R = 2\pi\sqrt{\frac{k_x^2}{g \cdot GM}}$$

From the above expression, substituting $0.4B$ (B: beam breadth) for "k_x" and 9.8m/sec^2 for "g",

$$T_R = \frac{0.8B}{\sqrt{GM}} \quad \cdots\cdots (1.2.7)$$

Where, "B" is the ship's beam breadth in meters, "T_R" is ship's rolling period in second. From "B" and "T_R", "GM" can be deduced. This deduction method is quite practical.

Volume 2

Principle of Ship Handling

Ⅰ. Maneuvering by Rudder

Ⅱ. Maneuvering by Main Engine and Propeller

Ⅲ. Maneuvering by Thruster and Tug

Ⅳ. Maneuvering under External Disturbances

I
Maneuvering by Rudder

Chapter 1 Rudder

1. Function of the Rudder

Ship handling is an operation for moving or stopping a ship in command as intended, controlling its position, posture and speed. Primarily it means that the person engaged in ship handling can use its rudder and main engine. The main engine is a prime mover to control a ship's speed and the rudder is an important device to control its moving direction, posture and position.

2. Performance of the Rudder

2.1 Forces Acting on the Rudder

To get a better understanding of the mechanism of forces to be generated on a ship's rudder, ***Bernoulli's theorem*** which is often used to illustrate how an airplane is kept aloft in the air would be most convenient. Bernoulli's theorem concerning steady flow of perfect fluid free from viscosity can be expressed as shown in the equation below, where "p" is pressure at an arbitrary point in fluid current, "z" is the depth at the point, "v" is current velocity, "ρ" is the density of water and "g" is gravitational acceleration.

$$\frac{p}{\rho g} + z + \frac{v^2}{2g} = const. \quad \cdots\cdots (2.1.1)$$

The above equation illustrates the principle that at an arbitrary point in flowing fluid if depth is not changed, the larger the current velocity "v", the smaller the pressure "p" becomes and vice versa.

In the case of the function of the wing of aircraft in the air, the mechanism in which air on the upper face of wing of airplane flows faster than that on its under face can be explained by ***circulation theorem***. And the mechanism that the difference of current speeds between upper face and under face creates lift can be explained by Bernoulli's theorem.

The effect of rudder is the same mechanism as the wing of aircraft. When a rudder is turned to $\delta°$ in water current, the current along the rudder face of the steered board slacks, by contrast the current along the opposite face of the rudder quickens. This phenomenon demonstrates Bernoulli's theorem that when current quickens pressure drops and when current slacks pressure rises. This illustrates the force which works from the front side (steered board) with higher pressure to the rear side of the rudder with lower pressure is generated.

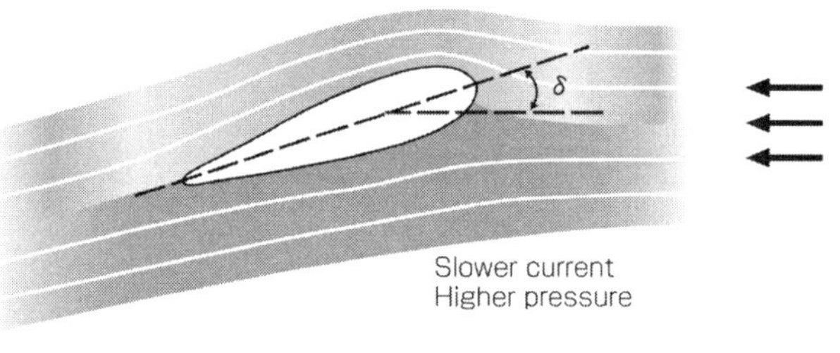

Fig. 2.1.1 Currents and pressures around rudder

2.2 Normal Force, Friction Force, Lift and Drag

As shown in Fig. 2.1.2, normal and friction forces are created on the face of the rudder simultaneously when it is turned in fluid current. The ***resultant force*** of these two forces is called ***rudder force***.

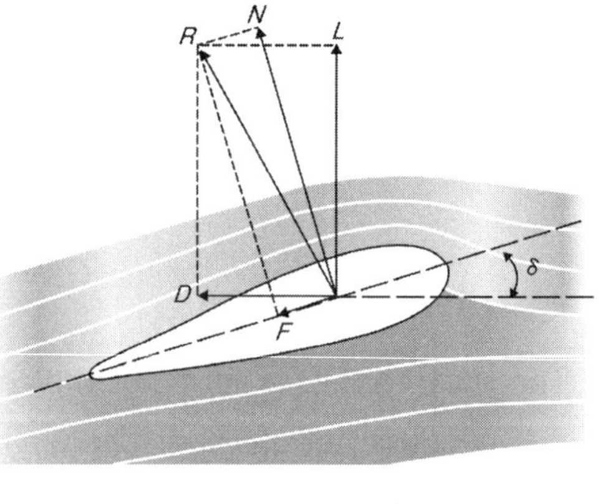

Fig. 2.1.2 Forces acting on the rudder

Further this force can be analyzed into a longitudinal component of ***drag*** which impedes the ship's speed acting as an headway resistance and transverse component of ***lift*** which creates not only a horizontal turning moment to turn the ship around the Z-axis centered on its center of gravity, but also a heeling moment to list the ship around the X-axis centered on its center of gravity pushing the center of gravity of the ship's outward opposite to the steered board.

Neglecting friction force "F" as it is too small compared with normal force "N", the relationship among normal forces "N", lift "L" and drag "D" when the rudder is turned to $\delta°$ can be shown in the following equations.

$$\text{Lift} \quad : L = N \cdot \cos\delta \quad \cdots\cdots (2.1.2)$$

$$\text{Drag} \quad : D = N \cdot \sin\delta \quad \cdots\cdots (2.1.3)$$

2.3 Beaufoy's Empirical Formula to Deduce Normal Force

To estimate the scale of normal force, Beaufoy's empirical formula in which normal force is expressed as a function of the rudder area, ship speed and helm angle, is available as shown below.

$$\text{Normal force} : N = 58.8 \cdot A \cdot V_p^2 \cdot \sin\delta \quad \cdots\cdots (2.1.4)$$

where

N : normal force (kgf)
A : rudder area (m^2)
V_p : current velocity at the rudder face (m/sec)
δ : helm angle (degree)

The above Beaufoy's empirical formula (2.1.4) implies that the broader the surface area of the rudder, the stronger the normal force grows proportionally to the squared current velocity. Besides, the relation between helm angle and normal force indicates that normal force increase simply following sine curve. However, it should be noted that an estimated value derived from Beaufoy's empirical formula is said to tend to be smaller than the actual value and it is incapable to make deduction reflecting the shape of rudder.

2.4 Stall Phenomenon

Fig. 2.1.3 illustrates how normal force increases corresponding to the angle of the rudder based on the results of model rudder experiments without hull. The vertical axis represents the ratio (nondimensionalized number) of measured normal forces at each helm angle to that at 90°.

In experiments, three types of rudders, which each ***aspect ratio*** (height/breadth of rudder) is 2.0 vertically oblong, 1.0 square and 0.5 horizontally oblong, were used. Fig. 2.1.3 shows the curves derived from Beaufoy's empirical formula (2.1.4) together with the results of the above experiments in comparison.

Flux around the rudder remains uniform along the face of the rudder until the angle of the rudder reaches a certain angle, but is peeled off from the face of rudder and forms eddies as the angle of the rudder increases, resulting in the reduction of normal force. An abrupt normal force reduction due to the above effect is called ***stall***. And the angle of the rudder at which stall occurs is called ***critical helm angle***.

Summarizing the characteristic relation between normal force and critical angle in view of the shapes of rudders, it can be said that with vertically oblong rudders larger normal force can be gained by comparatively smaller helm angle but stall occurs at earlier stage, and by contrast with horizontally oblong rudders it is possible to maintain rudder effect up to comparatively larger helm angle as stall

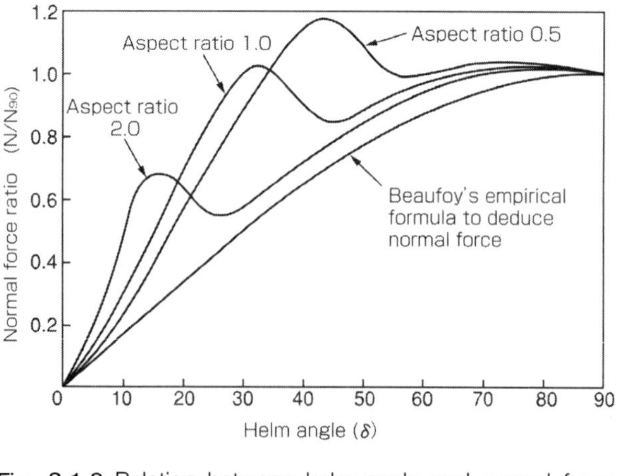

Fig. 2.1.3 Relation between helm angle and normal force

is delayed. The characteristic of square rudders lies in about the middle of the two above.

2.5 Joessel's Empirical Formula to Deduce Center of Normal Force

Although power required for the steering engine to drive a rudder may vary depending on the position where the rudder stock is placed, it must be larger than the moment around the rudder stock. For instance, when the rudder stock is mounted to the front end of the rudder, its moment will be equivalent to the multiplication of the normal force acting on the rudder and the distance from the front face of the rudder to the acting center of normal force.

To estimate the position of acting center of normal force, Joessel's empirical formula is generally applied as shown below.

$$\text{Center of normal force}: x = (0.195 + 0.305\sin\delta)b \quad \cdots\cdots (2.1.5)$$

where

x : distance from the front end of rudder to the center of normal force (m)
b : breadth of rudder (m)
δ : angle of rudder (degree)

According to Joessel's empirical formula (2.1.5), the center of normal force moves as the angle of the rudder increases, and comes to lie up to around 35% of the breadth of rudder behind the front end of the rudder at 30° of the helm angle.

2.6 Types of Rudders for Merchant Ships

(1) Aspect Ratio

In general, a rudder of a merchant ship is fixed to its hull and current from there will be sent out right astern. And the rudder to change the ship's heading is designed to be placed right behind the propeller so that it may catch accelerated propeller current effectively. Therefore, while a ship is running ahead, propeller current "V_p" is larger than ship speed "V_s" as shown in Fig. 2.1.4. And thus accelerated propeller current prompts the effect of rudder.

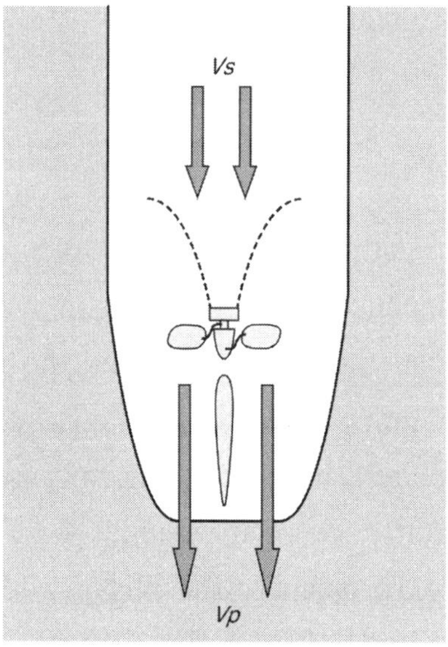

Fig. 2.1.4 Propeller current acting on the rudder (Ahead run)

In view of the characteristic of a rudder, generally larger lift and smaller drag may be favored. In contrast, in the view of maneuvering, a rudder that can create larger initial rudder effect by smaller helm angle to maintain the ship's heading effectively may be favored. Furthermore, in view of turning ability, such characteristic in which maximum helm angle free from stall can be maintained as long as possible would be favored.

According to the results of model rudder experiments, for the former purpose, a vertically oblong rudder is more advantageous, while for the latter purpose a horizontally oblong rudder is more advantageous.

In these conflicting advantages, vertically oblong rudders which aspect ratio ranges from 1.5 to 2.0 are more favored for general merchant ships. This is because comparatively large normal forces can be gained by small helm angle. Although a vertically oblong rudder to which stall phenomenon tends to appear at comparatively smaller angle in model tests, with the actual rudder it may seldom happen within the ordinary steering range as it can catch rectified and accelerated propeller current at the stern. In addition, as moment acting on vertically oblong rudders are smaller compared with that acting on horizontally oblong ones, the power necessary for the steering engine can also be reduced compared with horizontally oblong ones.

(2) Surface Area, Shape of Cross Section

Needless to say, the broader the area of rudder, the larger the effect of rudder can be expected. However, its size and shape will be restricted by the actual ship size and type depending on physical conditions of the shape of stern. Generally for merchant ships, in view of steerability and course keeping ability, rudders which area ranges around 1.5% to 2.0% of ship's lateral underwater area at its designed full load line are widely fitted.

For the shape of the cross section of the rudder, a streamlined shape that enables the current to flow smoothly along its face to minimize resistance to obtain maximum normal force is advantageous. Practically, streamlined shape which maximum thickness ranges 12% to 18% of its breadth is favored.

(3) Structure, Position of Rudder Stock

Most streamlined rudders are fabricated from steel frames and plates with their rudder stocks placed almost near to the center of normal force which ranges around 20% to 30% of each rudder breadth backward from the front end so that power to drive their steering engines can be minimized.

A rudder fabricated from steel frames and plates is called ***double plate rudder*** and a rudder made by a single plate is called ***single plate rudder***. A rudder which stock is fitted close to the center of normal force to counterbalance rudder moment is called ***balanced rudder***, while its rudder stock is fitted to the front end is called ***unbalanced rudder***.

3. Effect of the Rudder

3.1 Turning Moment by the Rudder Immediately after Rudder Action

The effect of a rudder appears as a developing rate-of-turn after it is turned. When the turning response of the ship is as swift as the person engaged in ship handling expected, its steerability is deemed as ***good steerability*** while if its turning response is sluggish contrary to their expectations, its steerability is deemed as ***poor steerability***.

Prompting the force of a ship's turning after its helm operation is a turning moment around the vertical axis (Z-axis) centered on the gravity of the ship created by normal force. Steerability just after steering depends on the scale of turning moment by the rudder force. The rudder moment is to be created immediately after steering is called ***initial rudder force moment.***

As shown in Fig. 2.1.5, when a ship is running straight and begins to turn by $\delta°$ of helm angle, its initial rudder force moment "M" can be expressed in the following equation (2.1.6)

$$M = N\left(\frac{1}{2}\ell \cdot \cos\delta\right)$$

$$= (N \cdot \cos\delta) \times \frac{1}{2}\ell \quad \cdots\cdots (2.1.6)$$

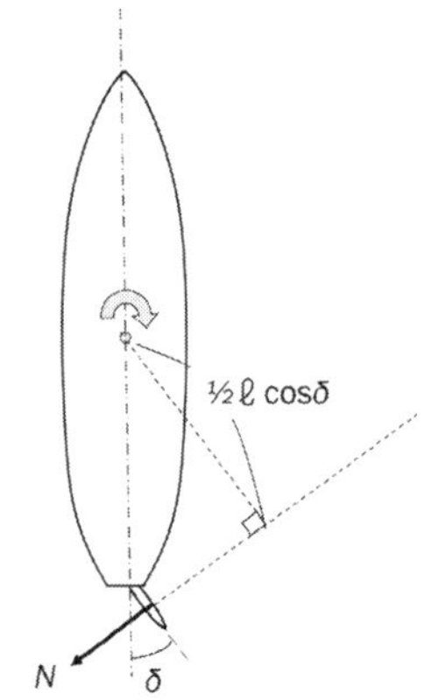

Fig. 2.1.5 Initial rudder force moment immediately after taking rudder

where the length of ship is shown as "ℓ".
Substitute equation (2.1.2) for the above $N \cdot \cos\delta$,

$$M = L \times \frac{1}{2}\ell \quad \cdots\cdots (2.1.7)$$

Substitute equation (2.1.4) for "N",

$$= (58.8 A V^2 \sin\delta)\cos\delta \times \frac{1}{2}\ell$$

$$= 14.7 \ell A V^2 \sin 2\delta \quad \cdots\cdots (2.1.8)$$

From the above equation (2.1.8), the initial rudder force moment reaches its maximum when the helm angle "δ" is 45°. However, the maximum helm angle for general merchant ships is settled on 35°. This is because in practice such a large helm angle is seldom used and requires a large capacity of steering engines and invites a large speed loss due to its increased resistance.

3.2 Reduction of Turning Moment by the Rudder in the Course of Turning Motion

(1) Effect of Fluid Drag Resistance

After the rudder is steered, the ship starts turning gradually by its initial rudder force moment overcoming the resistance of water. At the initial stage, the ship runs obliquely in water with its bow inward and its stern outward to its turning direction. Consequently the hull receives the pressure of the current on its outboard side. A part of outboard current forward from the center of gravity of the ship acts in the direction to turn her but the same current behind the ship's center of gravity acts in the opposite direction as shown in Fig. 2.1.6.

As the figure demonstrates, at the initial stage of turning motion, as the area behind the ship's center of gravity under the effect of oblique current is larger than its forward area and water pressure acts opposite to the direction of turning, the initial rudder force moment will be reduced.

(2) Effect of Angle of Flow to the Rudder

There is another reason to decrease the initial rudder force moment. That is the change of current angle to the rudder in the course of turning. While the turning motion progresses, the ship turns around the vertical axis centered on its pivoting point with its bow inward and stern outward to its turning circle. Under this situation lines of flow impinge on the hull with some angles which grow larger toward its stern.

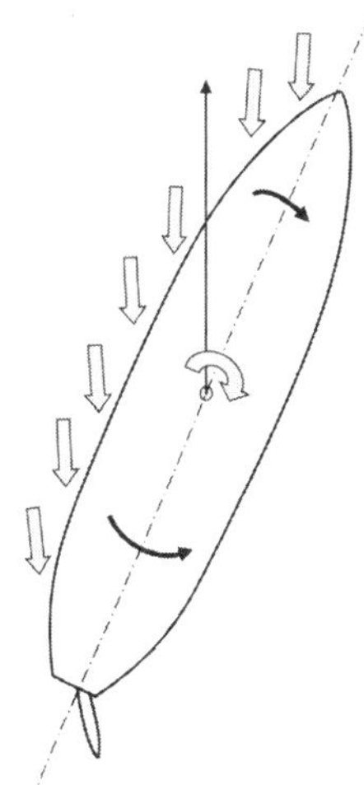

Fig. 2.1.6 Hydrodynamic force acting on the hull while in oblique running

As shown in Fig. 2.1.7, the effect of the rudder is counter-acted by an oblique flow to the rudder fitted at the stern because the angle of relative flow to the rudder becomes smaller than the actual helm angle taken. This resultant angle of flow to the rudder is called ***effective attack angle***.

Although the scale of effective attack angles are primarily considered to be a result of subtracted drift angle "β" from helm angle "δ", it is said that because of the effects of propeller current with a rectified and accelerated stream, an effective attack angle under steady turning by hard over rudder ranges around 2/3 of the actual helm angle taken.

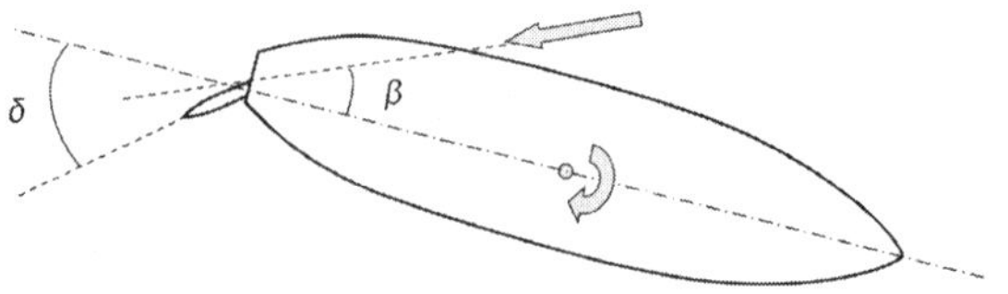

Fig. 2.1.7 Effective attack angle to the rudder

Chapter 2 Steering Operation

1. Steering Response Characteristics

As to a ship's responsive motion to steering, remarkable progression has been made by many continual studies in the field of ship maneuvering theory. On the one hand, as ship's responsive dynamics to steering have been understood subjectively and individually through the feelings of persons engaged in ship handling, it often occurs that the same phenomenon was talked about in different terms and jargons. To avoid such inconvenience, I would like to define the terms to be used as shown in the following Fig. 2.1.8.

Steering response characteristics (Maneuverability) (Steerability)		
	Characteristics of turning	Responsiveness to the helm (quick response to turn)
		Turning ability (turn in a small circle)
	Characteristics of running straight	Course stability (steady course keeping) (quick settling on course)

Turning: Rotating motion around ship's center of gravity in a fixed ship's coordinates after steering

Circling: Locus of ship's center of gravity in turning in fixed geometrical coordinates after steering

Fig. 2.1.8 Terms related to steering response characteristics

For steering response characteristics, the term ***maneuverability*** is used in ship maneuvering dynamics, while the term of ***steerability*** corresponds to it in practical ship handling. Steering response characteristics may be divided into the characteristics of turning and running straight in the sense of persons engaged in ship handling. By contrast, in ship maneuvering dynamics it can be expressed in three specific values of responsiveness to the helm, turning ability, and course stability. In the past, the difference between the characteristics of a simple turn to change heading and a turning to draw circle was rather ambiguous. In this book I have tried to describe those motions following the definition shown in Fig. 2.1.8 as much as possible.

One of the most primary operations for a person engaged in ship handling is to settle their ship in its intended direction. To control a ship's turning and circling motions properly by steering is a fundamental operation to keep and alter course and to avoid risk of collision.

What a person engaged in ship handling expects most is quick development of the rate-of-turn (responsiveness to the helm) and the ship's ability to turn in a small circle (turning ability). Thus abilities to "respond quickly to the helm" and "turn in the least locus" are important elements for persons engaged in ship handling to evaluate their ship's response characteristics.

In the ship maneuvering dynamics, response characteristics to steering are called ***maneuverability*** as a whole. The characteristics to change heading following steering, to draw locus in turning and to keep course and settle on an intended course are respectively called ***responsiveness to the helm, turning ability*** and ***course stability***.

And indices to substitute a ship's maneuvering characteristics are comprehensively called ***maneuverability indices*** and among them the indices of responsiveness and turning ability are respectively called ***index of responsiveness to the helm*** and ***index of turning ability***. The index for course stability is identical to that for responsiveness to the helm.

2. Maneuverability Indices

2.1 Linear Equations of Ship Maneuvering Motion

In the past, a ship's maneuverability was evaluated focusing on the scale of a turning circle drawn by the ship's center of gravity under steady turning. However, the data that explains the relation between the time and motion was not included; e.g., time lag from the time when the rudder is turned to the time when the ship's bow starts to turn, and total time required to complete a turn.

Now we have an approach based on the dynamics of ship handling that enables evaluation of turning characteristics using the time duration, in addition to the conventional method. For this purpose, the ***linear equations of ship maneuvering motion*** are well known as simple and effective equations to explain maneuvering motion. The concept and structure of these equations are explained below.

For a ship's maneuvering motion created by steering, we can focus on three coupling motions; longitudinal motion, lateral motion and turning motion centering on the center of gravity as the origin on the ship's body-fixed coordinate system. And by applying ***Newton's second law of motion*** ($F = m\alpha$) to these three motions, following equations of (2.1.9), (2.1.10) and (2.1.11) on the above three coupling motions can be obtained.

For longitudinal motion : $m(\dot{u} - vr) = X$ ······ (2.1.9)

For lateral motion : $m(\dot{v} + ur) = Y$ ······ (2.1.10)

For turning motion : $I_z \cdot r = N$ ······ (2.1.11)

where

m : ship's mass
u : speed parallel to ship's longitudinal axis (X-axis)
v : speed parallel to ship's transverse axis (Y-axis)
r : rate-of-turn around vertical axis (Z-axis) passing ship's center of gravity
I_z : ship's moment of inertia around Z-axis
X : force acting on hull parallel to X-axis
Y : force acting on hull parallel to Y-axis
N : moment around Z-axis

Among motions of longitudinal, lateral, and turning on the horizontal plane, longitudinal motion does not change largely in ordinary turning by rudder. And subsequent lateral motion is also negligible. Therefore, the longitudinal motion of equation (2.1.9) can be excluded from coupling motion. Assuming the forces of current acting on hull parallel to Y-axis "Y", the right side of equation (2.1.10) and moment around the Z-axis "N", the right side of equation (2.1.11) will make gradual change proportionally to the ship's speed and acceleration. And if linear-processing is applied to these simultaneous equations, the equation of rate-of-turn "r" response to helm angle "δ" can be obtained. Further simplifying is given to the coefficients of this equation, the following first order approximation equation of ship maneuvering motion can be derived.

$$T\dot{r}+r=K\delta \quad \cdots\cdots (2.1.12)$$

where

T : index of responsiveness to the helm
K : index of turning ability

2.2 Physical Implication of "T", "K"

Assuming a ship's turning motion as an object with turning moment of inertia "I" rotates centered on its center of gravity, and supposing that its rotation is created by the moment which is proportional to the helm angle "δ", and a resistance moment proportional to rate-of-turn "r" affects while turning, its turning motion equation will be the following, let gain coefficients "a" and "b".

$$I\dot{r}+br=a\delta \quad \cdots\cdots (2.1.13)$$

where

I : turning moment of inertia
r : rate-of-turn
a : gain coefficient of rudder force
b : gain coefficient of resistance moment

Dividing the both sides of equation (2.1.13) by "b",

$$\left(\frac{I}{b}\right)\dot{r}+r=\left(\frac{a}{b}\right)\delta \quad \cdots\cdots (2.1.14)$$

In comparing the coefficients of the above equation (2.1.14) with those of the linear equation of ship maneuvering motion expressed as an equation (2.1.12) derived from dynamics of ship maneuvering, the followings are obtained.

$$\left(\frac{I}{b}\right)=T \quad \cdots\cdots (2.1.15)$$

$$\left(\frac{a}{b}\right) = K \quad \cdots\cdots (2.1.16)$$

That is to say, the index of responsiveness to the helm "T" equals the ratio of turning moment of inertia "I" to gain coefficient of resistance moment "b". And index of turning ability "K" corresponds to the ratio of the gain coefficient of rudder force "a" to that of resistance moment.

In the equation (2.1.15), assuming resistance moment is unchanged, if the index of responsiveness to the helm "T" is large, it implies that a turning inertia of the object is also large and it is difficult to be turned and stopped. On the other hand, when "T" is small, its turning inertia is also small and it is easy to be turned and stopped. The value of index of responsiveness to the helm "T" reflects such physical characteristics as explained as above.

In the equation (2.1.16), assuming resistance moment is unchanged, when the value of "K" is large, capability to make the object turn is large, when "K" is small, its capability becomes small. The value of index of turning ability "K" reflects such physical characteristics as explained as above.

2.3 Acceleration of Rate-of-Turn after Steering

☞ **Refer to Vol. 4 Ship Handling Characteristics Data Base**

Reducing the differential equation (2.1.12) in time domain, the change of the turning motion "r" corresponding to the helm angle "δ" can be understood in the process of time "t". Reducing the linear equation of a ship maneuvering motion expressed as equation (2.1.12) to obtain "r",

Rewrite equation (2.1.12) to $T\dfrac{dr}{dt}+r=K\delta$,

$$\frac{dr}{dt}=\frac{1}{T}(K\delta-r)$$

Reducing the above differential equation,

$$\int \frac{1}{K\delta - r} dr = \int \frac{1}{T} dt$$

$$\log|K\delta - r| = \frac{t}{T} + C$$

$$K\delta - r = \pm e^{C} \cdot e^{\left(-\frac{t}{T}\right)}$$

$$r = K\delta \mp e^{C} \cdot e^{\left(-\frac{t}{T}\right)}$$

When $\mp e^{C} = C$,

$$r = K\delta + C \cdot e^{\left(-\frac{t}{T}\right)} \quad \cdots\cdots (2.1.17)$$

"C" in equation (2.1.17) means constant of integration. Assuming steering is started from the helm midship state, initial conditions are, $t=0$ and $r=0$.

$$\therefore \ C = -K\delta$$

Therefore

$$r = K\delta\left\{1 - e^{\left(-\frac{t}{T}\right)}\right\} \quad \cdots\cdots (2.1.18)$$

From the above equation (2.1.18), drawing a graph in relation between time "t" as the horizontal axis and rate-of-turn "r" as the vertical axis, it will be as shown in Fig. 2.1.9. Fig. 2.1.9 illustrates the acceleration of rate-of-turn as a duration of time after the rudder is turned.

When $T>0$, the right side 2nd member of equation (2.1.18) $e^{(-t/T)}$ will decrease as time elapses as shown in Fig. 2.1.10, and make $e^{(-t/T)} \Rightarrow 0$ and "r" will be settled to a constant value "$K\delta$". At around the time when "$K\delta$" is settled to a constant value, the ship starts to enter into a steady turning motion. Thereafter the ship continues steady turning motion.

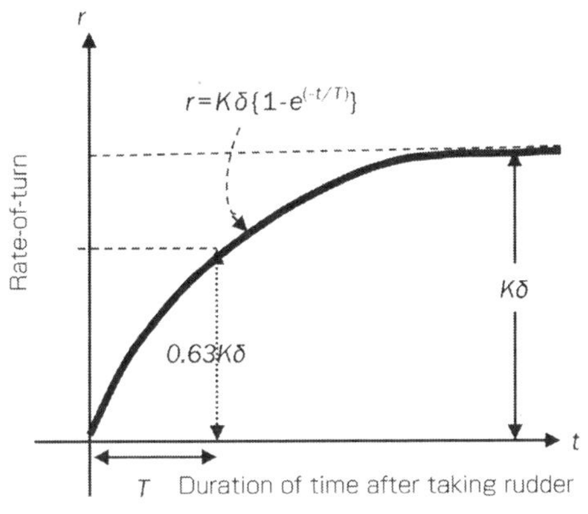

Fig. 2.1.9 Acceleration of rate-of-turn after rudder take

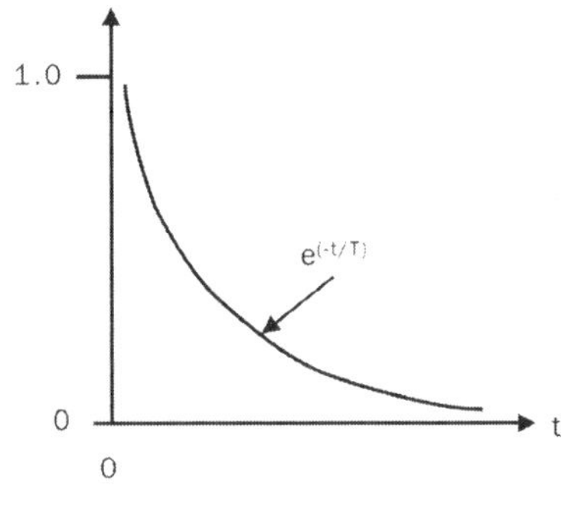

Fig. 2.1.10 Negative exponential function

2.4 Definition of Index of Responsiveness to the Helm, "T"

In equation (2.1.18), when $t=T$,

$$
\begin{aligned}
r &= K\delta\left\{1-e^{\left(-\frac{t}{T}\right)}\right\} \\
&= K\delta\{1-e^{(-1)}\} \\
&= K\delta(1-0.368) \\
&= 0.63K\delta \quad \cdots\cdots (2.1.19)
\end{aligned}
$$

From the above, you can understand that time required for rate-of-turn to reach 63% of that of steady turning corresponds to the index of responsiveness to the helm "T".

Therefore, index of responsiveness to the helm "T" is defined as time required for rate-of-turn to reach 63% of that of steady turning. Where, unit of "T" is [sec]. And non-dimensional value of "T" is expressed as $T'=T(V/L)$. "V" and "L" represent ship speed and ship length respectively.

The index of responsiveness to the helm "T" is an element to control quickness of follow up motions to turn a ship responding to steering operation and time required to attain steady turning. In equation (2.1.18), when "T" is small, ship can enter steady turning faster as $e^{(-t/T)}$ damps quickly. Fig. 2.1.11 illustrates the relation between rate-of-turn and time required to reach steady turning per the scales of "T". The smaller the index of responsiveness to the helm "T", the quicker the ship's turning response to the helm becomes. And as the ship can enter into steady turning faster, a person engaged in ship handling may feel that the ship has good steerability. Thus index of responsiveness to the helm is generally used to evaluate the quickness of a ship's turning response to the helm.

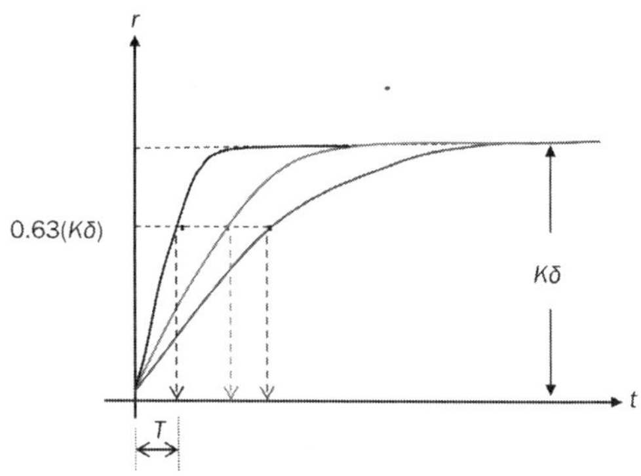

Fig. 2.1.11 Size of "T" and time to reach steady turning

2.5 Definition of Index of Turning Ability, "K"

In equation (2.1.18), when sufficient time "t" elapses, a ship enters into steady turning under a certain constant value of "r" as the value of $e^{(-t/T)}$ becomes zero. When rate-of-turn at this moment under steady turning is "r^*", then,

$$r^* = K\delta \qquad (2.1.20)$$

Therefore, index of turning ability is defined as below.

$$K = \frac{r^*}{\delta} \qquad (2.1.21)$$

From the above equation (2.1.21), the index of turning ability "K" is defined as rate-of-turn per unit of helm angle while ship is under steady turning.

Where, unit of "K" is [sec^{-1}]. And nondimensional value of "K" is expressed as $K' = K(L/V)$. "V" and "L" respectively represent ship speed and ship length.

Index of turning ability "K" is an element to control a ship's rate-of-turn while a ship is under steady turning after steering. Fig. 2.1.12 illustrates the relation between rate-of-turn and time elapsed per the scales of "K". The larger the index of turning ability "K", the stronger the turning moment of inertia becomes and subsequently a smaller turning radius is attained. In such a case, persons engaged in ship handling usually feel that the ship's steerability is good. Thus the index of turning ability "K" is also generally used to evaluate the strength of a ship's turning moment of inertia.

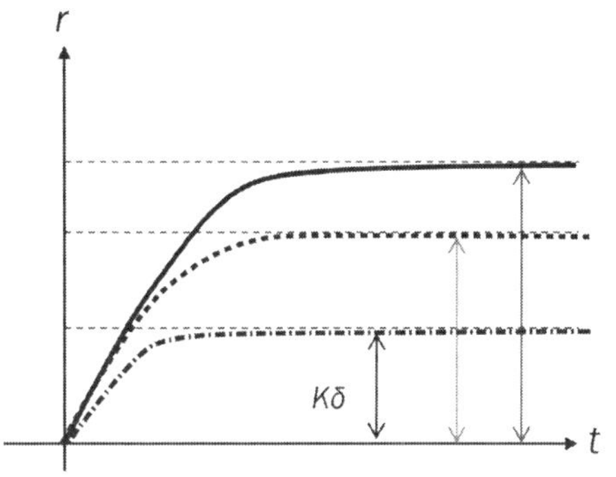

Fig. 2.1.12 Size of "K" and rate-of-turn at steady turning

2.6 Maneuverability Indices of Real Ships, "T" and "K"

☞ Refer to Vol. 4 Ship Handling Characteristics Data Base

Indices of responsiveness to the helm "T" and turning ability "K" vary depending on each ship and at each speed. Therefore, when you are going to evaluate the steering response characteristics of the ship from these values in comparison with those other ships, it would be better to use non-dimensionalized indices "T'" and "K'" instead of "T" and "K". Generally, both "T'" and "K'" range around 1.0, while those of fat ships range around 1.5 to 2.5 and 4 to 5 respectively.

As "T'" and "K'" can be obtained by analyzing the result of zigzag maneuver tests based on the linear equations of ship maneuvering motion, in a test done by small rudder angle the values of "T'" and "K'" tend to be larger, while the larger the helm angle in a test, the smaller they tend to be. Therefore, the values derived from zigzag tests which helm angles range 15° to 20° are generally taken as natural values. Besides, the larger the ship type, the larger the values of "T'" and "K'" tend to be.

3. Characteristics of Turning Motion

3.1 Relation between Time Lag of Responsiveness to the Helm and "T"

When a rudder is turned while a ship is running straight ahead, the ship begins to turn gradually after a certain time lag. When the rudder is turned to "δ_0" for a time span "τ" from the state of its helm midship, change of heading angle "ψ" to elapsed time "t" can be derived by reducing the linear equations of ship maneuvering motion expressed as equation (2.1.12).

$$\psi(t) = K\delta_0 \left\{ t - \left(T + \frac{\tau}{2}\right) + \frac{T^2}{\tau\left(e^{\frac{\tau}{T}} - 1\right)} e^{\left(-\frac{t}{T}\right)} \right\} \quad \cdots\cdots (2.1.22)$$

In the equation (2.1.22), if its maneuverability indices are "T" and "K", the helm angle taken "δ_0" and time required to turn rudder to be "τ" are defined, change of heading angle "ψ" to the duration of time "t" can be derived. A chain line in Fig. 2.1.13 illustrates the curve of "ψ".

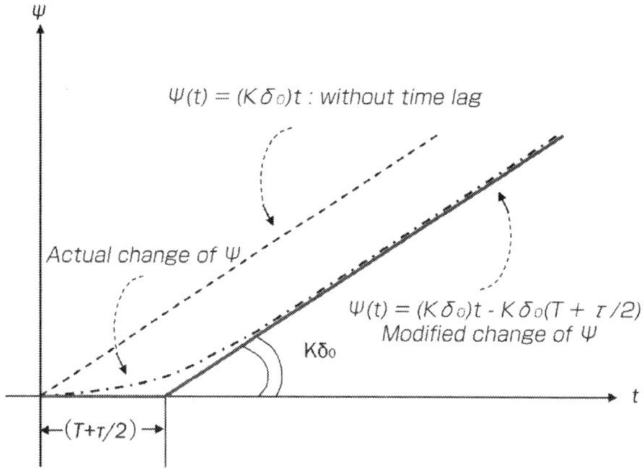

Fig. 2.1.13 Change of heading angle to the duration of time

As time elapses, $e^{(-t/T)}$ on the right side of the equation (2.1.22) becomes zero. And "$\psi(t)$" changes in a linear fashion to the duration of time "t" as shown in equation (2.1.23). The equation shown in (2.1.24) can be derived transforming the equation (2.1.23) to get a clear linear equation form like $Y=\langle a \rangle X+\langle b \rangle$.

$$\psi(t) = K\delta_0 \left\{ t - \left(T + \frac{\tau}{2}\right) \right\} \quad \cdots\cdots (2.1.23)$$

$$\psi(t) = \ll K\delta_0 \gg t - \ll K\delta_0 \left(T + \frac{\tau}{2}\right) \gg \quad \cdots\cdots (2.1.24)$$

Accordingly, change of the heading angle "ψ" to the duration of time is a solid line which the gradient is $\langle K\delta_0 \rangle$ as shown in Fig. 2.1.13. When a ship's heading changes following the above equation, the ship is making steady turning. In the equation (2.1.24) when reducing the intersection between this line and the horizontal axis by giving $\psi(t)=0$ in the equation (2.1.24), the intersection on the horizontal axis can be obtained as the time when $(T+\tau/2)$ elapsed.

$$t=\left(T+\frac{\tau}{2}\right) \qquad \cdots\cdots (2.1.25)$$

If the gradual change of ship's heading after steering is approximated to the thick bold solid line as shown in Fig. 2.1.13, it can be said that for the duration of time for $(T+\tau/2)$, the ship continues running straight, then immediately enters into steady turning following a line which gradient is $\langle K\delta_0 \rangle$.

The duration of time corresponding to $(T+\tau/2)$ means the ***time lag of responsiveness to the helm***, and distance run for which time span is called ***reach***. Reach can be derived from the following equation where "V" is ship speed.

$$\text{Reach}= V\left(T+\frac{\tau}{2}\right) \qquad \cdots\cdots (2.1.26)$$

Time lag to respond and reach derived from the equations of (2.1.25) and (2.1.26) largely depend on the index of responsiveness to the helm "T". Consequently, time lag to respond to steering and subsequent reach become shorter for ships in which "T" is small and responsiveness to the helm is good.

Fig. 2.1.14 represents a schematic diagram of turning circles composed of a straight running part and steady turning part, assuming that when the rudder of the ship running straight ahead along the vertical axis of the X-Y coordinate is turned at the original point in the coordinate to make a complete turn round, the ship will enter into a steady turning motion immediately after running forward for a definite time. Fig. 2.1.14 (a) represents an actual locus of a turning ship. Fig. 2.1.14 (b) represents the case where a ship enters into steady turning after a certain straight running distance. And Fig. 2.1.14 (c) represents the case where ship enters into steady turning as soon as the rudder is turned. In Fig. 2.1.14 (a), the change of ship's heading in relation to the duration of time corresponds to the chain line curve in Fig. 2.1.13. The case in Fig. 2.1.14 (b) corresponds to the solid line graph and the case in Fig. 2.1.14 (c) corresponds to the dotted line in Fig. 2.1.13.

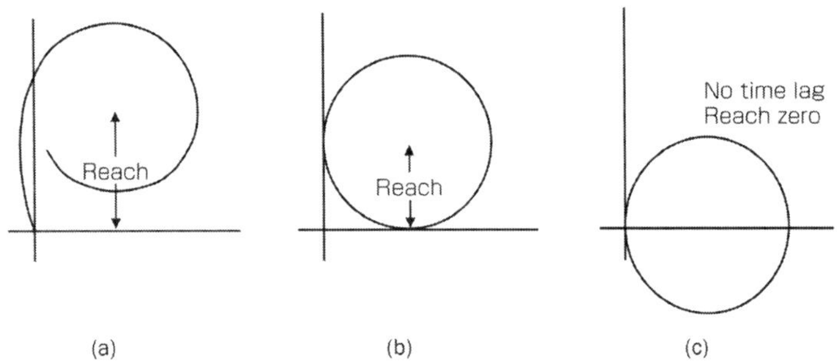

Fig. 2.1.14 Modeled turning loci

3.2 Relation between Radius of a Steady Turning Circle and "K"

As studied on the uniform circular motion in particle dynamics, when a particle moves on a circle which radius is "R" at a constant tangential velocity "V", the relation among angular velocity "ω", radius of circle "R" and tangential velocity "V" can be expressed as $\omega = V/R$ (refer to Fig. 2.1.15). Applying this equation to a ship's steady turning motion and substitute "$K\delta$" for a constant angular velocity "ω", the following equation is derived.

$$K\delta = \frac{V}{R} \quad \cdots\cdots (2.1.27)$$

Consequently, the radius of steady turning circle "R" will be,

$$R = \frac{V}{K\delta} \quad \cdots\cdots (2.1.28)$$

From the above, you can understand that radius "R" of a steady turning circle is defined by the index of turning ability "K". This implies that a ship which value of "K" is large and turning ability is good can turn in a small circle. And this rate-of-turn "$K\delta$" under steady turning motion corresponds to the gradient of the thick solid line in the graph, Fig. 2.1.13. A large gradient implies that rate-of-turn is large and the radius of the turning circle becomes small.

3.3 Effect of "T" and "K" to Locus of a Turning Circle

Sorting ship's turning characteristics, indices of responsiveness to the helm "T" and turning ability "K", into respectively two patterns of large "T" and small "T", and large "K" and small "K", then, four combinations of "T" and "K" from (a) to (d) described below and four characteristics of turning loci as shown in Fig. 2.1.16 can be made. Generally, high speed fine type ships which are superior in responsiveness to the helm but inferior in turning ability belong to pattern (b), while fat type ships which are inferior in responsiveness to the helm but superior in turning ability belong to Pattern (c).

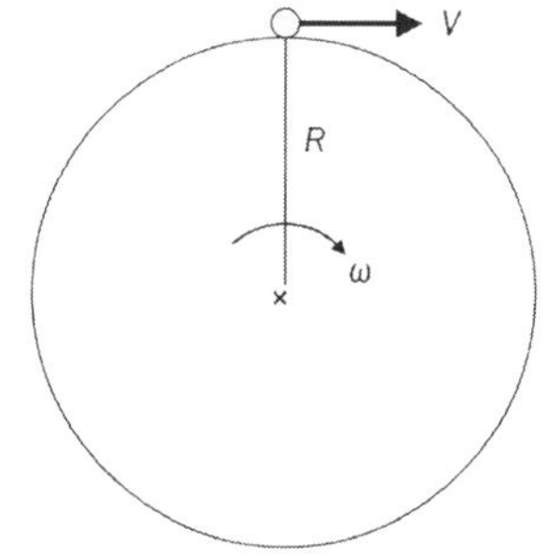

Fig. 2.1.15 Uniform circular motion of a particle

(a) small "T" (superior in responsiveness to helm) and large "K" (superior in turning ability)
: short straight running distance and small turning circle

(b) small "T" (superior in responsiveness to helm) and small "K" (inferior in turning ability)
: short straight running distance and large turning circle

(c) large "T" (inferior in responsiveness to helm) and large "K" (superior in turning ability)
: long straight running distance and small turning circle

(d) large "T" (inferior in responsiveness to helm) and small "K" (inferior in turning ability)
: long straight running distance and large turning circle

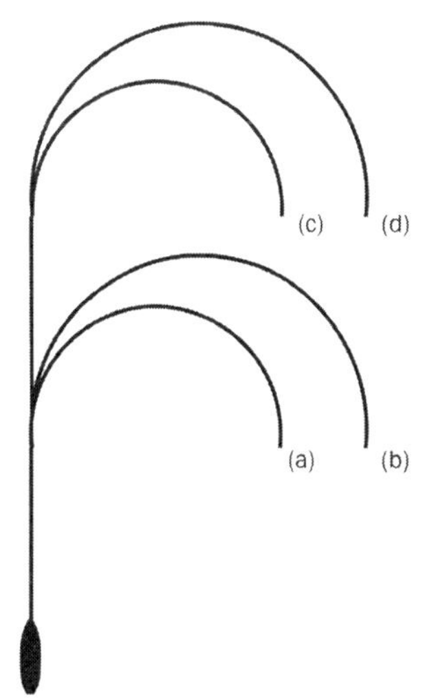

Fig. 2.1.16 Patterns of turning loci by combination of "T" and "K"

4. Characteristics of Running Straight

4.1 Steady Course-keeping

Fundamental steering operations are course-keeping, alteration of course and collision avoidance. Among these operations, for alteration of course and collision avoidance, it is important for a person engaged in ship handling to turn a ship as necessary to achieve operational purposes as easy as possible by reasonable turning ability. Another fundamental steering operation is course-keeping to maintain a ship's course straight as intended. For this purpose the ability to run straight is required.

For ships having small turning inertia which enables a ship to turn and stop easily, their headings are liable to be deflected by external disturbances. However, such deflection of heading can be easily offset by checking the rudder. These are the characteristics of ships having a small value of responsiveness to the helm "T" (refer to equation 2.1.15).

Therefore, it can be said that ships superior in responsiveness to the helm are easier in maintaining their courses. However, on the other hand, ships having small turning inertia are liable to falter by external disturbances, constant care to maintain course must be exerted on frequent deflection of course although they may be offset by small steering operations.

4.2 Quick Settling on Course

(1) Course Stability

In the operation to alter course or avoid a risk of collision, we usually return the rudder to midship once after giving a certain turning inertia to the ship and wait until its inertia dampens, then again get the ship to settle on a new intended course using or adjusting rudders. At this instance, time until turning inertia dampens differs depending on ships. Rather, there are rare cases to develop turning inertia. Thus the characteristic to make a ship run straight on a new course quickly after the operation to turn the ship depends on how quickly initial turning inertia created by steering which works as an external force on the ship can be offset.

The characteristic of how turning inertia dampens when the heading of a ship running straight is deflected due to external disturbances is called ***course stability***. When a ship can be settled quickly on the original course after external disturbances are removed, its course stability can be evaluated as good.

Assuming that a rate-of-turn is created on a ship in which rudder is placed midship, how its rate-of-turn dampens thereafter can be derived from the following equation substituting $\delta=0$ in equation (2.1.12).

$$T\dot{r}+r=0 \quad \cdots\cdots (2.1.29)$$

Reducing this differential equation (2.1.29) for "r",

$$r=r_0 e^{\left(-\frac{t}{T}\right)} \quad \cdots\cdots (2.1.30)$$

Where, "r_0" represents rate-of-turn created by external force. Fig. 2.1.17 shows graphs derived from equation (2.1.30). As you see in the equation (2.1.30) and Fig. 2.1.17, the smaller the value of "T", the faster the exponential part dampens and the faster the rate-of-turn approaches to zero. Conversely, the larger the value of "T", the slower the damping becomes. That is to say, the characteristics of course stability correspond to those of responsiveness to the helm.

Occasionally there are some ships which turning inertia develops even after external force was removed. In such cases, "T" bears a negative value and such ships are evaluated as inferior in course stability.

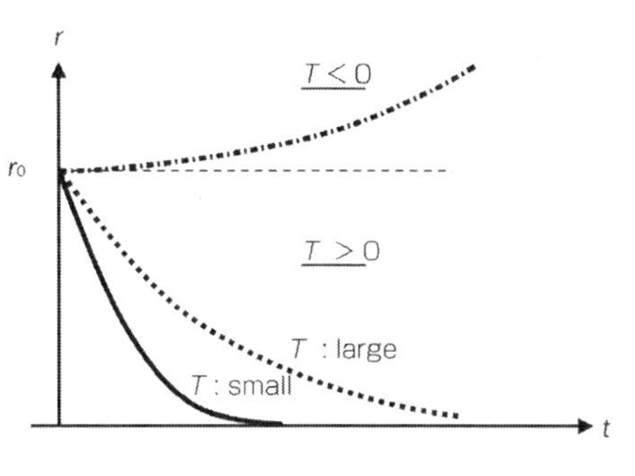

Fig. 2.1.17 Turning inertia damping behavior and course stability

(2) Overshoot Angle after Rudder Returns to Midship

☞ Refer to Vol. 4 Ship Handling Characteristics Data Base

Over shoot angle "ψ" after rudder is returned to midship when rate-of-turn is "r_0" can be derived from the equation (2.1.30) which represents dampening of turning inertia after the rudder is returned to midship.

$$\psi = r_0 T \left\{ 1 - e^{\left(-\frac{t}{T}\right)} \right\} \quad \cdots\cdots (2.1.31)$$

When "t" $\Rightarrow \infty$, equation (2.1.31) will be

$$\psi = r_0 T \quad \cdots\cdots (2.1.32)$$

Overshoot angle until turning stops after rudder is returned to midship can be derived from the equation (2.1.32). The scale of the overshoot angle after the rudder was returned to midship depends on rate-of-turn at the time. The superior the responsiveness to the helm, the smaller the overshoot angle becomes after the rudder is returned to midship. It is important for a person engaged in ship handling to confirm the characteristic of turning inertia of their own ship beforehand.

(3) Overshoot Angle after Checking Rudder

☞ Refer to Vol. 4 Ship Handling Characteristics Data Base

In the steering operation to alter course or to avoid the risk of collision, we often apply checking the rudder a little before the intended course to offset extra turning inertia trying to settle her on that course quickly. An overshoot angle at which turning stops after the checking rudder was applied can be obtained through damping of the turning inertia as per the duration of time until "$r(t)$" becomes zero in the equation (2.1.33). In other words, overshoot angle can be derived integrating the equation (2.1.33) from the time when checking the rudder is taken to the time when turning stops.

$$r(t) = r_0 e^{\left(-\frac{t}{T}\right)} - K\delta_0 \left\{1 - e^{\left(1-\frac{t}{T}\right)}\right\} \quad \cdots\cdots (2.1.33)$$

where

t : time duration after checking the rudder

r_0 : rate-of-turn at the time when checking rudder was taken

δ_0 : checking rudder angle

T : index of responsiveness to the helm

K : index of turning ability

In the coordinate of Fig. 2.1.18 showing the relation between overshoot angle and rate-of-turn after checking the rudder was applied, horizontal axis represents rate-of-turn at the time of checking the rudder was taken and vertical axis represents overshoot angle after checking the rudder. In the figure solid line represents ships with superior course stability and dotted line represents ships with inferior course stability. For instance, in the case that the rate-of-turn is 0.5°/sec at the time checking the rudder was taken, in a ship of superior course stability, overshoot angle is about 10° against 15° of checking the rudder, while in a ship of inferior course stability, the overshoot angle reaches about 25° against the same checking rudder angle. Although overshoot angle may vary depending on rate-of-turn at the time when checking the rudder is taken, a person engaged in ship handling should be conversant to the characteristic of overshoot angle after checking the rudder of their own ship is taken beforehand.

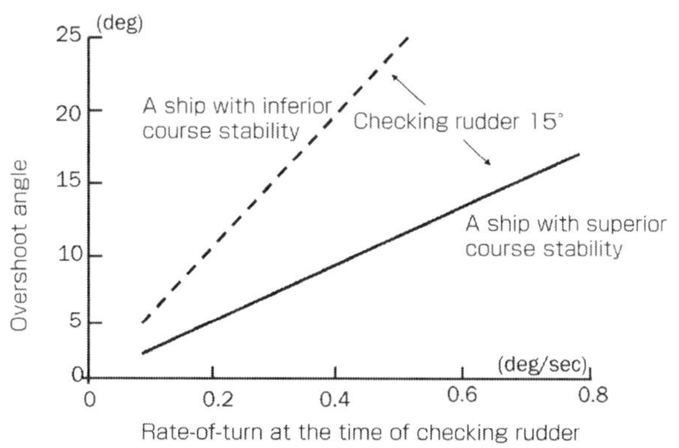

Fig. 2.1.18 Control of overshoot angle by checking rudder

Chapter 3 A Ship's Motions Accompanied by Rudder Action

1. Circling Motion after the Rudder has Turned

1.1 Turning Circle

☞ Refer to Vol. 4 Ship Handling Characteristics Data Base

Fig. 2.1.19 shows a locus drawn by the center of gravity of a ship when it is turned by a constant helm angle from its straight running state. After the rudder has turned, the hull starts circling changing its heading. And when ship's heading completes 360° turning, its hull also draws 360° locus. Such 360° circling locus of the center of gravity of turning ship is called ***turning circle***.

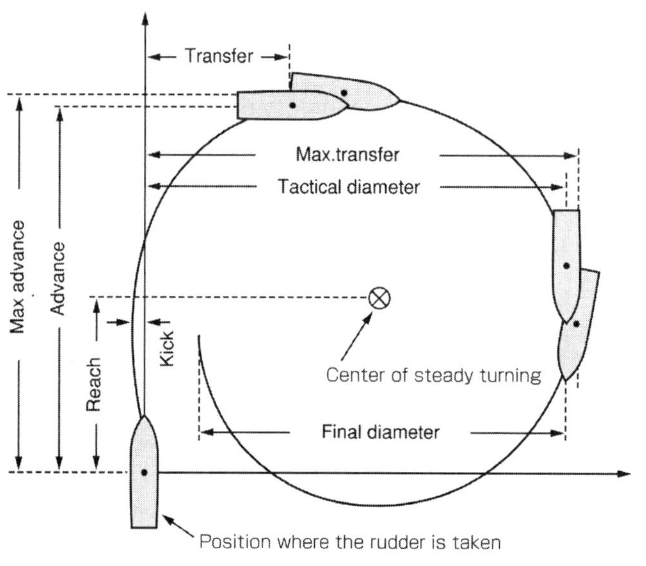

Fig. 2.1.19 Turning circle

1.2 Components of Turning Circle

A turning circle is composed of three elements, namely, ***advance, transfer*** and ***tactical diameter***. These are defined in relation to the turning angle of a ship's head as follows and they are important elements for persons engaged in ship handling to evaluate a ship's turning characteristics.

- Advance : longitudinal distance from the position where the rudder is taken to the position where the ship's head is turned 90°
- Transfer : transverse distance between the ship's original course and the ship's center of gravity where the ship's heading turned 90°
- Tactical Diameter : transverse distance between ship's original course and the ship's center of gravity where the ship's heading turned 180°

In addition to the above, ***maximum advance, maximum transfer*** and ***final diameter*** are defined as follows.

- Maximum Advance : longitudinal distance from the position where the rudder is taken to the position where ship's center of gravity has advanced most
- Maximum Transfer : transverse distance between the ship's original course and the ship's center of gravity where it is diverted most.
- Final Diameter : Diameter of ship's steady turning circle

Points for maximum advance and maximum transfer correspond to the positions where the ship's head turned further to ***drift angle*** in addition to 90° and 180°. Fig. 2.1.19 illustrates ***reach*** and ***kick*** together.

1.3 Elements Affecting Scale of the Turning Circle

The scale of the turning circle may vary depending on helm angle, ship type, ship's condition and the water depth (h) to ship's draft (d) ratio h/d. Focusing on the effects of these elements on tactical diameter, the followings can be said.

(1) Effect of Helm Angle

When a ship is turned by 35° of helm angle instead of 15° in the area where depth is enough to be free from shallow water effect, its tactical diameter decreases by about two thirds (2/3). The factor enabling turns in such small circles is the rise in index of turning ability "K" as rudder force "a" of $(a/b)= K$ in equation (2.1.16) is increased by larger helm angle.

(2) Effect of Ship Type and Size

When a ship is turned by 35° of the helm angle in deep water area, tactical diameter of a fat ship (VLCC with C_b=0.83) is about $3L$, while that of a fine ship (PCC with C_b=0.52) reaches $4.5L$ which is considerably larger compared with that of a fat type ship. This is because for a fine ship the index of turning ability "K" will decrease according to $(a/b)=K$ of equation (2.1.16) as the resistance of water "b" increase.

(3) Effect of Ship's Condition

In ballast conditions, turning becomes easier compared with a full load condition as the resistance of water decreases as ship's draft decreases. Although the extent of its effect may vary depending on helm angle, ship type and size, tactical diameter of VLCC in ballast condition decreases by 10 to 20% compared with a full load condition.

(4) Effect of the Ratio of Water Depth and the Ship's Draft

In shallow water areas where water depth to ship's draft ratio h/d is 1.3, tactical diameter increases to twice that in deep water area. This is because the resistance of water "b" in turning increases as depth decreases and index of turning ability "K" also decreases according to $(a/b)=K$ of equation (2.1.16).

(5) Effect of Trim

In the condition of trim by the stern, tactical diameter increases as the acting point of resistance in turning moves toward the stern and turning ability declines. Conversely, in the condition of trim by the head, it becomes smaller. The extent of the effect by trim may vary depending on helm angle and the scale of trim, tactical diameter may change by 10 to 20% in 1 to 2% of trim.

(6) Effect of Ship's Speed

For the effect of ship's speed, substituting non-dimensionalized "K'" $(=K\times L/V)$ for "K" in $R=V/K\delta$ of equation (2.1.28), $R=L/K'\delta$ is derived. This derived equation demonstrates that radius "R" in steady turning has no relation with ship's speed "V". From the foregoing fact, tactical diameter may not be affected significantly within normal ship's speed except in case of extremely low or high speed.

2. Kick of Ship's Stern

☞ Refer to Vol. 4 Ship Handling Characteristics Data Base

When a rudder is turned, the lift of the rudder acting on its center of gravity makes the ship drift. In the initial phase of steering, the ship starts to turn its heading and the hull drifts laterally due to the effects of lift. Lateral distance between the original course of the ship and its center of gravity is called ***kick***. When a ship's speed and helm angle are large, lift acting on the rudder grows larger and increases kick. When a ship's resistance to drift in water is small, the kick of ship's center of gravity also becomes larger. The kick of a ship's center of gravity generally ranges 2 to 4% of ship length.

Fig. 2.1.20 shows the relation between the original course of the ship and the ship's heading when kick of the ship's center of gravity reaches its maximum. ***Stern kick***, the outward deflection of a ship's stern from the original course, can be measured geometrically in relation to the scale of kick of center of gravity and the ship's heading at that moment. The scale of kick out at the stern generally ranges 10 to 13% (1/8L to 1/10L) of a ship's length.

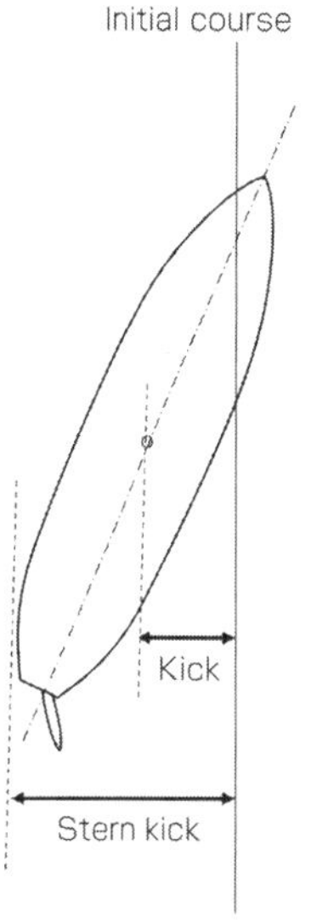

Fig. 2.1.20
Stern kick

Utmost caution should be exerted for a ship's stern kick when wharfs, buoys or any other obstacles exist on the opposite side to the side to be steered, to avoid contact with them. However, this effect can be favorably afforded in the case of man overboard to protect the person from being hit by the propeller, otherwise avoiding floating objects close to the ship's side by taking the rudder to the board of the person overboard or the object to be avoided.

It can be deemed that stern kick reaches its maximum when

a ship's heading is turned to 5° to 10°. Seeing the relation between the scale of stern kick and the duration of time after the rudder is turned, although it may vary depending on ship's speed at that time, its maximum may be attained 40 to 60 seconds after the rudder is turned in ships of good steerability and about 90 seconds in ships of poor steerability under S/B Full speed. Under slow speed, it will take 1.5 to 2 times longer in ships of good steerability and 2 to 3 times in ships of poor steerability. That is to say, to do optimum ship handling making best use of this effect, persons engaged in ship handling are required to be familiar with the data of relationship among their ship's heading deflection, time duration, and stern kick beforehand.

3. Drift Angle

☞ Refer to Vol. 4 Ship Handling Characteristics Data Base

Drift angle means an angle between a ship's longitudinal center line and the direction of current around its hull. When a ship is turning, lines of currents coming from the opposite directions to the speed vectors of each part of the hull impinge on the hull at varied angles depending on the ship's longitudinal part. In general, drift angle means the angle between the ship's heading and the direction of current at the ship's center of gravity.

In Fig. 2.1.21, let O, P, G and Q respectively be the center of a turning circle, the foot of the perpendicular drawn from the center of the turning circle O to the ship's longitudinal center line, the ship's center of gravity and a point close to the stern, and let drift angle at each point be "β_P", "β_G" and "β_Q", then the following equations can be derived.

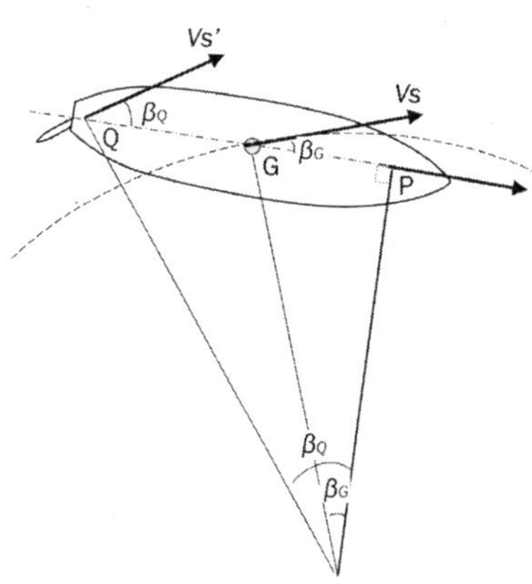

Fig. 2.1.21 Drift angle

As you can see from the equation (2.1.34), the angle of current impinging on the hull grows larger as an arbitrary point on the ship's longitudinal center line comes closer to the stern and

$$\left.\begin{aligned}\beta_P &= \tan^{-1}\left(\frac{0}{OP}\right) = 0 \\ \beta_G &= \tan^{-1}\left(\frac{GP}{OP}\right) \\ \beta_Q &= \tan^{-1}\left(\frac{QP}{OP}\right)\end{aligned}\right\} \cdots\cdots (2.1.34)$$

the angle achieves its maximum on a point of the stern. And the ship receives current from both sides, at any point from P to the bow, the current comes from inside and at any point from P to the stern, from outside, as it turns swinging its head inward and stern outward. Therefore, signs (±) of drift angle "β" will be opposite forward or backward of the point P.

Drift angle is larger on fat ships which have good turning ability. And their drift angle in turning by hard over rudder in deep water ranges 20° to 25°, while that of fine ships which turning ability is low range 10° to 15°. However, it will be drastically reduced to a few degrees in shallow water where the water depth "h" to ship's draft "d" ratio h/d is 1.3 due to the shallow water effects which causes high water resistance and invites difficulties in turning.

4. Reduction of Speed

☞ Refer to Vol. 4 Ship Handling Characteristics Data Base

When the rudder is turned, a ship's speed decreases as its turning motion develops. There are three reasons a ship's speed reduces in turning motion. The first one is due to the decline of propeller effect. Propeller effect decreases because of the reduction of propeller revo-

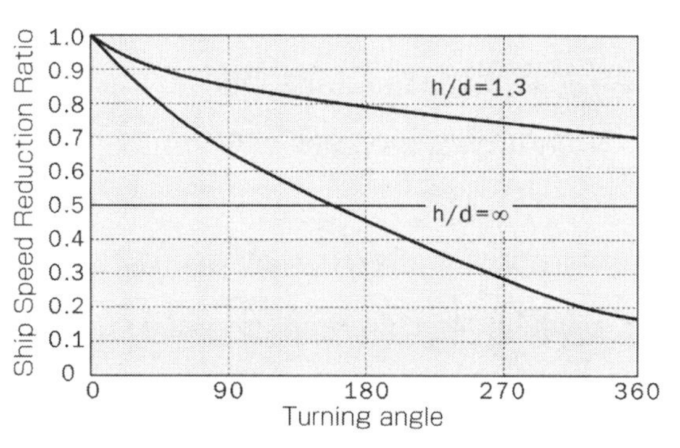

Fig. 2.1.22 Speed reduction due to turning motion

lution and the speed of propeller current due to increased angle of flow to the propeller as the turning of ship progresses. The second is the drag force acting on the face of rudder. And the third is an increased hull resistance in turning. In turning, the resistance of ship's hull increases as the ship sits obliquely in current increasing its form resistance compared with that when the ship is running straight ahead. In addition, a longitudinal component of centrifugal force of the turning motion acting on the ship's center of gravity constitutes a part of advance resistance.

Fig. 2.1.22 shows the general trend of speed reduction in turning motion by hard over rudder in deep and shallow water. Reduction of speed appears conspicuously as turning motion progresses. Average speed reduction in deep water when ships have completed a 90° turn is about 65% compared with the speed when the ship is running straight ahead regardless of ship size and type. And it increases to about 85% in shallow water in which the water depth "h" to ship's draft "d" ratio h/d is 1.3.

5. Heel

5.1 Inward Heel Immediately after Rudder Turns

When the rudder is turned to "δ" while the ship is going straight ahead, lift $N\cos\delta$ will be created by normal force "N" acting on the face of the rudder. Assuming the ship is listed to "θ", then a horizontal component of lift will be $N \cdot \cos\delta \cdot \cos\theta$ and let the correction of the horizontal component of force acting on the ship due to the heel angle "θ" be $\cos\theta \fallingdotseq 1$, deeming the heel angle is quite small, the resultant horizontal component of lift $N \cdot \cos\delta$ acts not only to give the ship turning moment but also to push the ship outward opposite to its steered board.

Immediately after the rudder is turned, the hull keeps running straight ahead for a while. At this moment inertia force $(M+M_y)\ \dot{v}$, where "M": ship's mass, "M_y": added mass of lateral motion and lateral force $N \cdot \cos\delta$ acting to push the hull outward are in equilibrium. From this fact the following equation is derived.

$$N \cdot \cos\delta = (M + M_y)\dot{v} \quad \cdots\cdots (2.1.35)$$

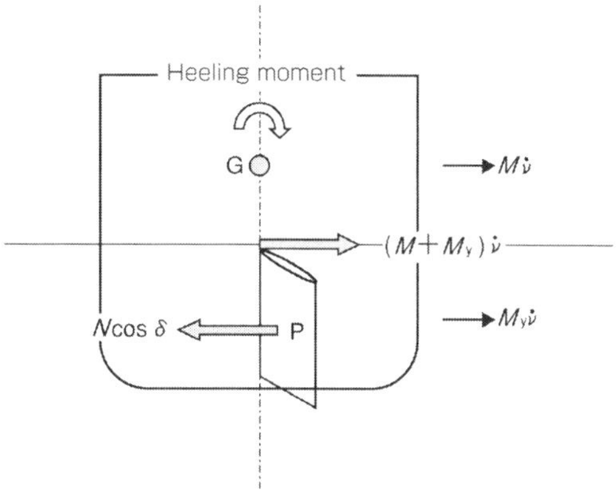

Fig. 2.1.23 Inward heel

As shown in Fig. 2.1.23, giving a positive sign to clockwise moment, moments around the ship's center of gravity can be expressed as below (refer to the relation between acting forces and moment levers shown in Table 2.1.1).

Table 2.1.1 Acting forces and moment levers (Inward heel)

Acting force	Acting point	Moment lever from G
Horizontal component of normal force of rudder	Almost center of rudder	GP
Horizontal component of inertia force including added mass	About the mid point between G and P	$\frac{1}{2}$GP

$$\begin{aligned}\text{Heeling moment } &= N \cdot \cos\delta \cdot GP - (M + M_y)\dot{v} \times \frac{1}{2} GP \\ &= N \cdot \cos\delta \left(GP - \frac{1}{2} GP \right) \\ &= \frac{1}{2} N \cdot \cos\delta \cdot GP \geq 0 \quad \cdots\cdots (2.1.36)\end{aligned}$$

As the sign of the equation (2.1.36) is positive, the ship will list to the same direction of turning immediately after its rudder was turned. This heel is called as ***inward heel.***

At this moment, the ship lists to an angle which heeling moment equals to righting moment $W \cdot GM \cdot \sin\theta$. These two moments in equilibrium can be expressed as in the equation (2.1.37) and the consequential amount of inward heel will be derived from the equation (2.1.38).

$$W \cdot GM \cdot \sin\theta = \frac{1}{2} N \cdot \cos\delta \cdot GP \quad \cdots\cdots (2.1.37)$$

$$\therefore \sin\theta = \frac{N \cdot \cos\delta \cdot GP}{2W \cdot GM} \quad \cdots\cdots (2.1.38)$$

The faster the ship speed, the larger the angle of rudder and the larger the trim by the stern in which condition the rudder is placed deeper in water, the larger the inward heel becomes. In addition, the higher the ship's center of gravity and the smaller the "GM", the larger the inward heel becomes. In general, inward heel of merchant ship is small and does not last long.

5.2 Outward Heel in the Progression of Turning

As turning progresses, the ship's center of gravity draws a circle keeping a constant drift angle "β". At this moment, on the face of rudder the horizontal component of lift $N \cdot \cos\delta$ and on the ship's center of gravity, centrifugal force "E" are acting. From Fig. 2.1.24, the lateral component of centrifugal force is $E \cdot \cos\beta$.

The lateral component of centrifugal force $E \cdot \cos\beta$ and the horizontal component of hydraulic pressure acting on the ship's side "F" affect it simultaneously. While steady turning, this hydraulic pressure "F" and the total of the horizontal component of lift $N \cdot \cos\delta$ and lateral component of centrifugal force $E \cdot \cos\beta$ are balanced. From the above the following equations will be derived.

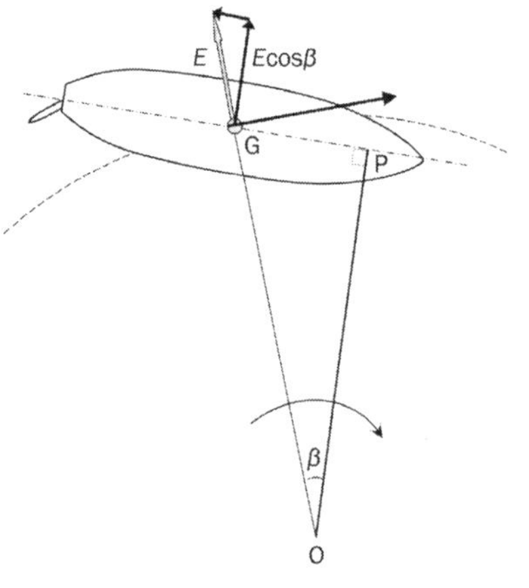

Fig. 2.1.24 Centrifugal force acting on ship's center of gravity

$$F = N \cdot \cos\delta + E \cdot \cos\beta$$
$$= N \cdot \cos\delta + \frac{M}{R} V^2 \cos\beta \quad \cdots\cdots (2.1.39)$$

where

M : ship's mass

R : radius of steady turning circle

V : ship speed while steady turning

From the moments around G in Fig. 2.1.25 where letting clockwise moments be positive and the relation between the acting forces and their acting points as shown in Table 2.1.2, the following equation can be derived.

$$\text{Heeling moment} = N \cdot \cos\delta \cdot GP - F \cdot GQ$$
$$= N \cdot \cos\delta \cdot GP - \left(N \cdot \cos\delta + \frac{M}{R} V^2 \cos\beta\right) GQ$$
$$= N \cdot \cos\delta(GQ + PQ) + \left(N \cdot \cos\delta + \frac{M}{R} V^2 \cos\beta\right) GQ$$
$$= -\left(\frac{M}{R} V^2 \cos\beta \cdot GQ - N \cdot \cos\delta \cdot PQ\right) \quad \cdots\cdots (2.1.40)$$

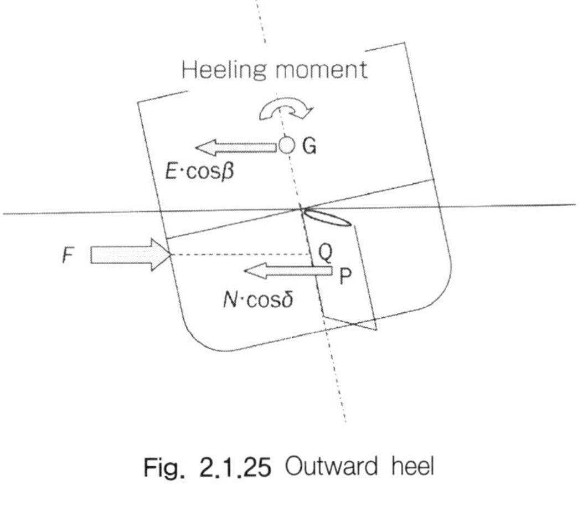

Fig. 2.1.25 Outward heel

Table 2.1.2 Acting forces and moment levers (Outward heel)

Acting force	Acting point	Moment lever from G
Lateral component of centrifugal force	Ship's center of gravity	0
Horizontal component of normal force of the rudder	Almost center of rudder	GP
Horizontal component of hydraulic pressure	About 1/2 the ship's draft	GQ

As the acting points of lift P and hydraulic pressure Q are very close to each other, it can be deemed as $PQ \fallingdotseq 0$. Then the sign of heeling moment in the equation (2.1.40) will be negative $(-)$. This means the hull lists outward opposite to the initial phase of steering, as centrifugal force acting on the hull increases in the progression of the turning motion. This heel is called ***outward heel.***

At this moment, the ship lists to an angle in which the heeling moment as shown in equation (2.1.40) equals to the righting moment $W \cdot GM \cdot \sin\theta$. These two moments in equilibrium can be expressed as shown in the equation (2.1.41) and the consequential amount of outward heel will be derived from the equation (2.1.42).

$$W \cdot GM \cdot \sin\theta = \frac{M}{R} V^2 \cos\beta \cdot GQ - N \cdot \cos\delta \cdot PQ \quad \cdots\cdots (2.1.41)$$

In the above equation, as the two points of P and Q are almost in the same place and $W = M/g$ (W: ship's weight, M: ship's mass and g: gravitational acceleration), therefore,

$$\therefore \sin\theta = \frac{V^2 \cos\beta}{g \cdot R \cdot GM} GP \quad \cdots\cdots (2.1.42)$$

This means that while steady turning as expressed by the above (2.1.42), the faster the ship speed and the larger the angle of the rudder, the larger the outward heel becomes. In addition, the higher the ship's center of gravity and the smaller the "GM", the larger the outward heel grows.

5.3 Maximum Outward Heel

Fig. 2.1.26 shows the transitional phase from inward heel to outward heel as time elapses after steering. In the transitional phase from inward heel to outward heel, caution is necessary as the outward heel angle grows larger compared with the normal heel angle. Such a large outward heel angle created by an inertial force is called ***maximum outward heel.***

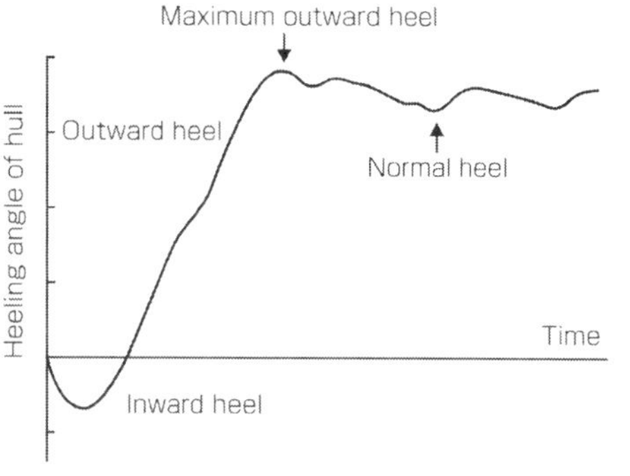

Fig. 2.1.26 Maximum outward heel

In turning, if the rudder should be returned abruptly to midship during such a large outward heel, the outward heel will be increased as the moment created by rudder force as shown in the second item of the right side of the equation (2.1.40) dissipates suddenly. And if the rudder should be turned to the opposite board, the outward heel will be aggravated further because initial inward heel will be added. To avoid such large heel due to inertial forces, it is important to ensure enough stability, to keep a slow speed and to turn the ship with small helm angle.

6. Pivoting Point

6.1 Definition of Pivoting Point

When a ship is turning, an intact point that is not affected by the dynamic change by its motion exists on some point of the longitudinal center line of the ship. Such a point is called the ***pivoting point***, the center on which the ship rotates. In a more physical expression, it can be defined as "a point on the longitudinal centerline or its extension of a ship in turning where the velocity of drift and consequential drift angle are zero".

6.2 Location of Pivoting Point

When any instantaneous lateral external forces "F" such as turning of rudder when a ship is running straight or a tug's push while it is dead in the water is given to a point C distant from the ship's center of gravity G, its impulse "S" will be given by the following equation.

$$S=\int_{t1}^{t2} F \cdot dt \quad \cdots\cdots (2.1.43)$$

As the dynamic instantaneous change of G by the above external force, "Δv" in the same direction of external force and "Δr" in the same direction of external force moment will be created. In addition, dynamical changes will be created all over the hull corresponding to the distance from G. However, one point where Δv is zero ($\Delta v=0$) is left free from such effect of dynamical change. Such a point is the pivoting point. Let the pivoting point be P, dynamical conditions about P can be expressed as follows.

$$\Delta v - GP \cdot \Delta r = 0 \quad \cdots\cdots (2.1.44)$$

On the conditions that,

$$\Delta v = \frac{S}{M} \quad \cdots\cdots (2.1.45)$$

$$\Delta r = \frac{S \cdot GC}{I} \quad \cdots\cdots (2.1.46)$$

where

$$I = M \cdot k^2 \quad \cdots\cdots (2.1.47)$$

In the above, "I", "M" and "k" respectively represent turning moment of inertia, ship's mass and radius of gyration of turning moment of inertia. And substituting expressions (2.1.45), (2.1.46) and (2.1.47) for equation (2.1.44),

$$\frac{S}{M} - GP\frac{S \cdot GC}{I} = 0 \quad \cdots\cdots (2.1.48)$$

$$GP \cdot GC = \frac{I}{M}$$

$$GP = \frac{k^2}{GC} \quad \cdots\cdots (2.1.49)$$

Accordingly, if we can estimate a point where instantaneous external force acts on the ship moving straight or lying dead in water, the location of the pivoting point can be identified from "GP" derived from the equation (2.1.49).

Radius of gyration of turning moment of inertia "k" is also called ***radius of gyration of yawing***. In cargo ships, it will be defined by longitudinal distribution of the cargo load and its values generally range $1/4L$ to $1/3L$ (L: ship length).

When this instantaneous force continues to affect thereafter, the ship begins to move laterally at a speed "v" and to turn around the ship's center of gravity at a rate-of-turn "r" simultaneously. As the ship's pivoting point while turning is defined that it exists on the ship's longitudinal center line where ship's lateral velocity is zero, its dynamical conditions can be expressed as follows.

$$v - GP \cdot r = 0 \quad \cdots\cdots (2.1.50)$$

$$\therefore\ GP = \frac{v}{r} \quad \cdots\cdots (2.1.51)$$

Conversely, the location of a ship's pivoting point in the transitional phase in the process that the ship is going to be drifted and turned from the state of going straight or lying dead in water can be derived from the equation (2.1.51), if speed of drift "v" and rate-of-turn "r" can be identified.

In addition, the equation (2.1.51) can be rewritten as (2.1.52) using the speed vector at the ship's center of gravity "V" and subsequent equation $v = V\sin\beta$ (β: drift angle at the ship's center of gravity).

$$GP = \frac{v}{r} = \frac{V\sin\beta}{r} \quad \cdots\cdots (2.1.52)$$

Further, while steady turning, it can be derived using radius of steady turning circle "R" and $R = V/r$ referring to Fig. 2.1.15. Then, the above equation (2.1.52) can be rewritten as (2.1.53) below.

$$GP = \frac{v}{r} = \frac{V\sin\beta}{r} = R \cdot \sin\beta \quad \cdots\cdots (2.1.53)$$

This also demonstrates that the pivoting point is the foot of the perpendicular on the ship's longitudinal center line drawn from its steady turning center as can be seen in Fig. 2.1.21.

6.3 Shift of Pivoting Point in the Transitional Period

As you know from the equations (2.1.49) and (2.1.51), the pivoting point does not exist while the ship keeps a straight course or is lying dead in the water. It appears somewhere all at once on the ship's longitudinal center line at the instance when an external force is added and moves on the ship's longitudinal centerline depending on the development of "v" and "r".

In the course of progression of motion, hydraulic pressure, centrifugal force and other external force and moment will be added to the ship. Impulse and moment of impulse due to the effect of each external force along with the duration of time can be expressed as follows.

$$\text{Impulse} = \sum_i \int_{t_1}^{t_2} F_i dt \quad \cdots\cdots (2.1.54)$$

$$\text{Moment of impulse} = \sum_i GC_i \int_{t_1}^{t_2} F_i dt \quad \cdots\cdots (2.1.55)$$

As the ratio of impulse to moment of impulse represents the acting point of the resultant force of various external forces affecting to the ship, let the distance from the ship's center of gravity be "GC'",

$$GC' = \frac{\sum_i GC_i \int_{t_1}^{t_2} F_i dt}{\sum_i \int_{t_1}^{t_2} F_i dt} \quad \cdots\cdots (2.1.56)$$

As the change of momentum for this period equals to the impulse of external forces affecting the ship for the period,

$$M \cdot v = \sum_i \int_{t_1}^{t_2} F_i dt \quad \cdots\cdots (2.1.57)$$

Further, the change of angular momentum equals to the moment of impulse of external forces affecting the ship for this period,

$$I \cdot r = \sum_i GC_i \int_{t_1}^{t_2} F_i dt \quad \cdots\cdots (2.1.58)$$

From the ratio of the equations (2.1.57) and (2.1.58), and the equation (2.1.56), the following equation can be derived.

$$\frac{I \cdot r}{M \cdot v} = \frac{\sum_i GC_i \int_{t_1}^{t_2} F_i dt}{\sum_i \int_{t_1}^{t_2} F_i dt} \quad \cdots\cdots (2.1.59)$$

Substituting equation $I = Mk^2$ of equation (2.1.47) for the above left side,

$$\frac{k^2}{GC'} = \frac{v}{r} = GP \quad \cdots\cdots (2.1.60)$$

Thus, when external forces continue to affect, the pivoting point "GP" changes depending on the acting point of the resulting force "GC'".

6.4 Example to Deduce Location of the Pivoting Point after Steering

(1) Shift of the Pivoting Point in Ahead Turning

Table 2.1.3 Specifications of container ship used for estimation

Ship's type		2,800 TEU container ship	
Lpp		230.0m	
Breadth		32.2m	
		Full Load	Ballast
Draft		11.5m	6.34m
Trim		even keel	3.5m B/S
Center of gravity		-5.55m	-11.9m
Displacement		53,875ton	25,200ton
Spped	Nav. Full	21.67kts	23.74kts
	S/B Full	10.62kts	11.74kts

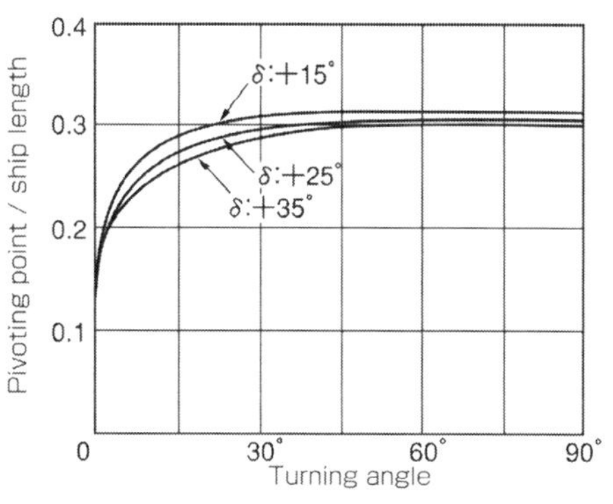

Fig. 2.1.27 Estimated pivoting point (2,800 TEU container ship in ahead turning by helm angles 15°, 25° and 35° in full load condition at Nav. Full speed 11.7 kts in deep water)

Fig. 2.1.27 shows the shift of the pivoting point of a container ship in full load condition derived from the equations (2.1.49) and (2.1.51) responding to helm angles 15°, 25° and 35° until it is turned to 90° in deep water at navigation full speed, which specifications are given in Table 2.1.3.

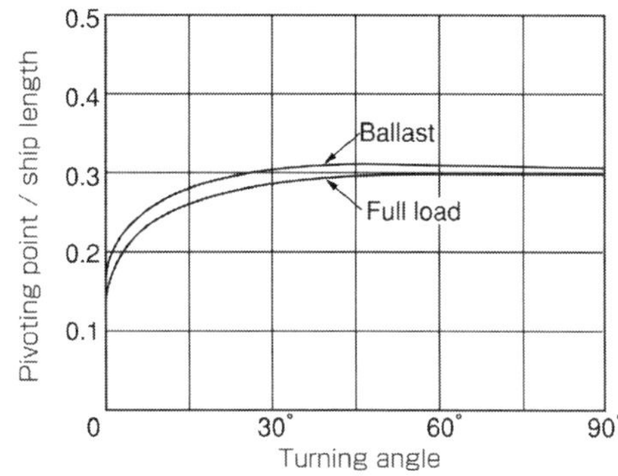

Fig. 2.1.28 Estimated pivoting point (2,800 TEU container ship in ahead turning by helm angle 35° in full load condition at S/B Full speed 10.6 kts in deep water and in ballast condition at S/B Full speed 11.7 kts in deep water)

In estimating the location of the pivoting point, the equation (2.1.49) was used when the ship was going straight ahead while equation (2.1.51) was used instead after the ship began to turn. In estimation, the radius of gyration for turning moment of inertia is assumed $1/4L$ (L: ship length) uniformly.

Fig. 2.1.27 illustrates that the pivoting point appears in a place close to $GP=0.13L$ immediately after steering, and moves forward as turning progresses and settles on a place close to $GP=0.3L$. Fig. 2.1.28 shows the shift of the pivoting point of the same container ship as the above in full load condition and in ballast condition, in deep water, at S/B Full speed and helm angle 35°.

The curves in Fig. 2.1.27 and Fig. 2.1.28 indicate that the change of the pivoting point is not much affected by helm angle, ship speed, and the conditions of the ship.

(2) Shift of the Pivoting Point in Astern Turning

Fig. 2.1.29 shows the deduced results of the shift of the pivoting point when the above container ship in full load condition is turned astern in deep water at speed 3.5 knots by helm angle 15°, 25° and 35°. Fig. 2.1.29 illustrates that the pivoting point appears in a place close to $GP=0.13L$ immediately after steering, and moves towards the stern as turning progresses and it retards close to the ship's center of gravity when the ship's heading turns to 15° and settles on a position at about $GP = -0.1L$.

This is because the direction of drift speed at the ship's center of gravity which is the same as steered board immediately after the rudder is turned will be reversed as turning progresses. By this reason, while turning astern the ship circles swinging its bow largely outward. In the case of astern turning, it also can be said that the shift of the pivoting point is not much affected by helm angle, ship speed and the conditions of the ship.

(3) VLCC Case

Here, we will see a different type of ship from the foregoing container ship. Fig. 2.1.30 shows the deducted result of the shift of the pivoting point of a 230,000 DWT VLCC in full load condition until it is turned to 90° at full navigation speed of 15.4 knots in deep water by helm angle 15°, 25° and 35°, which specifications are given in Table 2.1.4. Fig. 2.1.30 illustrates that a pivoting point appears immediately after steering and locates at around $GP=0.12L$ like the foregoing container ship. It moves forward as turning progresses. In a VLCC it takes longer time until its pivoting point settles. At around the time when the ship starts steady turning, it comes close to $GP=0.4L$ which is a little forward compared with the container ship.

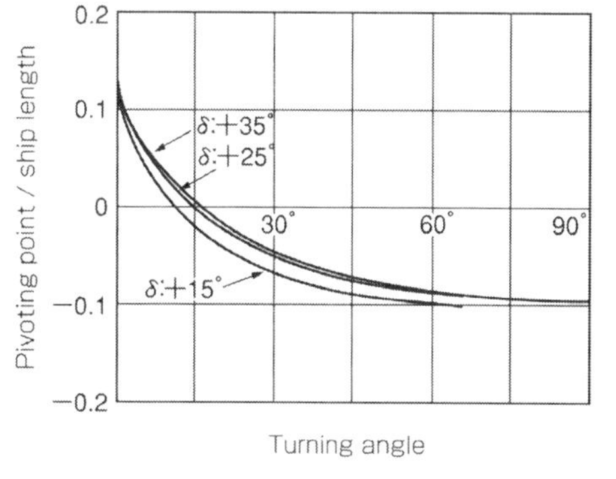

Fig. 2.1.29 Estimated pivot point (2,800 TEU container ship in astern turning by helm angles 15°, 25° and 35° in full load condition at 3.5 kts in deep water)

From the difference of the characteristics between the shift of the pivoting point of a container ship and those of a VLCC during steady turning and also from the time required for the pivoting point to settle on a certain position, you may deepen your understanding of the fact that high speed type container ships are generally high in responsiveness to the helm but low in turning ability while fat type VLCCs are generally low in responsiveness to the helm but high in turning ability.

Table 2.1.4 Specifications of VLCC used for estimation

Ship's type		230,000 DWT VLCC	
Lpp		305.0m	
Breadth		58.9m	
		Full Load	Ballast
Draft		19.33m	8.17m
Trim		0.03m B/S	5.38m B/S
Center of gravity		+8.91m	+11.64m
Displacement		269,799ton	104,255ton
Speed	Nav. Full	15.4kts	17.2kts
	S/B Full	12.1kts	13.5kts

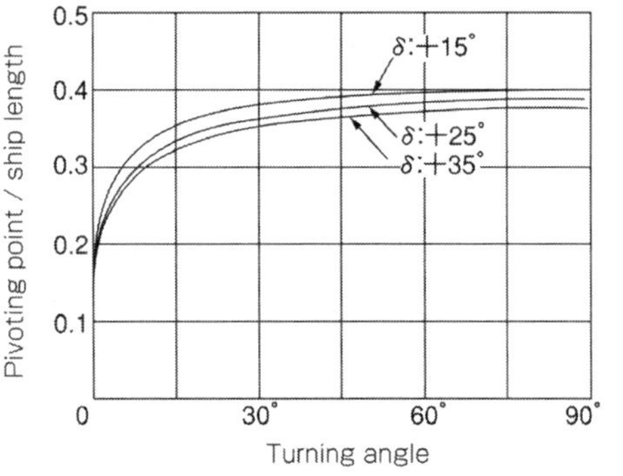

Fig. 2.1.30 Estimation of pivoting point (230,000 DWT VLCC in ahead turning by helm angles 15°, 25° and 35° in full load condition at Nav. Full speed 15.4 kts in deep water)

6.5 General Characteristics of Pivoting Point

In relation to equation (2.1.49),

a) The pivoting point does not exist while a ship is going straight ahead or lying dead in the water. It appears somewhere on the ship's longitudinal center line at the moment when an external force is added to the hull and thereafter moves depending on the progress of a ship's motion.

b) The location of a pivoting point is defined depending on the point where an external force is added. And the farther the acting point of it from the ship's center of gravity, the closer the pivoting point to the ship's center of gravity locates.

c) When an external force is added forward of the ship's center of gravity, a pivoting point at first locates behind the ship's center of gravity. Conversely, when an external force is added behind the ship's center of gravity, a pivoting point at first locates forward of the ship's center of gravity.

In relation to equation (2.1.51)

d) When a ship turns slightly making large lateral movements, sometimes its pivoting point exists somewhere remote point on the extended ship's longitudinal center line.

e) The shift of the pivoting point in ahead turning, it appears once close forward to the ship's center of gravity, thereafter gradually moves to forward. In this instance, in ships with high turning ability a pivoting point exists farther forward.

f) In astern turning, a pivoting point once appears forward close to the ship's center of gravity, thereafter moves aft as turning progresses and finally settles behind the ship's center of gravity.

Generally,

g) The shift of the pivoting point is not affected by the helm angle taken, ship speed, the conditions of the ship and the thrust power of tugs.

Chapter 4 Maneuverability Tests

1. Zigzag Maneuver Test

☞ Refer to Vol. 4 Ship Handling Characteristics Data Base

The indices of responsiveness to the helm "T" and turning ability "K" are derived from the result of ***zigzag maneuver test***. The test method and process to deduce the values of "T" and "K" are as discussed below.

1.1 Test Method

i) Run the ship straight ahead at a constant speed.

ii) Order starboard (or port optionally) rudder $\delta_0°$ and have the helmsman keep $\delta_0°$.

iii) When the ship's heading has turned to starboard where $\psi=\delta_0$, $(t=t_2)$, order port rudder $\delta_0°$ and have the helmsman keep $\delta_0°$.

iv) Although the ship's head continues to turn starboard for a while after the rudder has been reversed, when the ship's heading has turned to port $\delta_0°$ from its original course where $\psi=\delta_0$, $(t=t_4)$, order starboard rudder $\delta_0°$ again and have the helmsman keep $\delta_0°$.

v) When the ship's heading has turned to starboard where $\psi=\delta_0$, $(t=t_6)$, order port rudder $\delta_0°$ and have the helmsman keep $\delta_0°$.

vi) The test will complete when the ship's heading returns to the original course.

From the above test, the chronological relation between helm angle "δ" and turning angle of the ship's head "ψ" can be obtained as shown in Fig. 2.1.31.

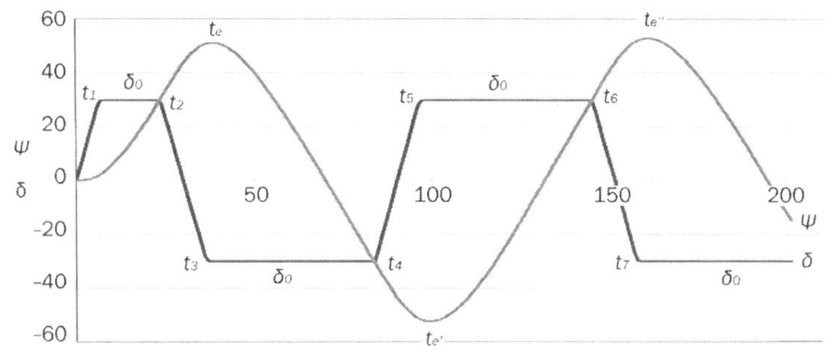

Fig. 2.1.31 Zigzag maneuver test

1.2 Deduction of "T", "K"

As there are some ships that couldn't run straight even when their rudders were set to midship, for them the linear equation of ship maneuvering motion, the equation (2.1.61) in which checking helm angle "δ_r" is taken into account to keep ships running straight is used to deduce "T" and "K".

$$T\dot{r}+r=K(\delta+\delta_r) \quad \cdots\cdots (2.1.61)$$

i) Integrating the equation (2.1.61) from "α" to "β" in respect of time, the equation (2.1.62) is obtained.

$$T\{r(\beta)-r(\alpha)\}+\{\psi(\beta)-\psi(\alpha)\}=K\int_{\alpha}^{\beta}\delta dt+K\delta_r(\beta-\alpha) \quad \cdots\cdots (2.1.62)$$

In the equation (2.1.62), substituting $t=0$ for "α" and the times $t=te'$, $t=te''$ corresponding to $r=0$ for "β" respectively,

$$\psi(te')=K\int_{0}^{te'}\delta dt+K\delta_r \cdot te' \quad \cdots\cdots (2.1.63)$$

$$\psi(te')=K\int_{0}^{te''}\delta dt+K\delta_r \cdot te'' \quad \cdots\cdots (2.1.64)$$

In the above integrations of $\int_{0}^{te'}\delta dt$ and $\int_{0}^{te''}\delta dt$, chronological data of helm angle "δ" can be derived by trapezoidal approximation. For instance,

$$\int_0^{te'} \delta dt = \delta_1 \left\{ \frac{(t_3+t_2)}{2} + \frac{t_1}{2} \right\} + \delta_2 \left\{ \frac{(t_5+t_4)}{2} - \frac{(t_3+t_2)}{2} \right\} + \delta_3 \left\{ te' - \frac{(t_5+t_4)}{2} \right\} \quad \cdots\cdots (2.1.65)$$

Reducing the simultaneous equations of (2.1.63) and (2.1.64), "K" and "δ_r" are obtained. Let "K" obtained here be "K_{68}".

ii) In the equation (2.1.62), paying attention to other time spans of integration, ($t_4 \sim te'$) and ($t_6 \sim te''$), "T" will be obtained by using "K" and "δ_r" obtained above.

When $\alpha = t_4,\ t_6$ and $\beta = te',\ te''$, equation (2.1.62) will be as follows.

$$-T \cdot r(t_4) - \{\psi(t_4) - \psi(te')\} = K\int_{t_4}^{te'} \delta dt + K\delta_r (te' - t_4) \quad \cdots\cdots (2.1.66)$$

$$-T \cdot r(t_6) - \{\psi(t_6) - \psi(te'')\} = K\int_{t_6}^{te''} \delta dt + K\delta_r (te'' - t_6) \quad \cdots\cdots (2.1.67)$$

Reducing the simultaneous equation of (2.1.66) and (2.1.67) substituting "K_{68}" and "δ_r" obtained previously, two "T" values are obtained. Let the mean of these values be "T_{68}"

iii) Further, paying attention to other time spans of integration, ($0 \sim te$) and ($t_2 \sim te$), "T" and "K" are obtained. When $\alpha = 0$ and $\beta = te$, equation (2.1.62) can be rewritten as follows:

$$\psi(te) = K\int_0^{te} \delta dt + K\delta_r \cdot te \quad \cdots\cdots (2.1.68)$$

Substituting "δ_r" obtained in i) for the equation (2.1.68), "K" is derived. Let this "K" be "K_4".

When $\alpha = t_2$ and $\beta = te$, the equation (2.1.62) will be as follows:

$$-T \cdot r(t_2) - \{\psi(t_2) - \psi(te)\} = K\int_{t_2}^{te} \delta dt + K\delta_r (te - t_2) \quad \cdots\cdots (2.1.69)$$

Substituting "δ_r" obtained in i) and "K_4" obtained in the above for the equation (2.1.69), "T" is derived. Let this "T" be "T_4"

iv) "K_4" and "T_4" represent the values of "T" and "K" at the first extreme and "K_{68}" and "T_{68}" represent the second extreme in the chronological data in zigzag maneuver test. Although the mean of these two extremes are often used, "K_{68}" and "T_{68}" seem to represent a ship's characteristics the most.

2. Spiral Test

☞ Refer to Vol. 4 Ship Handling Characteristics Data Base

For a ship with good course stability, turning motion can be dampened once its rudder has been returned to midship. By contrast, for a ship with poor course stability turning motion cannot be easily stopped only by returning its rudder midship. Moreover, even when checking rudder is applied, it takes some time to be settled on; otherwise the turning motion will not stop. To evaluate such characteristics, ***spiral test*** is conducted. Method of spiral test is discussed below.

2.1 Test Method

i) Make the ship run straight ahead at a constant speed.

ii) Order starboard (or port optionally) rudder $\delta_0°$ and have the helmsman keep $\delta_0°$.

iii) Take measurement of the rate-of-turn when the ship starts steady turning.

iv) Decrease the helm angle and keep a decreased angle. Then, when the ship starts steady turning, take measurement of the rate-of-turn at that time.

v) Decrease helm angle step by step and repeat the same procedure as iv).

vi) When the above procedures from starboard to port have been completed, repeat the same procedures from port to starboard.

2.2 Evaluation of Course Stability

Draw δ-r curve plotting helm angle "δ" and corresponding rate-of-turn "r"

from the above test result on the coordinate which vertical axis represents for rate-of-turn and horizontal axis represents for helm angle.

As shown in Fig. 2.1.32 (a), when δ-r curve passes the origin; its course stability is evaluated as good. By contrast, the course stability of a ship, even after reverse rudder is taken, the previous turning inertia is slow to diminish and shows characteristics as in Fig. 2.1.32 (b), is evaluated as poor course stability. In Fig. 2.1.32 (b), the range between "δ_1" and "δ_2" is called the width of an unstable loop. The broader the width of the unstable loop, the poorer the course stability becomes.

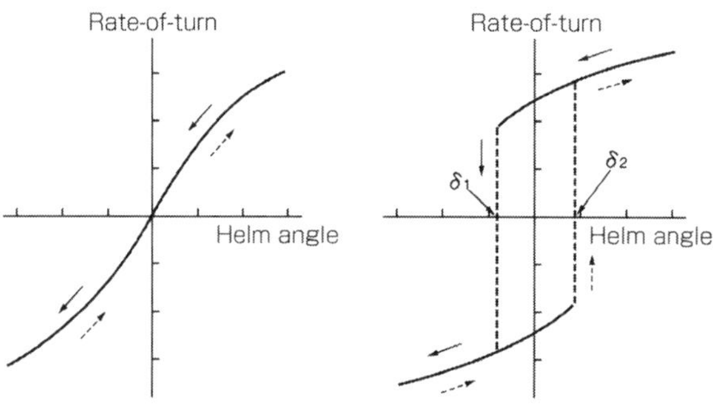

(a) Good course stability (b) Poor course stability

Fig. 2.1.32 Course stability test

3. Turning Test

☞ Refer to Vol. 4 Ship Handling Characteristics Data Base

Turning test is a test to turn a ship by a certain constant helm angle to confirm the circling locus of the ship's center of gravity. In general, to measure a ship's ultimate turning ability, the test is conducted by its maximum available helm angle until the ship makes 360° turn. The method of testing is quite simple so that the test may be made in various combinations of helm angles and ship speeds.

In the past, measurement of the locus of the ship's center of gravity in turning tests was made relying on visual observation of land marks or temporary buoys otherwise numerical integration based on the observed data for ship speeds and turning angles. However, today continuous and precise position fixing is available using the GPS.

From turning tests, we can obtain various important data for ship maneuvering, not only static data such as ***advance, transfer, tactical diameter, maximum advance, maximum transfer*** and ***final diameter*** but also dynamical data such as the development or change of turning motion, drift angle, kick and reduction of speed.

Ⅱ
Maneuvering by Main Engine and Propeller

Chapter 1 Main Engine

1. Type of Main Engine

1.1 Diesel Engine

Diesel engines which are driven by low price heavy oil are widely used for merchant ships. Diesel engines are internal combustion engines that use the heat of compression to initiate ignition burning the fuel injected into the combustion chamber. Then, the pressure of burnt fuel moves the piston inside the cylinder. By repeating the cycle of induction, compression, ignition and exhaust inside the cylinder, the piston continues to move reciprocally. Such motion of the pistons is converted to turning motion via connecting rod and crank connected to the propeller shaft and finally makes the propeller turn.

To a certain extent, better propulsive efficiency can be gained by the lower propeller revolution; recently larger ships fitted with larger diameter propellers connected to low speed diesel engines are increasing. Engines which revolutions ranging from 100 to 300 per minute are sorted into the category of low speed diesel engines. Today, for larger container ships low speed diesel engines which power per unit exceeds 50,000 HP are fitted.

Engines which revolution ranging from 300 to 1,000 per minute are sorted into the category of medium speed diesel engine. As the revolution of this type of engine is lowered by reduction gears, optimum propulsive efficiency can be gained depending on the selection of it. Furthermore, because of their lighter weight and smaller volume in addition to good combustion efficiency, they are widely used for ships of medium and small types, passenger, ferry and RORO (Roll on Roll off) ships.

For example, for a 6,000 TEU container ship which has sea speed of 23 knots at 125rpm is fitted with a main engine of 42,000 HP while for a 260,000 DWT crude oil tanker, 5 times that of the foregoing container ship in dead weight, which has sea speed of 14 knots at 70 rpm is fitted with a main engine of 28,000 HP. The fact that a container ship which has a dead weight not more than one fifth that of a VLCC is fitted with 1.5 times the engine power of a VLCC demonstrates that a ship's required engine power increases proportionally to the two thirds (2/3) power of its displacement and the third (3) power of its speed. Therefore, container ships requiring higher speed need larger engine power than VLCCs.

For your reference, today as a unit of power "watt" is universally used instead of "HP" (horsepower) or "PS" (pferdestärke : German) according to the International System of Unit "SI" (Le System International d'Unites : French). One HP or PS means a unit of power to move 75 kgf at a rate of one meter per one second, 75 kgf · m/s, and it equals to 735.49875 watt. One horsepower roughly equals to 735.5 watt.

1.2 Steam Turbine

In view of fuel consumption, a steam turbine is inferior by about 30% to 40% compared to diesel engines. However, it is capable not only to reduce its revolution to the same level as a low speed diesel engine intervening reduction gears to gain optimum propulsive efficiency, but also to build an engine which power reaches about 70,000 HP per unit. For these reasons, it had been often fitted to larger ships such as VLCC, high speed container ships, LNG carriers and larger passenger ships. However, due to soaring fuel price, many of turbine engines once fitted to these ships have been or are being retrofitted to diesel engines.

The mechanism of a steam turbine, driven by the injection of high temperature and high pressure steam generated in a boiler, is to turn the ship's propeller by delivering the power to the propeller shaft through the revolution of rotating blades and a turbine rotor via reduction gears. To make the best use of energy, steam generated in the boiler is at first sent to a high pressure turbine revolving at 5,000 to 7,000 rpm then sent out to a low pressure turbine revolving 3,000 to 5,000 rpm.

For a steam turbine, a dedicated turbine for reverse turning of the rotor when a ship is going astern is provided. A turbine for astern engine is provided in the low pressure turbine casing and it is left rotating idle until the astern engine is ordered and steam will be injected.

1.3 Electric Propulsion

The ***electric propulsion drive*** is a system to turn propellers by electrical energy generated by a dynamo. Although this system has been adopted for bow/stern thrusters for general merchant ships and some special types of ships like ice breakers and cable layers due to its good torque characteristic and easy control of revolutions of electric motors, it is seldom replaced with a conventional propulsion systems for general ships because of its high cost and low energy efficiency.

Despite this, following the aforementioned special types of ships ferry boats and cruise ships were the next to adopt the system making use of good maneuverability combined with ***pod type propulsion***, quietness and space saving design. Recently however, other types of ships such as survey ships, training ships and coastal ships have started using it as well in the effort to keep pace with the improvement of fuel consumption. Pod type propulsion is, as shown in Fig. 2.2.1, a system to turn propeller by a contained electric motor inside a cocoon shaped vessel. Pod type propulsion can be fitted to the ship's stern independently suspended enabling itself to rotate 360°. Therefore, not only can the ship be handled without the rudder, but good maneuverability can be achieved as well.

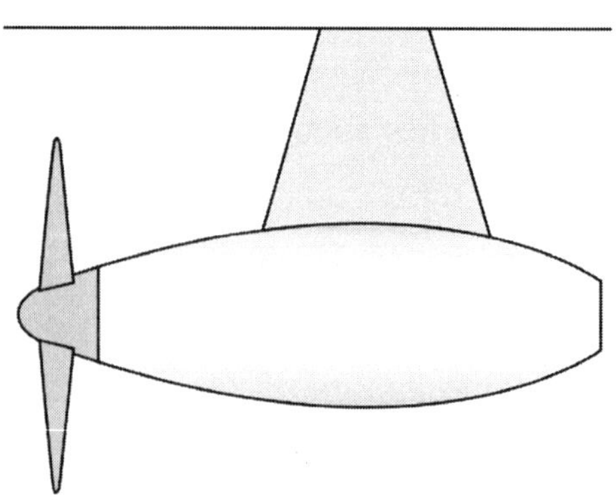

Fig. 2.2.1 Pod propelling unit

2. Characteristics of Engine Performance and Maneuvering

2.1 Diesel Engine

(1) Switchover of Ahead or Astern

In the case of diesel engines, switchover of ahead and astern is made by reversing the rotating direction of crank shaft. For this purpose, the engine needs to be stopped once to change the cam for reversing and starting the engine again. Generally, a switchover should be made after the revolution of the crank shaft slows down to a certain revolution. As this type of engine is able to control at comparatively high revolution state by injecting compressed control air to dampen revolutions, the time required to switchover and consequential running distance for that time will be shorter.

(2) Thrust Power of Astern Engine

Diesel engines can respond comparatively faster to a reversing order and are quick to increase their revolutions. In addition, strong astern thrust can be gained. However, when a ship is going astern, the same thrust as that when the ship is going ahead cannot be obtained as propeller currents are interrupted by its hull.

(3) Durability under Low Revolution

In the case of diesel engines, continuous low revolution navigation, either ahead or astern is not possible. In view of the possible lowest speed by an average of dead slow ahead, generally, dead slow ahead ranges 5.5 to 6.5 knots. When a speed lower than the above speed range is necessary, it needs to repeat stops and starts of the engine as applicable.

(4) Critical Revolution

Violent vibration on the hull will be created due to resonance of the natural frequency of ***shafting system*** connecting main engine, propeller and shaft as a whole when propeller revolution reaches a certain revolution. The revolution at this moment is called ***critical revolution*** and the operation of engine at this revolution is impossible. The range of critical revolution is generally displayed on the wall of the navigation bridge.

(5) Starting Air

In the operation of diesel engines, at every switchover of ahead/astern, restarting is necessary. And for restarting, compressed (starting) air is necessary. Caution should be taken as sometimes frequent operations may cause the exaust of starting air.

2.2 Steam Turbine

(1) Switchover of Ahead or Astern

In the case of steam turbines, when the ahead engine is changed to the astern engine, steam to the ahead turbine must be cut once and the supply of steam to the astern engine will not be started until turning inertia of the rotor dampens. Since the inertia of the rotor is large corresponding to the high revolution of steam turbines, it will take a longer time until the revolution of the rotor dampens so that the supply of steam to the astern engine may be possible. Thus ships fitted with steam turbines need a longer time to switchover engines compared with ships fitted with diesel engines. Consequently, the ship's running distance for that period grows larger.

(2) Thrust Power of Astern Engine

A dedicated turbine for astern is so downsized as to minimize the loss of energy while idling when the ship is going ahead. It would be better to think that its thrust may range about a half of that of forward thrust.

(3) Durability under Low Revolution

In steam turbines, the revolution and consequential thrust of both ahead and astern do not increase quickly as it increases gradually according to the supply of steam. In addition, since it is capable of maintaining low revolution, continuous navigation under exceptionally low revolution is possible. Moreover, for ships fitted with a steam turbine it is unnecessary to worry about the exhaustion of starting air and the critical revolution as with diesel fitted ships.

2.3 Electric Propulsion

An electric motor possesses quite good torque characteristics both to increase and decrease revolutions, even during low speed operation. And as it is easy to control revolution and change its polarity, ships with electric propulsion have good maneuvering characteristics in switchover of ahead or astern, the strength of astern thrust, and durability under low revolution.

Chapter 2 Propeller

1. Shape of Propeller

For general merchant ships screw type propellers (***screw propeller***) are fitted to their stern for propulsion.

The structure of propeller, as shown in Fig. 2.2.2, comprises ***propeller shaft*** and combined casting of ***blades*** and ***boss*** fitted into it. For blades, sometimes slanted blades are used in relation to the shape of the ship's stern. This aft slant is called ***rake***.

However the extent of rake ranges from 10° to 15°, to ensure the strength of blades with 10° rake, the thickness of blades needs to be increased more than 10%. In addition, rake itself does not affect propulsive efficiency, propellers without rake are widely used as far as there is no restriction by the shape of ship's stern. Blades of many propellers are curved back about 10° relative to the direction of rotation to mitigate impact pressure between the ship's hull plates. This is called ***skew back***.

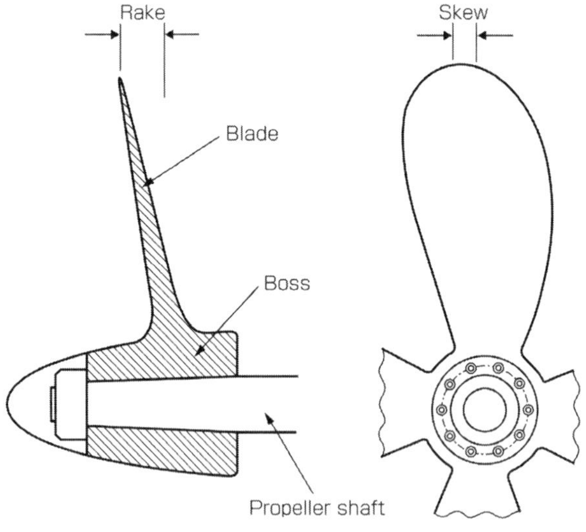

Fig. 2.2.2 Structure of propeller

The size of a propeller is measured by the diameter (***propeller diameter***) of the circle that its ***propeller tip*** draws. Propeller blades are embedded in the boss at a certain angle like a screw which is called ***pitch angle***. The advanced distance made during one rotation of the propeller is called ***pitch***. The diameter "D" and pitch of propeller "P" are principal elements to affect its performance. The ratios of propeller pitch to its diameter P/D (***pitch ratio***) of low speed ships and high speed ships range 0.6 to 0.8 and 1.0 to 1.5 respectively.

2. Type of Propeller

2.1 Fixed Pitch Propeller

Propellers in which the blades are embedded in the boss at a fixed pitch angle (***fixed pitch propeller : FPP***) are widely used. Fig. 2.2.3 shows an example of a 5-bladed propeller. The number of blades will be 4 to 6 blades taking into consideration the thrust power and designed speed of the ship. Usually, right turn (clockwise) propellers as seen from the stern are widely used.

Fig. 2.2.3 Example of 5-bladed propeller

Most merchant ships are fitted with one propeller to one shaft but there are some fitted with two propellers to two shafts.

When two propellers are to be fitted, there are two rotation combinations, one is ***outward turning***, the starboard propeller turning clockwise and port propeller turning counterclockwise and the other is ***inward turning***, the starboard propeller turning counterclockwise, and port propeller turning clockwise.

2.2 Controllable Pitch Propeller

Propellers with which the pitch of blades is changeable are called ***controllable pitch propeller*** (***CPP***). Control of propeller pitch is remotely made by a hydraulic system. In CPP ships, propeller thrust can be quickly changed as intended without changing propeller revolution. This enables long hours of running under extremely slow speed and switchover from ahead to astern can be made by reversing propeller's pitch without reversing the propeller's revolution.

With diesel engines or steam turbines, reliability for restarting and time lag in operation whenever stop, start or switchover from ahead to astern are made give persons engaged in ship handling considerable psychological pressure. By contrast, CPP ships emancipate persons engaged in ship handling from their anxiety about engine motion, as the restarting operation of engines and the operation of switchover from ahead to astern sometimes become unnecessary.

However, it should be noted that when the propeller pitch angle is zero, the rudder effect of a CPP will be abruptly lost due to the ***disc effect of CPP*** and sometimes the ship's heading will be swung in an unexpected direction. Disc effect is a phenomenon that occurs when the propeller which has a pitch angle of zero rotates as if it is one piece of a disc and blocks rectified current to the rudder and forms eddies around the propeller, depriving steerability of the rudder. As a result the ship's course keeping ability will be lost.

In CPP fitted ships, their propellers are used keeping their rotation throughout berthing operation, care should be necessary lest mooring ropes should be fouled in them. In addition, preparedness and responsiveness to disengage the clutch immediately in an emergency such as a person overboard close to the ship's stern, if it should happen, are necessary.

2.3 Highly Skewed Propeller

When you are going to increase propulsive efficiency using a low revolution engine or a large diameter propeller, it is necessary to take countermeasures to prevent vibration and noise due to pressure fluctuation which may be created when the upper tip of the propeller blade passes close to the ship's bottom. Although an ordinary propeller blade is formed at about 10° skew back from its center, in a highly skewed propeller, it is further enhanced as shown in Fig. 2.2.4.

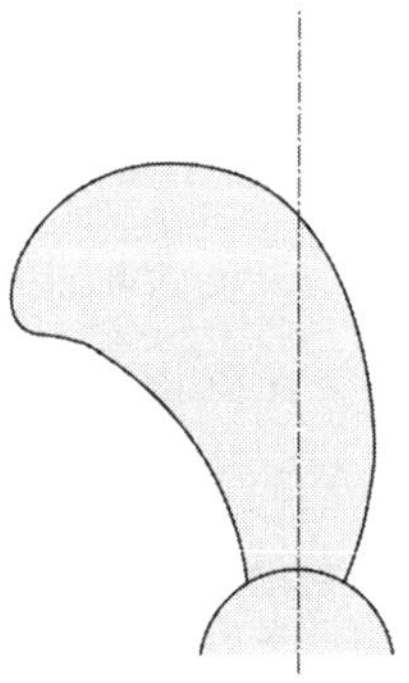

Fig. 2.2.4 Highly skewed propeller

The purpose of the skew of the propeller blade is to reduce vibration and noise by delaying the approach of the propeller blade to the ship's bottom to disperse pressure fluctuation. Highly skewed propellers can reduce vibration and noise without giving adverse effects to the ship's maneuverability. In addition, the effect of ***cavitation*** in which air bubbles generated by the propeller erode its face can be reduced.

2.4 Contra-rotating Propeller

A ***contra-rotating propeller*** (*CRP*) is a propeller that has two sets of blades arranged one behind the other in which pitch and rotation are totally reversed. As shown in Fig. 2.2.5, the purpose of this type propeller is to offset the rotation of current created by the forward propeller by the propeller behind it to save propulsive energy.

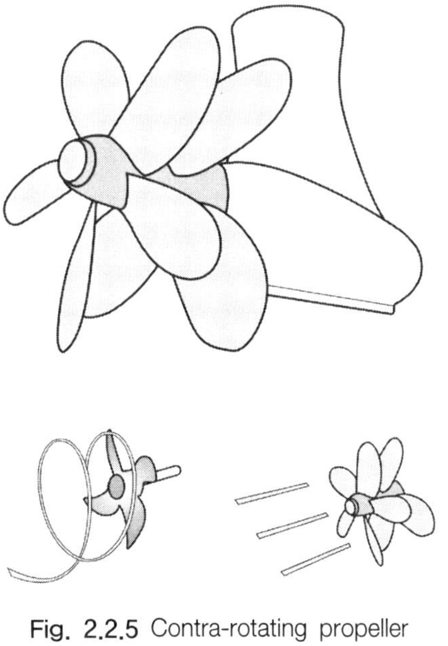

Fig. 2.2.5 Contra-rotating propeller

3. Thrust Power of Propeller

3.1 Ahead Thrust and Astern Thrust

Generally, propellers fitted to a ship's stern rotate clockwise when ships are going ahead as seen from the stern. In a ship fitted with a ***right handed propeller***, when the propeller blade rotates in water, a reactionary force is created by hydraulic pressure to the face of the propeller blade as shown in Fig. 2.2.6. A fore and aft component of this force makes ***ahead thrust.*** A lateral component of hydraulic pressure which acts to press the ship stern laterally is called ***sidewise pressure.*** When the rotation of a propeller is reversed to go astern, a fore and aft component of hydraulic pressure on the face of the propeller blade makes ***astern thrust*** as shown in Fig. 2.2.7. And the direction of sidewise pressure acts opposite to that created when the ship is going ahead.

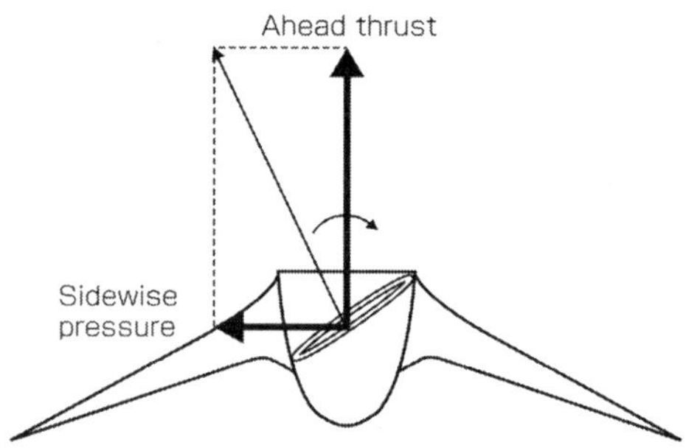

Fig. 2.2.6 Forces acting on propeller blade (Ahead thrust)

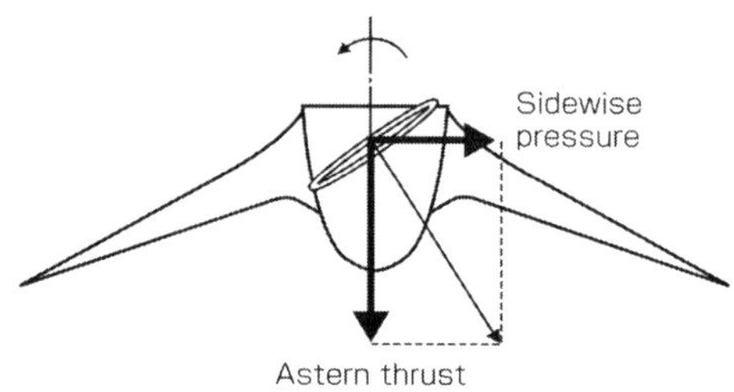

Fig. 2.2.7 Forces acting on propeller blade (Astern thrust)

3.2 Slip

When the pitch of a propeller is "p" meters, the ship moves "p" meters for every one rotation of the propeller. And if it rotates "n" times per minute, theoretically its speed will be $p \cdot n$ (meter /minute). However, as a propeller rotates in water, a difference between the ***theoretical propeller moving speed*** and the ***actual propeller moving speed*** is created. This difference is called ***slip.***

If the ship's actual propeller moving speed is "V_s" (meters / minute), its slip will be $p \cdot n - V_s$, which is called ***apparent slip.*** The ratio "V_s" to $p \cdot n$ is called ***apparent slip ratio*** and it can be written as follows.

$$s = 1 - \frac{V_s}{(p \cdot n)} \quad \cdots\cdots (2.2.1)$$

When a ship moves in water, the mass of water around the hull moves together in the same direction as if it is dragged by the hull. This is called ***wake.*** As the wake also exists around its propeller, propeller speed against the water will be reduced by wake speed. When the wake speed is "V_w", a ***propeller speed against***

the water "V_p" can be written as $V_p = V_s - V_w$. And $p \cdot n - V_p$ is called ***real slip*** and the ***real slip ratio*** can be written as follows.

$$s = 1 - \frac{V_p}{(p \cdot n)} \quad \cdots\cdots (2.2.2)$$

Fig. 2.2.8 shows the relationship among ship speed, propeller speed against the water, wake, apparent slip and real slip.

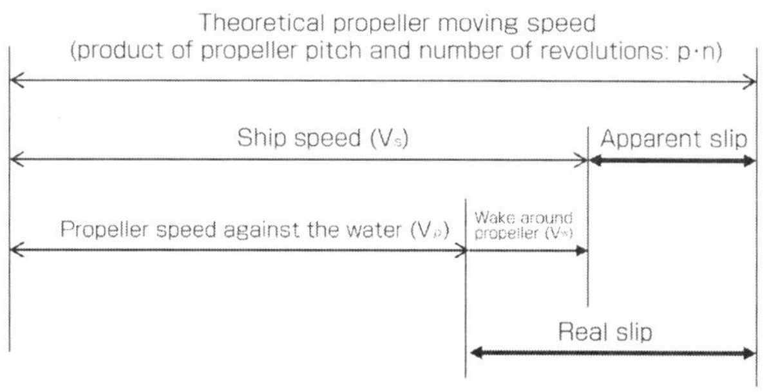

Fig. 2.2.8 Apparent slip and real slip

4. Effect of Propeller Current (Right-handed Single Propeller)

4.1 Propeller Current

Fig. 2.2.9 shows the features of current in front of and rear of the propellers. The currents flowing into and discharging out from the propeller are collectively called ***propeller current***. And the currents in front of and to the rear of the propeller are respectively called ***suction current*** and ***discharge current***.

As suction current flows parallel to the ship's hull when the ship is running straight ahead, the effect of suction current to the hull is negligible. However, when the ship is running astern, as current comes from behind the rudder and suction current impinges on the rudder first, its rudder effect will be increased.

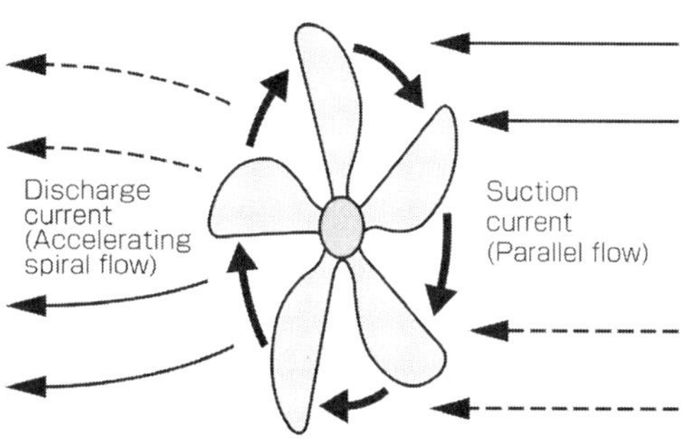

Fig. 2.2.9 Propeller current

As discharge current forms an accelerated spiral current by the rotation of the propeller, it gives asymmetrical force to the rudder behind the propeller when the ship is running ahead. By contrast, when the ship is running astern it gives the same effect to the ship's stern. As a result, the turning motion of the ship will be created without operating its rudder.

4.2 Deflection of Heading Due to Propeller Discharge Current

(1) Under Ahead Engine

When a ship fitted with a right-handed single propeller is running ahead, the spiral discharge current from the propeller impinges on the upper left face and the lower right face of the rudder forming clockwise spiral current as seen from the stern.

Fig. 2.2.10 shows the effect of spiral current from the propeller and the difference of the ***angle of incident*** between the right and the left face of the rudder.

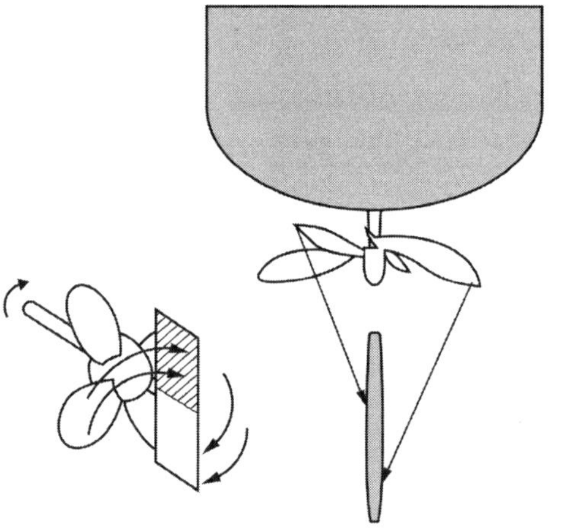

Fig. 2.2.10 Incident angle of propeller discharge current to rudder face

As shown in the above Fig. 2.2.10, the angle of incident of the propeller discharge current that impinges on the lower right face of the rudder from the upper right is larger than that to the upper left face from the lower left, pressure from the right to the left acting on the rudder face is superior to that from the left to the right. Consequently, the ship's stern is pressed to the left making the ship's heading turn to the right even if the helm is kept midship.

When ships are in light or ballast conditions in which rudder submergence is not deep enough, right turning tendency will be enhanced as the upper hydraulic pressure area of the rudder decreases.

(2) Under Astern Engine (Effect of Impingement Due to Propeller Discharge Current at the Stern)

When a ship fitted with a single right-handed propeller is running astern, current discharged from the propeller flows into the ship's stern rotating counterclockwise. A part of the counterclockwise spiral current on the left side flows passing under the ship's bottom at the stern, but on the right side it impinges on the ship's side plate directly at a large angle of incident.

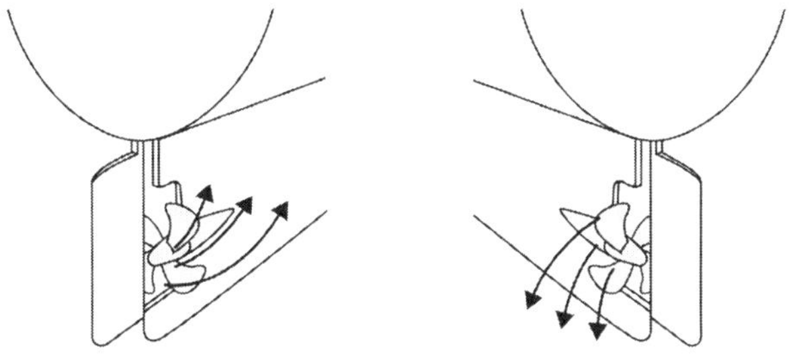

Fig. 2.2.11 Effect of impingement due to propeller discharge current at the stern

Due to this pressure on the side plate of the starboard side at the stern, the ship's heading turns to the right. This phenomenon is called the ***effect of impingement due to propeller discharge current at the stern***. This effect appears conspicuously when a ship fitted with a single right-handed propeller is about to run astern as its propeller rotates counterclockwise.

4.3 Deflection of Heading due to Sidewise Pressure

When a propeller rotates, the reactionary force of water acts at a perpendicular to the face of the propeller blade; fore and aft components of this force form the ship's ahead or astern thrust respectively and lateral components form ***sidewise pressure***. As shown in Fig. 2.2.12, this sidewise pressure works toward the left on the upper blade and toward the

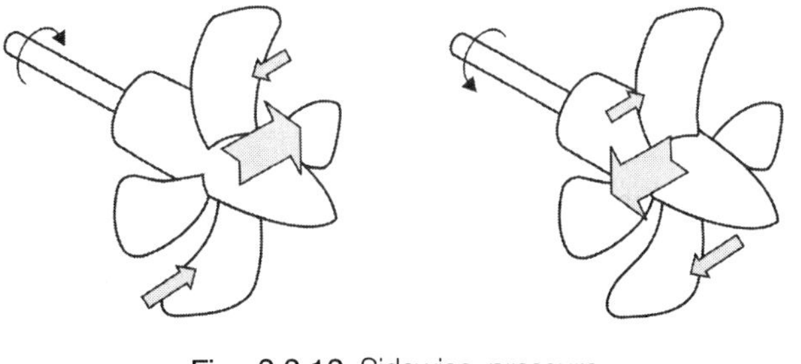

Fig. 2.2.12 Sidewise pressure

right on the lower blade when the ship is running ahead and its propeller is rotating clockwise. When the ship is running astern the foregoing relations are reversed.

When a propeller is close to the water surface while a ship is running ahead, as the reactionary force of water to the upper blade decreases due to aeration and the reactionary force to the lower blade increases comparatively, a reactionary force pushing the stern to the right is superior and the ship's heading turns to the left. Conversely, when the ship is running astern in this condition, the ship's heading turns to the right as reactionary force pushing the stern to the left is superior. This phenomenon is called the ***deflection of the ship's heading due to sidewise pressure***.

If the propeller is turned for a very short time when the ship is dead in water or running at low speed, which is called ***boosting***, this phenomenon is likely to occur. Furthermore when the rudder's submergence is not deep enough, this tendency will be more enhanced.

5. Integrated Effects of Rudder and Propeller (Right-handed Single Propeller)

5.1 Vector Analysis of Integrated Effects

While a ship is running ahead the ship's inertia to run straight ahead is so large that the deflection of the heading due to the two conflicting effects, propeller discharge current which works to turn ship's heading to the right and sidewise pressure which works to turn the ship's heading to the left, would not affect it substantially. However, such effects appear conspicuously when the ship is going to move from the state of dead in the water or proceeding at low speed.

Here, we will verify how the forces which act on the stern of a ship such as ahead and astern thrusts "T", rudder force "F", sidewise pressure "P_S", and lateral force created by the impingement due to propeller discharge current at the stern "P_L" affect the deflection of heading through the analysis of the affecting vectors of force. However, when the ship is running ahead, as the effect of the propeller discharging current is small, in this analysis, only the deflection of heading due to the effect of sidewise pressure is taken up.

In analysis, the integrated vector of all affecting forces is at first identified then it is broken down into longitudinal and lateral components.

Then, their sizes can be defined by comparing the ship's ahead and astern inertia. Fig. 2.2.13 shows the analysis of integrated vectors in the various combinations of helm angle and propeller thrust.

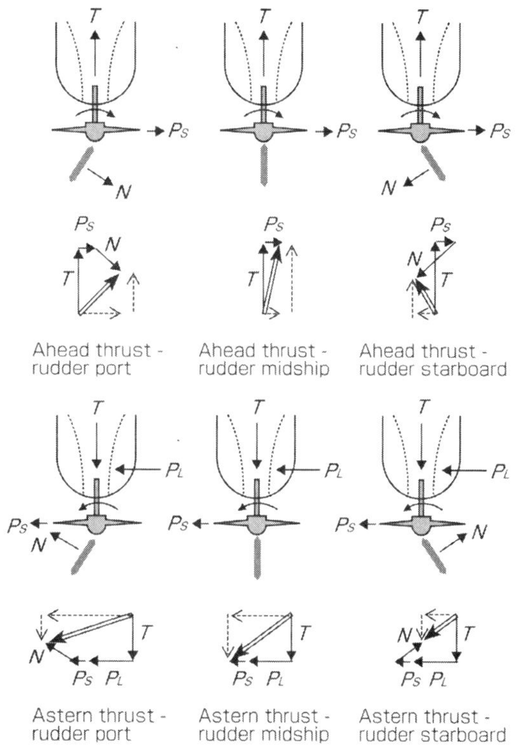

Fig. 2.2.13 Integrated effects of rudder and propeller

(1) When Ahead Engine is Started

When the propeller is turned under small ahead inertia with its helm midship, the stern is pushed to the right due to the effect of sidewise pressure. Consequently, the heading turns to the left. And if the helm turns to the left, left turning will be enhanced. On the contrary, if the helm is turned to the right, the rudder effect will be reduced or delayed a little. This effect is likely to appear conspicuously when the propeller is close to the water surface. On the contrary, if the submergence of the propeller is adequate, or ahead inertia is large, its effect is small.

(2) When Astern Engine is Started

When the propeller is turned astern, both the sidewise pressure of the propeller and the lateral pressure created by impingement due to the propeller discharge current at the stern act to push the stern to the left. Above all, with the effect of the lateral pressure created by the impingement due to the propeller discharge current at the stern, this phenomenon appears conspicuously. Therefore, ship's heading tends to turn to the right regardless the direction of rudder.

However, when the ship's astern inertia is small, even if the starboard helm is taken its effect is small, but the rudder effect will be enhanced as the ship's astern speed increases.

5.2 Examples of Ship Handling Considering Propeller Current

(1) Straight Astern

When a ship lying dead in the water is going to move astern with its helm midship, the ship's heading turns starboard due to the duplicated effect of the sidewise pressure of the propeller and the lateral pressure created by impingement due to the propeller discharge current at the stern. To make a ship move straight astern offsetting these effects, it would be better to take larger starboard helm beforehand as shown in Fig. 2.2.14 and thereafter making adjustments to the helm angle appropriately to utilize the effect of suction current to the rudder face in addition to the rudder effect by stern way.

Whenever the ship's heading is going to deflect, it is necessary to ease the helm a little or slow down sternway appropriately so that the deflecting motion may be cancelled. For this purpose, small and sensitive adjustments of the helm and engine will be necessary.

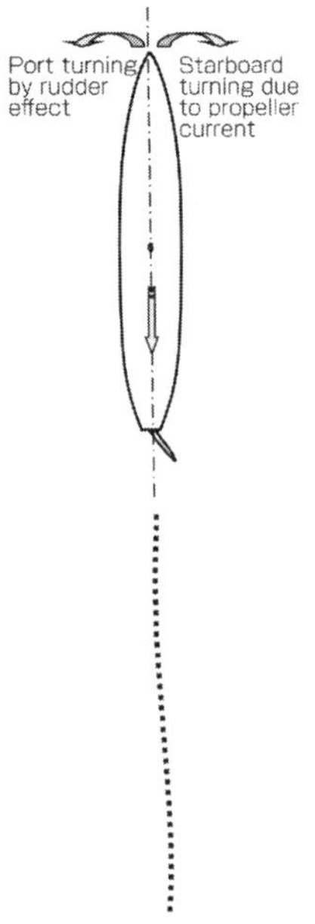

Fig. 2.2.14
Straight astern

(2) Starboard Turnaround on the Spot

When a ship is obliged to make a 180°turn within a restricted water area it would be reasonable to try to turn starboard because the ship can be turned easier as its turning motion is enhanced by the strong effects of sidewise pressures of the propeller and the lateral pressure created by impingement due to the propeller discharge current at the stern from ② to ③ as shown in Fig. 2.2.15.

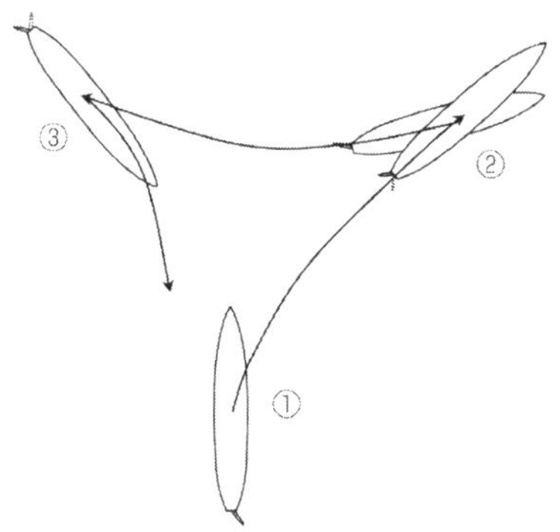

Fig. 2.2.15 Starboard turnaround on the spot

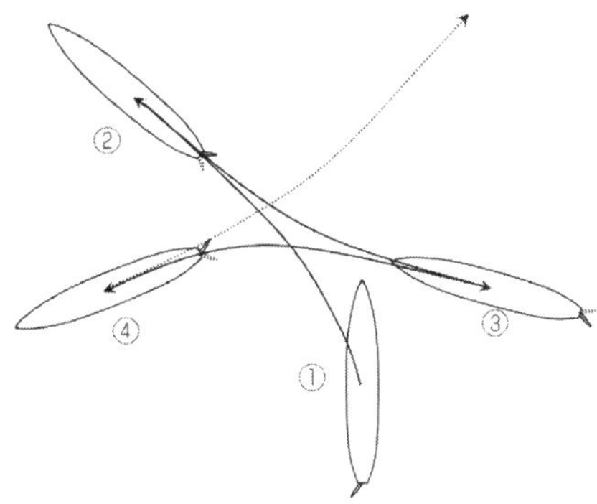

Fig. 2.2.16 Port turnaround on the spot

(3) Port Turnaround on the Spot

Conversely, if a ship is going to make a port turnaround, the effects of the sidewise pressure of the propeller and the lateral pressure created by impingement due to the propeller discharge current at the stern work totally opposite. As the ship retards almost straight or sometimes its heading turns starboard as shown in Fig. 2.2.16, it is difficult for a ship to make a port turnaround on the spot.

(4) Approach to Port-side-to Berth

When a ship is going to a port-side-to berth, approaching the berth on its port forward is reasonable. Approach the berth keeping 10 to 20°to its face, then, if the engine is put astern at the point ①, the ship will be able to be stopped almost parallel to the berth due to the heading deflection effect of the sidewise pressure of the propeller and the lateral pressure created by impingement due to the propeller discharge current at the stern as shown in Fig. 2.2.17.

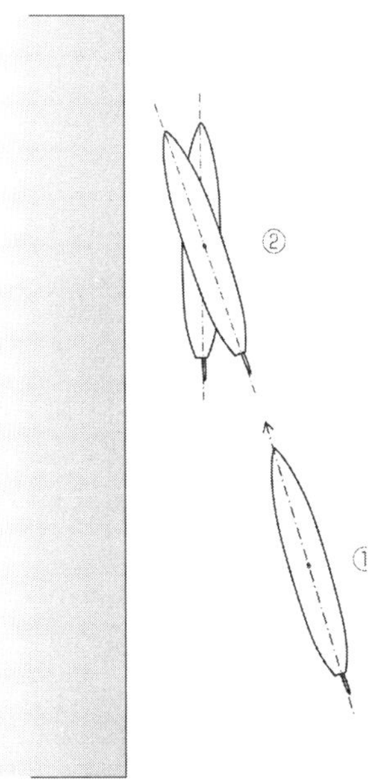

Fig. 2.2.17 Approach to port-side-to berth

(5) Approach to Starboard-side-to Berth

When a ship is going starboard-side-to berth, caution should be taken as the ship's heading turns to starboard toward the berth and its stern part moves from the berth by the heading deflection effects due to the sidewise pressure of the propeller and the lateral pressure created by impingement due to the propeller discharge current at the stern if the engine is put astern at the point ① as shown in Fig. 2.2.18. To avoid such adverse effect, it would be better to bring the ship as parallel to the berth face as possible keeping a speed which may be stopped by weak astern motion if possible.

(6) Approach to Mooring Buoy

A ship will be able to be stopped seeing the buoy below the forecastle, when a ship is approaching a buoy keeping a bearing on her starboard bow and if the engine is put astern at the point ① as shown in Fig. 2.2.19 by the heading deflection effects due to the sidewise pressure of the propeller and the lateral pressure created by impingement due to the propeller discharge current at the stern.

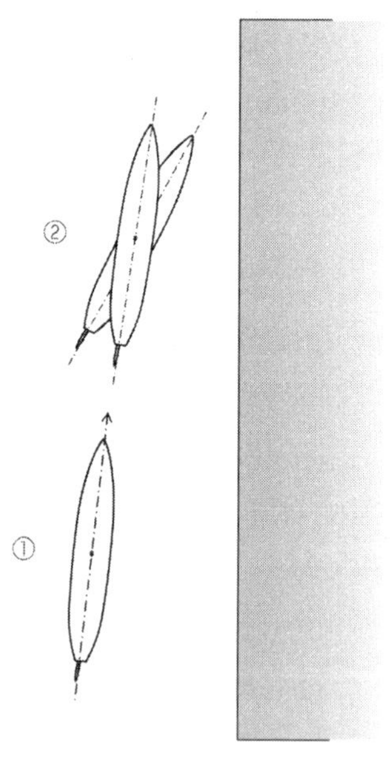

Fig. 2.2.18 Approach to starboard-side-to berth

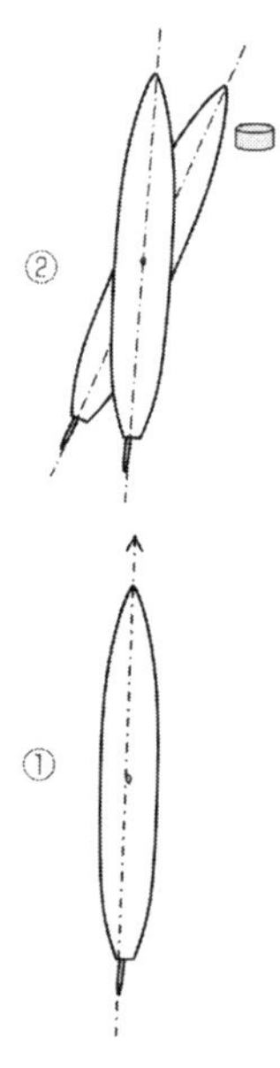

Fig. 2.2.19 Approach to mooring buoy

6. Effects of Propeller Currents of a Twin Propeller Ship

6.1 One Propeller Ahead while the Other is Running Astern

A ship fitted with twin propellers can move ahead or astern in a straight line as the heading deflection is effected due to the sidewise pressure and the lateral pressure created by impingement due to the propeller discharge current at the stern which can offset each other. In the case of twin propellers, ***outward turning*** is more widely accepted because it can prevent propellers from catching floating objects and it is advantageous to utilize the effect of propeller currents.

For instance, if the port propeller is put ahead and the starboard propeller is put astern in turning to starboard, the sidewise pressure and the lateral pressure created by impingement due to the propeller discharge current at the stern by the counterclockwise rotation of the starboard propeller work to push the ship's stern to the left. In addition the sidewise pressure of port propeller also acts to push the ship stern to the left. As a result, starboard turning of the ship will be enhanced.

The same effect can be available in turning to port by reversing the revolution of these two propellers. However, it should be noted that in the case of ***inward turning*** twin propellers, the same marked effect will be lost.

6.2 Examples of Ship Handling of Twin Propeller Ships

(1) Turnaround on the Spot

If two propellers are turned in different directions to each other, the ship can be turned to either board on the spot. This maneuvering characteristic is a marked difference from single propeller ships.

(2) Approach to Berth

For twin propeller ships, approach to the berth can be easily made regardless of berthing side. First, approach the berth keeping 10 to 20°to the berth face at slow speed, and put both propellers astern at an appropriate distance before the berth. If some angle remains against the berth, put the offshore side propeller astern or put the inshore side propeller ahead until the ship becomes parallel to the berth.

(3) Utilization of Differentiated Thrusts of Two Propellers

Two propellers are fitted symmetrically to the ship's longitudinal centerline. Therefore, the thrust of either propeller creates turning moments if they are independently used. When the thrust of two propellers driven are equal to each other, moments can be offset by each other and the ship can go straight ahead or astern. When the two thrusts are differentiated, the ship will start turning on either side depending on the difference of thrust.

For instance, this characteristic can be afforded to cancel the effect of drift by wind or current otherwise to avoid an impending risk such as a trouble of steering system or others.

6.3 Twin Propellers and Two Rudders, Twin Propellers and One Rudder

Ships fitted with twin propellers and twin rudders right behind each propeller are called ***twin propeller-twin rudder*** ships while ships fitted with twin propellers and a single rudder on the ship's longitudinal centerline are called ***twin propeller-single rudder*** ships. In twin propeller-twin rudder ships, good rudder effect can be gained because each propeller current can be directly sent to each rudder. By contrast, in twin propeller-single rudder ships, as each accelerated propeller current does not directly strike the rudder face, the effect of the rudder will be inferior to that of twin propeller-twin rudder ships.

Chapter 3 Speed and Inertia

1. Speed

1.1 Thrust

(1) Predictive Equation for Thrust

A ship is moved in reaction to the force created by the rotation of propeller blades and accelerated water is discharged behind. To move the ship, its speed needs to be increased until its thrust equals to hydraulic resistance affecting the ship's hull. On the other hand, if a ship's thrust is reduced while it is running, its speed gradually decreases to a point that the ship's thrust and hydraulic resistance to the ship's hull will be balanced.

A ship's ahead thrust created by a propeller can be expressed as a function of propeller diameter "D" and the number of propeller revolution "n", and deduced from the equation (2.2.3), letting water density be "ρ" and thrust coefficient be "K_T".

$$T = K_T \rho n^2 D^4 \qquad (2.2.3)$$

In the above, ***thrust coefficient*** is derived from the tank test of model propellers and a dynamometer. In the test, propeller thrusts are measured for varying speeds keeping the revolution of the propeller constant. From the test results, thrust coefficient is derived according to $K_T = T/\rho n^2 D^4$ and a propeller characteristic curve can be drawn based on propeller model tests in open water. In addition to the propeller characteristic curve, thrust coefficient, torque coefficient and propeller efficiency curves can also be drawn together with slip ratio on the coordinate in which the horizontal axis is the advance constant.

Although thrust coefficient may vary depending on the size, pitch and the number of blades of a propeller, its approximate value ranges 0.15 to 0.2 for slip ratio 0.2, 0.2 to 0.35 for slip ratio 0.4, 0.3 to 0.35 for slip ratio 0.6 and 0.35 to 0.4 for slip ratio 0.8. When a propeller is turned in a state where a ship is dead in the water, the so-called boosting state, the thrust coefficient ranges 0.4 to 0.5 for slip ratio 1.0.

(2) Examples Obtained from Predictive Equation

Let's look at the thrust of a ship which has a length of 100m and breadth of 15m and is fitted with a propeller which has a diameter of 4.5m and runs at 120 rpm. From equation (2.2.3),

Propeller Thrust : $T = K_T \rho n^2 D^4$

where

Water density : ρ=0.104 ton · sec^2/m^4
Revolution of propeller : n=2 rps
Propeller diameter : D=4.5 m

Applying an average value from the propeller characteristic curve for thrust coefficient "K_T",

for slip ratio 0.2, K_T=0.2 : T=34.1 tonf
for slip ratio 1.0, K_T=0.5 : T=85.3 tonf

1.2 Resistance

(1) Forms of Resistance

Resistance affecting the ship's hull by water can be sorted into ***frictional resistance*** due to the viscosity of water, ***wave making resistance*** by which energy will be lost, eddy ***resistance*** due to the ship's shape under water and ***air resistance*** due to friction with air.

Frictional resistance: is the resistance between a ship's hull under water and water viscosity including the effect of the 3 dimensional form of the hull under water, its roughness and the foulness of its surface. This type of resistance becomes lowest after bottom cleaning or painting at dry-dock.

Wave making resistance: is the resistance which equals consumed energy due to the build up of waves accompanied by the ship's running. A bulbous bow is a device used to minimize waves building up.

Eddy resistance: is the dragging resistance due to eddies generated around the discontinuous parts of a ship's hull under water or the stern part of the ship. A ship's streamlined shape is one way to reduce this type of resistance.

Air resistance: is the resistance generated by friction and the eddy of the air to the ship's structure above water. Although it is by far smaller compared with that of water as its density is 1/836 that of water, special consideration for ship handling as in the case of PCC, LNG carrier, and large container and cruise ships which have bulky structures above water is necessary.

Although it is necessary to carry out model tank tests for each type of ship to know precise values of the above resistances, in practice, various predictive equations proposed for typical ship types based on systematically measured values through the respective experiments can be used.

Frictional resistance varies depending on ship type, amounting to about 70 to 80% of the ***total resistance*** and wave making resistance follows with 10 to 30%. The other resistances are only a few percent. By the above fact, the whole resistance is sometimes divided into two categories, frictional resistance and ***residual resistance*** (the other remaining resistances). That is, Total resistance = Frictional resistance + Residual resistance. It can be safely said that residual resistance is almost equal to wave making resistance.

(2) Frictional Resistance

Frictional resistance "R_f" can be deduced from ship speed "V", water density "ρ" and the ***area of wetted surface*** of a ship's hull "S" according to equation (2.2.4). Frictional resistance increases proportionally to the wetted surface and ship speed squared.

$$R_f = \frac{1}{2} \cdot C_f \rho S V^2 \quad \cdots\cdots (2.2.4)$$

In the above equation, "C_f" means ***frictional resistance coefficient*** for which a few predictive equations are proposed as functions of ***Reynolds number***, $R_n = VL/\nu$ (ν: coefficient of kinematic viscosity). For instance, when the value of "ν" at 15℃ is taken in ***Schoenherr equation***, then, "C_f" will be 0.0013 at $R_n = 4\times10^9$, 0.0014 at $R_n = 2\times10^9$, 0.0015 at $R_n = 1\times10^9$, and 0.0017 at $R_n = 5\times10^8$.

(3) Residual Resistance

Residual resistance "R_r" can be derived by equation (2.2.5) according to the above equation.

$$R_r = \frac{1}{2} \cdot C_r \rho S V^2 \quad \cdots\cdots (2.2.5)$$

Where, "C_r" means ***residual resistance coefficient*** which can be derived from the Chart of Tailor Tank with parameters being, ∇/L^3 (∇: displacement, L: water line length), ***prismatic coefficient*** "C_p", ratio of breadth "B" to draft "d" B/d and ***Froude number*** $V/\sqrt{L \cdot g}$. Froude number "F_n" is the nondimensional expression of a ship's speed where "V" is the ship's speed and "g" is gravitational acceleration.

(4) Examples of Deduced Resistance Values

In the following, let's try to calculate "R_f" and "R_r" of a ship which ***block coefficient*** $C_b = 0.64$, ***midship coefficient*** $C_{mid} = 0.98$ and displacement $\nabla = 4{,}800\text{m}^3$, and let ship length, breadth, draft and ship speed respectively be 100m, 15m, 5m and 12.5 knots. From equation (2.2.4),

$$\text{Frictional resistance}: \ R_f = \frac{1}{2} \cdot C_f \rho S V^2$$

where

Water density $\rho = 0.104$ ton · sec^2/m^4

Wetted surface from ***Denny's simplified formula***,

$S = (1.7d + C_b B)L = 1{,}810\text{m}^2$

Ship speed 12 knots=6.57m/s, C_f=0.0015 from Schoenherr's formula,

$\therefore\ R_f$ =6.1 tonf

Residual resistance "R_r" from equation (2.2.5), $R_r = \frac{1}{2} \cdot C_r \rho S V^2$, "$C_r$" based on the following parameters,

∇ / L^3=0.0048

Prismatic coefficient : $C_p = C_b / C_{mid}$ =0.65

$\frac{B}{d}$=3

Froude number (F_n)=0.20

From Tailor's tank chart, C_r=0.616

$\therefore\ R_r$=2.5 tonf

As you can see in the above case, frictional resistance amounts to 71% while the residual is 29% to the total resistance, (6.1+2.5)=8.6 tonf.

1.3 Telegraph Speed

☞ Refer to Vol. 4 Ship Handling Characteristics Data Bas

A ship's speed cannot be gained until the ship's thrust supersedes its total resistance. Each speed level indicated on the telegraph shows steady speed gained at each level of thrust where it is balanced to its corresponding total resistance.

Generally, navigation at full speed is set on around 85 to 90% of the ship's ***MCR*** (***Maximum Continuous Rating***) taking an appropriate ***sea margin*** in safety and economical efficiency into account.

Stand by full speed and dead slow speed which are maneuvering speeds required for entering into or leaving from a port and transiting narrow channels and passages are set on around 12 knots regardless to a ship's MCR and are 4 to 6 knots respectively. However, dead slow for high speed container ships fitted with high powered engines is set much higher out of this range.

As seen from the relations among S/B Full, Half, Slow and Dead Slow, they are 85% (4/5), 55 to 65% (3/5) and 35 to 45% (2/5) of S/B Full respectively. The fractions in round brackets are given for the easy reference for persons engaged in ship handling to learn telegraph speed by heart. For your further information, the speed for astern will be about a half of its respective ahead speed level.

2. Inertia

2.1 Types of Inertia

It takes some time for a ship to increase or decrease speed or to stop as intended adjusting its thrust due to its inertia. For these periods to change speed, the ship runs corresponding to the previous speed and inertia. The elapsed time to reach an intended speed corresponding to an ordered thrust after it is changed is called inertia. The scale of inertia is substituted by the required time or distance run for that period. For ship handling, the data for inertia is indispensable. Inertia is categorized as follows.

(1) Driving Inertia

Time required to move a ship lying dead in water, or distance run until a ship reaches a speed corresponding to a given propeller thrust.

(2) Acceleration Inertia

When a ship is running at a steady speed, time elapsed or distance run until the ship reaches a speed corresponding to an increased propeller thrust.

(3) Deceleration Inertia

When a ship is running at a steady speed, time elapsed or distance run until the ship reaches a speed corresponding to a decreased propeller thrust.

(4) Stopping Inertia

When the propeller thrust is stopped in steady running state, time elapsed or distance run until the ship comes to a complete stop in water. However it takes too long until the ship comes to a complete stop in water, practically, elapsed time or distance run until the ship's speed comes down to 2 knots is deemed as stopping inertia.

(5) Emergency Stopping Inertia

When the propeller thrust is reversed in a steady running state, time elapsed or distance run until the ship comes to a complete stop in the water. The distance run when the ship running in full ahead is stopped by full astern, so-called crash astern is called short stopping distance and it is an important datum for ship handling in an emergency.

2.2 Estimated Inertia Based on the Law of Motion

(1) Driving and Acceleration Inertia

When a ship which has a mass of "m" moves in vacuum, its speed increases at a certain acceleration "α" proportionally, following the added force "F" according to Newton's law of motion ($F=m\alpha$). Actually, as a ship floats in water, it needs to be accelerated in the existence of hydraulic resistance "R". A ship starts to move when its thrust "T" breaks "R", then its steady speed "V" will be gained when its thrust "T" is balanced with resistance "R". This relation can be expressed in the following equation (2.2.6).

$$(m+m_x)\frac{dV}{dt} = T - R \quad \cdots\cdots (2.2.6)$$

Thrust power "T" can be derived from the equation (2.2.3) and resistance can be derived from the equations of (2.2.4) and (2.2.5). "m_x" in the above equation represents a loss of thrust required to accelerate a mass of water around the ship and is called ***added mass*** as it seems as if the mass of the ship itself increased. The quantity of added mass is about 5 to 15% in

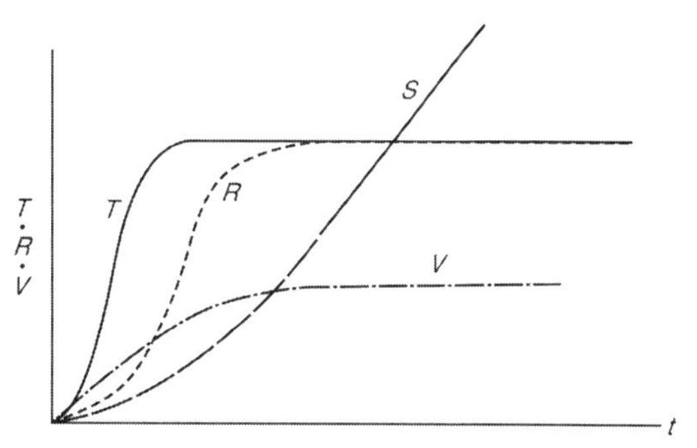

Fig. 2.2.20 Driving inertia

respect to ahead or astern motion of the ship.

Solving the equation of motion (2.2.6), let a purposed speed when thrust "T" and resistance "R" are in equilibrium be "V_0" and a speed at an arbitrary time be "V". Then, the equation of motion can be written as follows as the resistance to the hull "R" is proportional to the ship speed squared.

$$(m+m_x)\frac{dV}{dt}=T-R\left(\frac{V}{V_0}\right)^2 \quad \cdots\cdots (2.2.7)$$

As "T" and "R" are equal in a steady speed, equation (2.2.7) will be,

$$(m+m_x)\frac{dV}{dt}=R-R\left(\frac{V}{V_0}\right)^2 \quad \cdots\cdots (2.2.8)$$

$$dt=(m+m_x)\frac{V_0^2}{R}\cdot\frac{1}{V_0^2-V^2}dV \quad \cdots\cdots (2.2.9)$$

Integrating (2.2.9),

$$t=(m+m_x)\frac{V_0^2}{R}\cdot\frac{1}{2V_0}\log_e\left(\frac{V_0+V}{V_0-V}\right)+C \quad \cdots\cdots (2.2.10)$$

$$=\frac{1}{2}(m+m_x)\frac{V_0}{R}\log_e\left(\frac{V_0+V}{V_0-V}\right)+C \quad \cdots\cdots (2.2.11)$$

The initial condition when a ship lying dead in the water is started to move by giving propeller thrust is, $t=0$ and $V=0$. Consequently, the constant of integration is $C=0$.

Equation (2.2.11) will be,

$$t=\frac{1}{2}(m+m_x)\frac{V_0}{R}\log_e\left(\frac{V_0+V}{V_0-V}\right) \quad \cdots\cdots (2.2.12)$$

From the above, distance run "s" until thrust "T" and resistance "R" are balanced and "V_0" is gained,

$$s=\frac{1}{2}(m+m_x)\frac{V_0^2}{R}\log_e\left(\frac{V_0^2}{V_0^2-V^2}\right) \cdots\cdots (2.2.13)$$

(2) Deceleration and Stopping Inertia

When a ship's engine is stopped in a steady running state, speed decreases gradually and the ship finally stops as shown in Fig. 2.2.21.

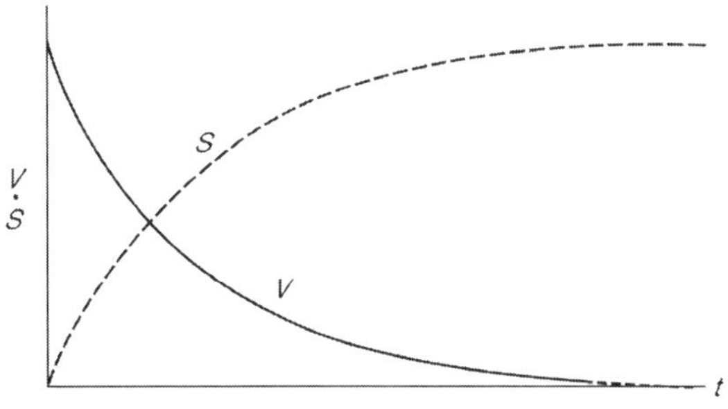

Fig. 2.2.21 Stopping inertia

Immediately after the engine stops, the ship's speed decreases drastically as its resistance is large proportionally to ship's speed. The rate of deceleration is slowed as resistance decreases. For such deceleration, stopping inertia corresponds to the equation (2.2.14) when zero is substituted for thrust power $T=0$ in equation (2.2.6).

$$(m+m_x)\frac{dV}{dt}=-R \cdots\cdots (2.2.14)$$

Solving the equation (2.2.14) based on the same principle of driving and acceleration inertia, let the initial speed be "V_0" and speed be at an arbitrary point "V" until a given thrust and speed are balanced. Then, the equation of motion will be,

$$(m+m_x)\frac{dV}{dt}=-R\left(\frac{V}{V_0}\right)^2 \cdots\cdots (2.2.15)$$

Rewriting the equation (2.2.15) about dt,

$$dt=-(m+m_x)\frac{V_0^2}{R}\cdot\frac{1}{V^2}dV \cdots\cdots (2.2.16)$$

Integrating the equation (2.2.16)

$$t = -(m+m_x)\frac{V_0^2}{R}\int\frac{1}{V^2}dV + C$$

$$= (m+m_x)\frac{V_0^2}{R}\cdot\frac{1}{V} + C \quad \cdots\cdots (2.2.17)$$

Since the initial conditions of deceleration and stopping inertia when an engine is stopped in steady running state are $t=0$ and $V=V_0$, the constant of integration,

$$C = -(m+m_x)\frac{V_0^2}{R}\cdot\frac{1}{V_0}$$

$$\therefore t = (m+m_x)\frac{V_0^2}{R}\left(\frac{1}{V}-\frac{1}{V_0}\right) \quad \cdots\cdots (2.2.18)$$

Further, finding the distance run "s" until the ship comes to a complete stop according to the equation (2.2.18),

$$s = (m+m_x)\frac{V_0^2}{V}\log_e\left(\frac{V_0}{V}\right) \quad \cdots\cdots (2.2.19)$$

(3) Emergency Stopping Inertia

For inertia when a propeller thrust is reversed in a steady running state, the time elapsed or distance run until the ship comes to a complete stop in water, the following equation (2.2.20) can be applied, letting astern thrust be "T_{ast}",

$$(m+m_x)\frac{dV}{dt} = -T_{ast} - R \quad \cdots\cdots (2.2.20)$$

As shown in Fig. 2.2.22, ahead thrust "T" reduces gradually after the propeller thrust is reversed, then, before long, astern thrust "T_{ast}" takes over and the ship's speed "V" drops abruptly. In that process, the relation among thrust "T", ship's speed "V" and resistance "R" varies intricately to each other until the ship comes to a complete stop in the water.

The equations of (2.2.21) and (2.2.22) are simplified formulae to deduce time elapsed and distance run of emergency stopping inertia derived from the equation (2.2.20). Equation (2.2.21) is to deduce ***short stopping time*** and equation (2.2.22) is to deduce ***short stopping distance***, from the time when the propeller thrust is reversed until a ship comes to a complete stop.

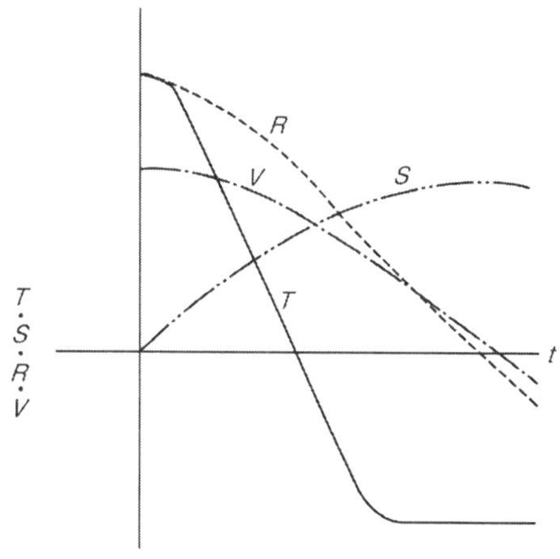

Fig. 2.2.22 Emergency stopping inertia

Short stopping time from the time when the propeller thrust is reversed to the time when the ship comes to a complete stop, where initial speed is "V_0" and hull resistance $R = kV^2$, will be,

$$t = \frac{1}{k}(m+m_x)\sqrt{\frac{k}{T_{ast}}}\tan^{-1}\left(V_0\sqrt{\frac{k}{T_{ast}}}\right) \cdots\cdots (2.2.21)$$

Short stopping distance from the time when a propeller thrust is reversed to the time when the ship comes to a complete stop is,

$$s = \frac{1}{2k}(m+m_x)\log_e\left(1+\frac{kV_0^2}{T_{ast}}\right) \cdots\cdots (2.2.22)$$

Aside from the above, the simplified formulae derived from the principle of conservation of energy according to which the change of kinematic energy is equal to the work done for that time are presented as follows.

Based on the fact that the kinematic energy of a ship with initial speed "V_0" is consumed by the work done by the astern thrust "T_{ast}" for the corresponding period while the ship runs a distance of "S",

$$\frac{1}{2}(m+m_x)V_0^2 = T_{ast} \cdot S \quad \cdots\cdots (2.2.23)$$

$$\therefore s = \frac{1}{2}(m+m_x)\frac{V_0^2}{T_{ast}} \quad \cdots\cdots (2.2.24)$$

As the change of momentum for the elapsed time until the ship comes to a complete stop in water is equal to the impulse of astern thrust "T_{ast}" and time "t",

$$(m+m_x)V_0 = T_{ast} \cdot t \quad \cdots\cdots (2.2.25)$$

$$\therefore t = (m+m_x)\frac{V_0}{T_{ast}} \quad \cdots\cdots (2.2.26)$$

Although the equations (2.2.24) and (2.2.26) seem simple compared with the equations (2.2.21) and (2.2.22), they are useful for knowing the approximate values of "t" and "s"; the characterized index of emergency stopping inertia when thrust is reversed in low speed free from consequential abrupt speed change.

(4) Speed Necessary to Stop a Ship in its Length

In applying knowledge about emergency stopping inertia in practical ship handling, particularly when approaching the berth in port, it is important to know beforehand the ship's speed necessary to stop the ship within its length. For this purpose, the following simplified formula is presented to deduce the necessary speed.

To stop a ship which length is "L" by astern thrust "T_{ast}" running at a speed "V_0" within its length can be obtained substituting "L" for "s" in equation (2.2.24).

$$V_0 = \sqrt{\frac{2T_{ast} \cdot L}{(m+m_x)}} \quad \cdots\cdots (2.2.27)$$

For your information, when you intend to stop a ship within half its length, substitute $s = 1/2\ L$ for "s" in equation (2.2.24).

For instance, when a ship which has length and displacement of 100m and 20,000 tonf respectively is approaching the berth, the speed necessary to stop the ship within its length by 50 tonf astern thrust, assuming its added mass 10%, will be,

$$V_0 = \sqrt{\frac{2 \times 50\,\text{tonf} \times 100\text{m}}{1.1 \times \left(\frac{20{,}000\,\text{tonf}}{9.8\text{m/sec}^2} \right)}}$$

$$= 2.1\text{m/sec}$$

$$\fallingdotseq 4.1\text{knots}$$

2.3 Estimation of Inertia by Simulation

☞ Refer to Vol. 4 Ship Handling Characteristics Data Base

(1) Driving and Acceleration Maneuvers

According to numerical calculations by simulation, the change of speed after engine motion can be obtained chronologically. In addition, as inertia can be visualized different from the deduction of inertia based on the analysis of the equations of motion, it seems more advantageous to understand and deduce inertia.

Simulation calculations of driving, acceleration, deceleration, stopping and an emergency stopping inertia in deep water of a LNG carrier which specifications are listed in Table 2.2.1 were carried out.

Fig. 2.2.23 shows the increase of speed from the time when the engine was driven to S/B Full Ahead in the state that the ship is lying dead in the water in the ***driving inertia test***. You will see that the speed increased gradually after the

Table 2.2.1 Ship's specification of simulation calculation
(Calculating conditions for simulation : h/d = ∞)

Ship type	Condition	Lpp	B	df	da	C_b
130,000 m^3 LNG carrier	Full loaded	276	46.0	10.8	10.8	0.72
ME HP	ME revolutions per min (rpm)					
		N/F	S/BF	Half	Slow	DS
43,000 PS	Ah'd	86	54	46	36	25
	Ast'n	· ·	54	46	36	25

engine was set in motion and reached its steady state about 30 minutes later, and that the distance run for that period is 8,760m, about 32 times of the ship's length.

You will see that the speed increases rapidly at the initial stage after its engine was driven to S/B Full Ahead, but the rate of speed increase slows down as it nears its steady speed. In this process, the time and distance run required to change speed one knot are called ***driving inertia factors*** at the initial stage of starting the engine.

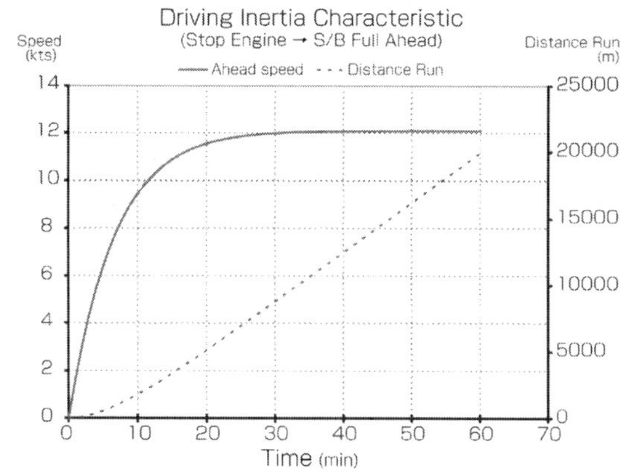

Fig. 2.2.23 Driving inertia test result (130,000m^3 type LNG carrier in full load, h/d = ∞)

As shown in Fig. 2.2.23, driving inertia factors vary depending on the place where measurements were taken. Paying attention to the driving inertia factors at the initial stage starting the engine of this ship, they are 0.8 min/knot, 86m/knot and 0.3L/knot.

Table 2.2.2 shows time required and distance run until steady speed is achieved and the driving inertia factors at the initial stage of starting the engine.

Table 2.2.2 Driving Inertia and driving inertia factors (130,000 m^3 type LNG carrier in full load, h/d = ∞)

Speed	(Driving Inertia) Stop→ S/B Full 0 knot → 12.0 knot		
Time and distance run until steady speed	Time	Distance	Distance/L
	30 min	8,760 m	31.7L
Average driving inertia factor	2.5 min/knot	730 m/knot	2.6L/knot
Driving inertia factor at the initial stage of starting engine	Time	Distance	Distance/L
	0.8 min/knot	86 m/knot	0.3L/knot

Fig. 2.2.24 shows the test results of speed increase when speed is increased step by step from Stop to Navigation Full Ahead (Stop → Dead Slow Ahead →

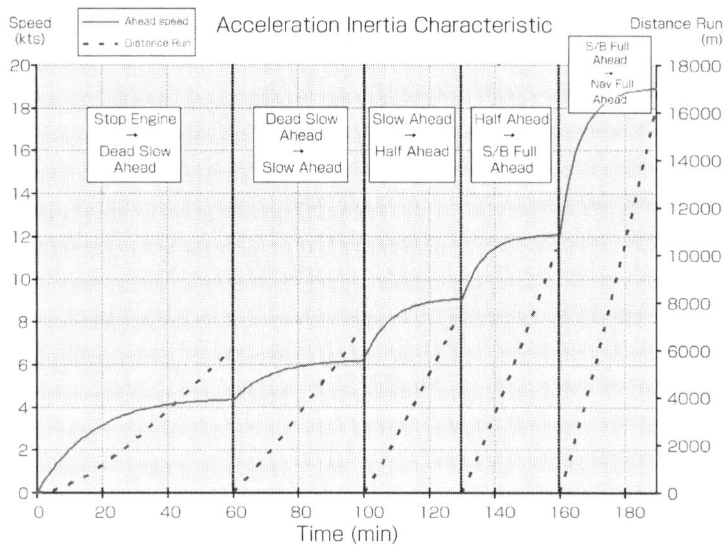

Fig. 2.2.24 Acceleration inertia test result
(130,000 m^3 type LNG carrier in full load, h/d = ∞)

Slow Ahead → Half Ahead → S/B Full Ahead → Navigation Full Ahead). Each time required and distance run to the next telegraph speed from the previous steady telegraph speed obtained from this figure is shown in Table 2.2.3 at every telegraph notch together with the ***acceleration inertia factors*** at each initial acceleration stage. The acceleration inertia factor means the time and distance run required to change speed one knot in the process of acceleration.

Table 2.2.3 Acceleration inertia and acceleration inertia factors
(130,000 m^3 type LNG carrier in full load, h/d = ∞)

	Stop → D. Slow	D. Slow → Slow	Slow → Half	Half → S/B Full	S/B Full → N. Full
Speed (knots)	0→4.3	4.3→6.2	6.2→9.0	9.0→12.0	12.0→18.9
Time required to next step steady speed (mins)	59	37	29	28	28
Distance run (m)	5952	6459	7489	9838	15309
Dist./Lpp ratio	21.6	23.4	27.1	35.6	55.5
Acceleration inertia factor at initial acceleration stage (Time)	5 min/knot	8.8 min/knot	3.9 min/knot	2.9 min/knot	1.3 min/knot
(Distance)	168 m/knot	134 m/knot	834 m/knot	883 m/knot	564 m/knot
(Dist./Lpp ratio)	0.61 L/knot	4.85 L/knot	3.0 L/knot	3.2 L/knot	2.0 L/knot

(2) Deceleration and Stopping Maneuvers

The results of deceleration inertia tests of the same ship from Navigation Full Ahead down to Stop step by step, Navigation Full Ahead→ S/B Full Ahead→ Half Ahead→ Slow Ahead→ D.Slow Ahead→ Stop, are shown in Fig. 2.2.25. Each time required and distance run per one knot reduction at every initial deceleration stage are identified. Table 2.2.4 shows these at every telegraph notch together with the ***deceleration inertia factors*** at each initial deceleration stage.

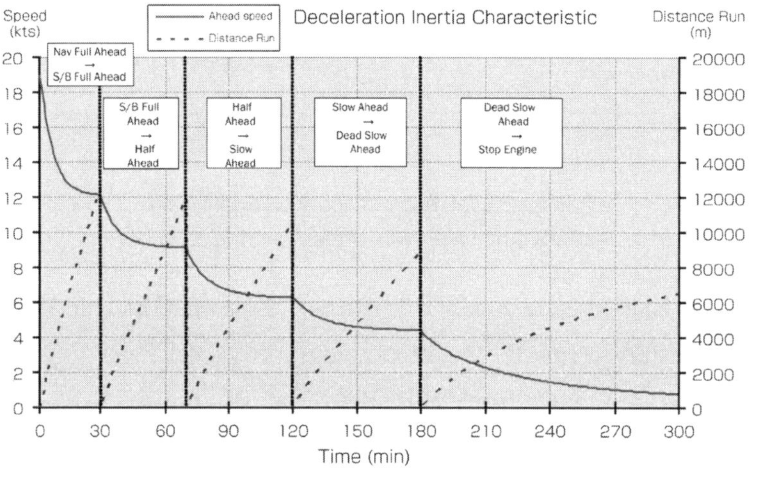

Fig. 2.2.25 Deceleration inertia test results
(130,000 m^3 type LNG carrier in full load, h/d = ∞)

Table 2.2.4 Deceleration inertia and deceleration inertia factors
(130,000 m^3 type LNG carrier in full load, h/d = ∞)

	N. Full → S/B Full	S/B Full → Half	Half → Slow	Slow → D. Slow	D. Slow → Stop
Speed (knots)	18.9→12.1	12.1→9.1	9.1→6.3	6.3→4.4	4.4→0.5
Time required to next step steady speed (mins)	28	33	40	60	180
Distance run (m)	11787	9967	8589	8916	7656
Dist./Lpp ratio	42.7	36.1	31.1	32.3	27.7
Deceleration inertia factor at the initial deceleration stage / (Time)	14 min/knot	3 min/knot	4.7 min/knot	8.3 min/knot	9.3 min/knot
(Distance)	733 m/knot	1079 m/knot	1226 m/knot	1538 m/knot	1148 m/knot
(Dist./Lpp ratio)	2.65 L/knot	3.9 L/knot	4.4 L/knot	5.6 L/knot	4.2 L/knot

The deceleration inertia factor is defined as the time and distance run required to change the speed by one knot in the process of deceleration.

Fig 2.2.26 shows the test result of stopping inertia to verify the relation between time required and speed reduction when the ship's engine is stopped in the state of running at a steady speed of S/B Full Ahead.

The results indicate that the ship's speed decreases rapidly immediately after Stop Engine but its gradient loosens as time passes due to the reduction of resistance to the hull.

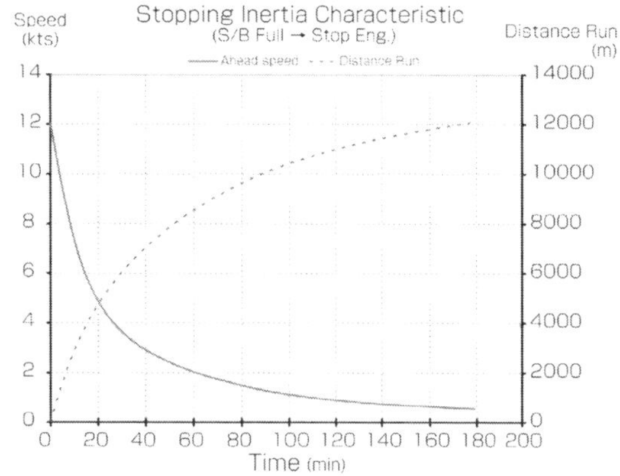

Fig. 2.2.26 Results of stopping inertia test (130,000 m^3 type LNG carrier in full load, h/d = ∞)

Generally, the characteristic of stopping inertia is evaluated until the ship's speed downs to 2 knots. In this simulation, time is measured until the ship's speed downs to 0.5 knots. In this ship, it took about 180 minutes and the distance run for that period is 12,114m (44L) until ship's speed decreases to 0.5 knots.

In this case, paying attention to the initial stage of stopping inertia where the ship's speed decreases abruptly, the time required to change the ship's speed by one knot is 2.1 min/knot, the distance run for that period is 622m and the ratio to ship's length is 2.3L/knot. These values are listed in Table 2.2.5.

Table 2.2.5 Stopping inertia and stopping inertia factors (130,000 m^3 type LNG carrier in full load, h/d = ∞)

Ship speed	S/B Full Ahead → Stop Eng. 12 knot → 0.5 knot		
Time required and distance run until stop	Time	Distance run	Distance run/L
	180 min	12,114 m	43.9 L
Average stopping inertia factor for the relevant period	15.7 min/ knot	1,053 m/knot	3.8 L/knot
Initial stopping inertia factor at the initial stage of stopping the engines	Time	Distance run	Distance run/L
	2.1 min/knot	622 m/knot	2.3 L/knot

(3) Crash Astern Maneuver

Ship maneuvers to stop a ship running at a steady rate of Full Ahead by reversing its engine to Full Astern is called ***crash astern maneuver***. Time required and distance run in crash astern maneuver are important characteristics for maneuvering in an emergency.

Fig. 2.2.27 shows the change of ship speed in crash astern maneuver of an LNG carrier in deep water for which specifications are listed in Table 2.2.1. In this case, the minimum time required to stop the ship and short stopping distance are 12 minutes and 3,148 m (11.4L) respectively. In crash astern, as strong astern thrust is given while a ship is running straight ahead, the deflection of the ship heading to starboard appears noticeably due to the multiplied effect of lateral force created by impingement due to propeller discharge current at the stern and sidewise pressure of the propeller. The deflection of the ship's heading to the starboard will be enhanced as ship speed decreases.

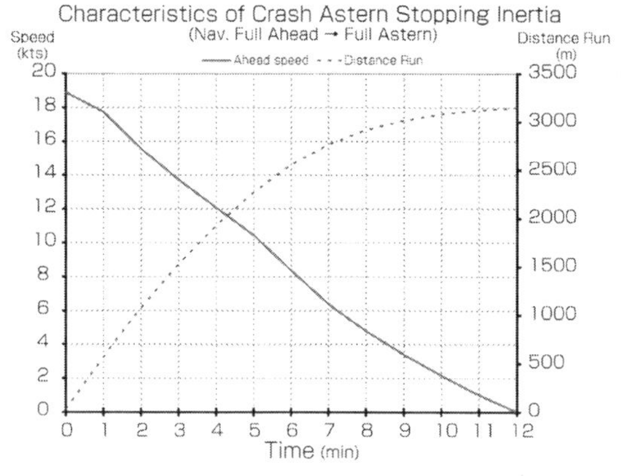

Fig. 2.2.27 Change of ship speed in crash astern maneuvers (130,000 m^3 type LNG carrier in full load, h/d = ∞)

In this case, the ship continues to go straight ahead while enough inertia remains, but deflection of the heading becomes increasingly apparent as its speed decreases. As shown in Fig. 2.2.28, at the time when the ship stops dead in the water, its rate-of-turn reaches 7.9°/min and its heading turns to 81°starboard.

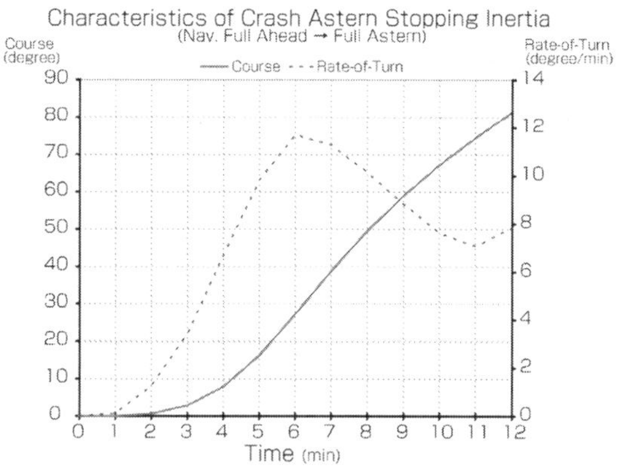

Fig. 2.2.28 Deflection of ship heading in crash astern maneuvers (130,000 m^3 type LNG carrier in full load, h/d = ∞)

3. Emergency Stop by Means of Crash Astern

☞ Refer to Vol. 4 Ship Handling Characteristics Data Base

3.1 Short Stopping Distance

It is not practical to carry out crash astern tests at sea to confirm the characteristics of crash astern maneuvers ship by ship. Instead, numerical and descriptive data can be derived more easily from simulation calculations by ship type. In addition, the data obtained through simulation calculations can be evaluated in comparison.

Fig. 2.2.29 shows the results of simulation calculations of crash astern maneuvers for the six typical ships above in Table 2.2.6, results for deep water are shown. Fig. 2.2.29 (a) shows the tracking of these ships from Full Astern in the steady running state at navigation Full Ahead to a complete stop in the water. While Fig. 2.2.29 (b) shows the tracks of these ships from Full Astern in the steady running state at S/B Full Ahead to a complete stop in the water.

Table 2.2.7 shows the short stopping distance, position where a ship comes to a complete stop in the water, direction of heading and rate-of-turn for each ship. Table 2.2.7 (a) and (b) respectively show data from Navigation Full Ahead to Full Astern and from S/B Full Ahead to Full Astern.

Table 2.2.6 Simulated calculations for crash astern maneuvers based on ship specifications

Ship type	Condition	Lpp	B	df	da	C_b
300,000 DWT VLCC (full)	Full loaded	324	60.0	20.5	20.5	0.86
10,000 DWT Bulker (full)	Full loaded	107	19.4	8.0	8.0	0.75
135,000 m^3 LNG carrier (full)	Full loaded	276	46.0	10.8	10.8	0.72
6,000 TEU container ship (full)	Full loaded	302.8	42.8	13.6	13.6	0.63
50,000 GT large cruise ship	Full loaded	209	29.6	7.5	7.5	0.62
4,500 unit PCC (full)	Full loaded	170	32.3	8.8	8.8	0.58

Table 2.2.7 Short Stopping Distance, Direction of Heading and Rate-of-Turn in Crash Astern Maneuvers

(a) Nav. Full Ahead → Full Astern, (h/d = ∞)

Ship type	Time to complete stop		When ship is dead in water				Ratio of y or x to L	
	Initial speed knots	T (min) t (sec)	Heading ψ (deg)	Rate-of-turn (deg/min)	Dist run y (m)	Drift x (m)	y/L	x/L
300,000 DWT VLCC (full)	15.2	(21.3) 1275	109	6.9	3973	1210	12.3	3.7
10,000 DWT bulker (full)	13.6	(5.3) 318	33	13.9	1126	51	10.5	0.5
135,000 m^3 LNG carrier (full)	18.9	(11.9) 716	81	7.9	3148	525	11.4	1.9
6,000 TEU container ship (full)	24.7	(8.3) 500	15	7.7	3555	22	11.7	0.1
50,000 GT large cruise ship	22.2	(2.9) 171	0	0	822	0	3.9	0
4,500 unit PCC (full)	19.9	(7.8) 470	35	10.5	2342	62	13.8	0.4

(b) S/B Full Ahead → Full Astern, (h/d = ∞)

Ship type	Time to complete stop		When ship is dead in water				Ratio of y or x to L	
	Initial speed knots	T (min) t (sec)	Heading ψ (deg)	Rate-of-turn (deg/min)	Dist run y (m)	Drift x (m)	y/L	x/L
300,000 DWT VLCC (full)	10.5	(16.8) 1010	77	6.8	2399	429	7.4	1.3
10,000 DWT bulker (full)	9.6	(3.9) 231	26	13.5	567	20	5.3	0.2
135,000 m^3 LNG carrier (full)	12.0	(8.9) 532	50	7.9	1548	112	5.6	0.4
6,000 TEU container ship (full)	18.1	(5.0) 299	16	7.6	1424	29	4.7	0.1
50,000 GT large cruise ship	11.0	(1.9) 114	0	0	318	0	1.5	0
4,500 unit PCC (full)	12.2	(4.6) 277	31	10.6	814	40	4.8	0.2

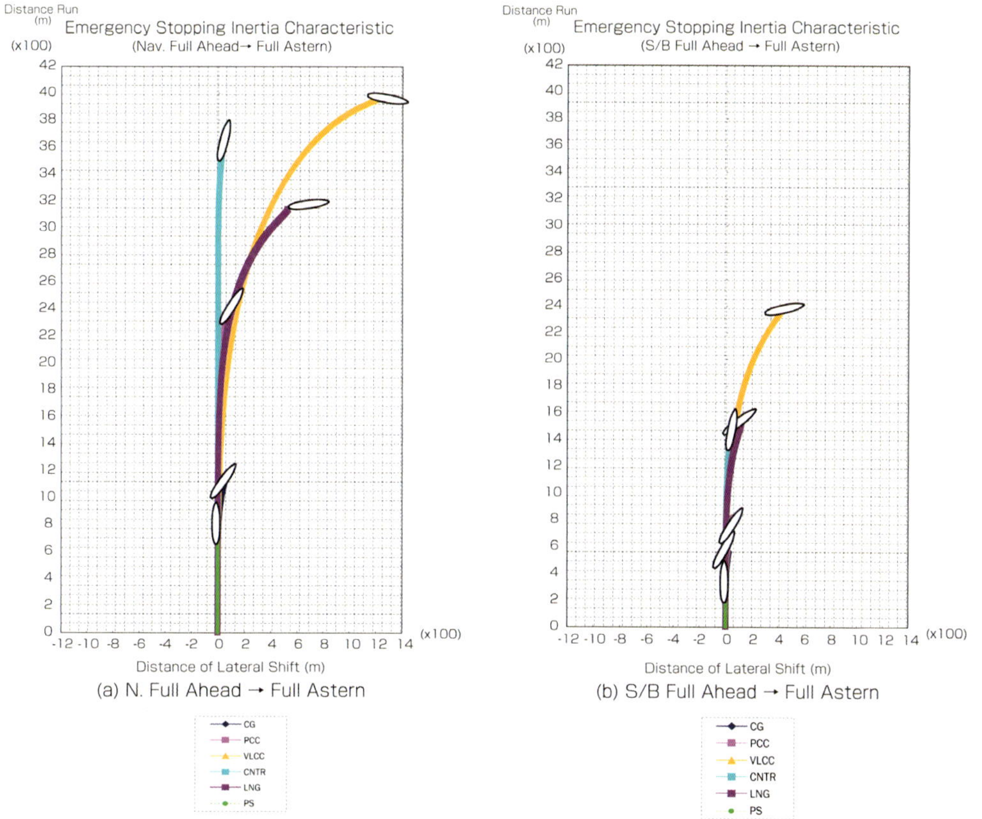

Fig. 2.2.29 Tracking of crash astern maneuver (h/d = ∞)

Referring to the data of typical types of ships in Fig. 2.2.29, you can understand that ships cannot be stopped easily even if a crash astern maneuver is done. For instance, the short stopping distances of these ships are about 4 km for a heavy VLCC, 3.5 km for a high speed container ship, 2 to 3 km for PCC and LNG carrier, and 1 km for a 10,000 DWT bulker. If these figures are compared to the ratio of ship length, it ranges from 10 to $14L$. However, for large cruise ship it is $4L$, fairly shorter compared with the other types of ship.

In respect to lateral shift distances from their original course until the ships come to a complete stop in the water, the range is from about $4L$ for VLCC and about $2L$ for LNG carrier toward the starboard side and for the other types of ships it is quite small. Particularly for container and large cruise ships provided with twin screws, drift is almost negligible because the effects of propeller discharge currents offset each other.

By contrast, the short stopping distances when the engine is reversed from S/B Full Ahead to Full Astern are about $7L$ for VLCC and about $5L$ for the other types of ships. The short stopping distance of large cruise ships fitted with pod type electric propulsion units is about $1.5L$, fairly shorter compared with other types of ships.

3.2 Deflection of Heading

Deflection of a ship's heading following crash astern maneuvers is a peculiar phenomenon to ships fitted with a right handed single propeller mainly due to the lateral force created by impingement due to propeller discharge current at the stern and the sidewise pressure of the propeller. Among ship types in Table 2.2.7, by referring to Fig. 2.2.29, this phenomenon emerges distinctly for VLCCs and LNG carriers in which headings turn almost completely to a right angle from their original course until they come to complete stop in the water, while the deflection of headings of the other types of ships is about 30°. At this stage, 8 to 10°/min of rate-of-turn remains constant regardless of ship type.

In the case of S/B Full Ahead to Full Astern, the deflection of heading is smaller proportionally to the time required until a ship comes to a complete stop. In this case, however, 8 to 10°/min rate-of-turn still remains. That is to say, as deflection of a ship's heading which is caused by the lateral force created by impingement due to propeller discharge current at the stern and sidewise pressure of the propeller continues as long as the astern engine is kept running, the longer the time required to stop the ship, the greater the defection of the ship's heading becomes.

However, it should be noted that as external disturbances such as wind and currents are not included in simulation calculations, the data does not necessarily reflect realistic ship motion at sea. A ship's behavior after its engine is reversed may vary greatly depending on the ship's own condition such as fouling of bottom and trim, the ship's inertia, the prior handling process, external disturbances of wind or current at sea and geographical or environmental conditions such as water depth and contour of the seabed at that point. Therefore persons engaged in ship handling should always pay attention to be able to judge correctly whether deflection of a ship's heading will actually occur, and if it does, how much and to which direction the ship's heading will be deflected.

3.3 Elements Affecting Short Stopping Distance

(1) Ship Type and Size

Generally, the heavier the weight of a ship, the longer its stopping distance becomes proportionally to its inertia. In other words, stopping distance of full load conditions are longer than that of ballast conditions, and the larger the block coefficient in which virtual mass is large, the longer the stopping distance becomes.

(2) Astern Maneuvering

The faster the ship's speed just before the reversion of the engine and the weaker the reversing thrust, the longer the stopping distance becomes. If it takes a long time to reverse the engine, stopping distance also becomes longer as the ship continues to go ahead for that period. In this sense, stopping distances of ships fitted with a turbine engine or a fixed pitch propeller become longer compared with ships fitted with a diesel engine or a controllable pitch propeller as they require longer time to maneuver.

(3) Effect of External Disturbances

When a ship receives wind from dead ahead, the stopping distance decreases as wind pressure increases proportionally to its surface area. Likewise, where the depth is shallow, it decreases as hydraulic resistance due to the shallow water effect increases. If a ship's bottom is fouled with marine growth or creatures, stopping distance decreases as hydraulic resistance increases. On the contrary, caution is necessary when tidal currents come from behind as stopping distance over the ground extends.

Ⅲ Maneuvering by Thruster and Tug

Chapter 1 Maneuvering by Thruster

1. Installation of Thrusters

1.1 Function of Thrusters

A ***side thruster*** is a propeller fitted inside a trunk hollowed out side to side at a ship's bow and/or stern below the water surface to give a ship lateral thrust with its discharge current. A side thruster fitted at the bow is called ***bow thruster*** while one fitted at the stern is called ***stern thruster***.

Recently the number of ships fitted with thruster as an auxiliary device to move ships laterally or turn around, in addition to the conventional ship handling means of the primary engine and rudder, have been increasing. The reason why their installation has been increasing is because their effectiveness in both safety and economy has been recognized. As an auxiliary means of ship handling such as supporting tugs for PCCs, LNG carriers and container ships which have large surface areas thrusters can be useful, especially in berthing or un-berthing operations.

For merchant ships, exclusive installation of a bow thruster is common. In contrast, some larger passenger ships are fitted with a stern thruster along with a bow thruster.

1.2 Output of Thrusters

Although the output of thruster per unit installed on general merchant ships may vary depending on a ship type, they range from 1,000 to 1,500 PS for PCCs, 1,500 to 2,000 PS for container ships and 2,000 to 2,200 PS for LNG carriers.

(Today according to the International System of Unit, watt is used instead of horse power. 1 Horse Power (PS) = 0.7355 kW)

When a thruster is fitted far enough below the water surface and the ship is lying dead in the water, the discharge current from the thruster can create a reactionary force effectively. Under such conditions, the relation between the rated thrust power and actual thrust power of the thruster can be expressed as in the equation (2.3.1) below.

$$\left.\begin{aligned} T &\fallingdotseq 1\ \text{ton}/100\ \text{PS} \\ T &\fallingdotseq 1.5\ \text{ton}/100\ \text{kW} \end{aligned}\right\} \quad \cdots\cdots (2.3.1)$$

2. A Ship's Motion Due to Bow Thrusters

2.1 Turning of Heading and Lateral Shift of Hull

Thrust generated by a bow thruster contributes to a ship's motion in two ways. As shown in Fig. 2.3.1, one works to turn the ship around its center of gravity and the other works directly on its center of gravity to move the ship laterally.

In particular, the latter effect of a bow thruster is created by the force acting on the ship's center of gravity, of which scale and direction are equal to the thrust power of the bow thruster as explained in the characteristics of vector in Fig. 2.3.2. It should be noted that motion created by the force of bow

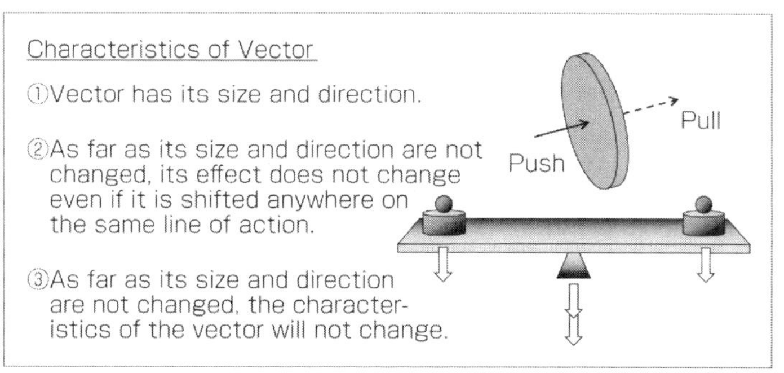

Fig. 2.3.2 Characteristics of vector

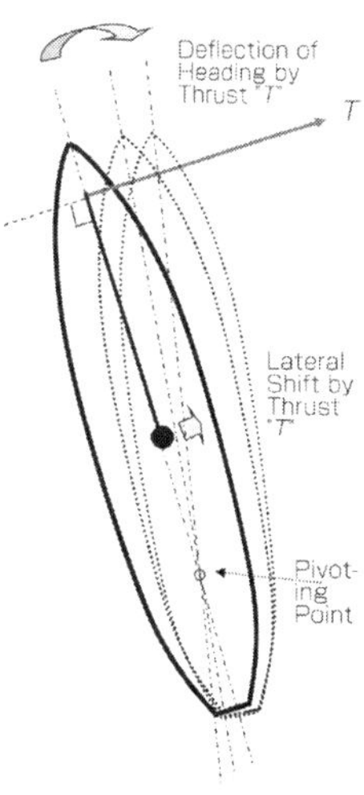

Fig. 2.3.1 Motions created by bow thruster

thruster works not only to turn the ship's heading but also to move the ship laterally.

2.2 Turning Moment Generated by Bow Thrusters

The main purpose of a bow thruster is to create a turning moment to the ship. From Fig. 2.3.1, turning moment "M" to the ship can be expressed as follows where thrust power of the bow thruster is "T" and distance between the ship's center of gravity and the bow thruster is "X".

$$M = T \times X \qquad (2.3.2)$$

You can easily understand from the above equation that the larger the thrust power "T" or the longer the lever of moment "X", the larger the effect of the bow thruster grows. Because of this, the bow thruster is usually, as far as structurally possible, fitted closer to the bow. From the statistics of PCCs, LNG carriers and container ships fitted with bow thrusters, their position range from around 3% of Lpp for LNG carriers, 3 to 7.5% Lpp for container ships and 5 to 10% Lpp for PCCs respectively from their fore perpendiculars. (Lpp: Length between perpendiculars)

2.3 Turning by Bow Thrusters and Location of the Pivoting Point

A ship is turned and simultaneously makes lateral movement by thrust power generated by bow thrusters. However, due to large hydraulic resistance perpendicular to the hull at the initial stage, lateral motion is quite slow compared with turning motion. At first, the ship starts to turn around a point closer to its stern on the ship's longitudinal center line. This point is a pivoting point as explained previously and can be derived from the equation (2.1.49) if the position of the bow thruster from the ship's center of gravity "GC" and radius of gyration "k" can be identified.

The value of "k" is defined by the longitudinal weight distribution of the ship. For instance, if its weight is concentrated around the center of the ship, "k" becomes smaller. By contrast if its weight is apportioned to the bow and stern,

"k" grows larger. Usually the value of "k" can be deemed as $1/4L$ to $1/3L$ (L: ship length). Therefore, if the bow thruster is fitted closer to the bow, location of the pivoting point can be deduced as 0.1 to $0.2L$ after the ship's longitudinal center of gravity from equation (2.1.49).

Rate-of-turn "r" grows larger and the velocity of lateral motion of the center of gravity "v" emerges as time elapses. The location of the pivoting point at this moment can be derived from the equation (2.1.51). As "v" gradually increases compared with the growth of the rate-of-turn "r" as time passes, the pivoting point also moves largely toward the stern.

Finally the pivoting point rests on a position 0.4 to $0.45L$ aft from the center of gravity and the ship turns around at a point closer to the stern. Then, we can come back to equations (2.1.49) and (2.1.51) to explain a ship's motion as shown in Fig. 2.3.3.

$$GP = \frac{k^2}{GC} \quad \cdots\cdots (2.1.49)\ \text{(reproduction)}$$

$$GP = \frac{v}{r} \quad \cdots\cdots (2.1.51)\ \text{(reproduction)}$$

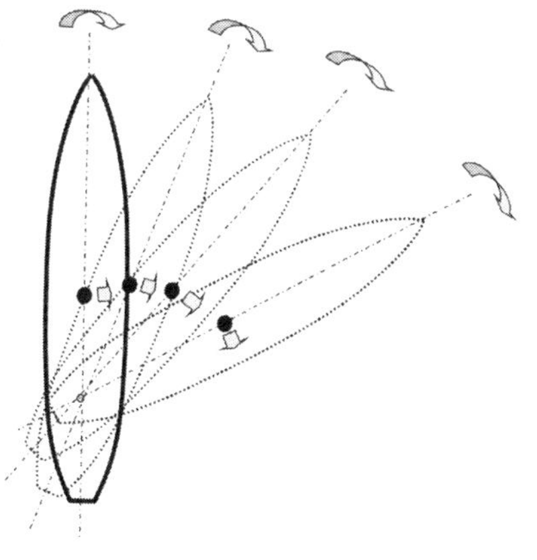

Fig. 2.3.3 Turning by bow thruster and location of the pivoting point

2.4 Turning Effect by Bow Thrusters

Fig. 2.3.4 shows the simulation results of an LNG carrier in Table 2.2.1 using the bow thruster to turn in deep water. In Fig. 2.3.4, to evaluate the effects of the bow thruster comparatively, simulations were made in the following three cases, when the ship has headway, when the ship is dead in the water and when the ship has sternway.

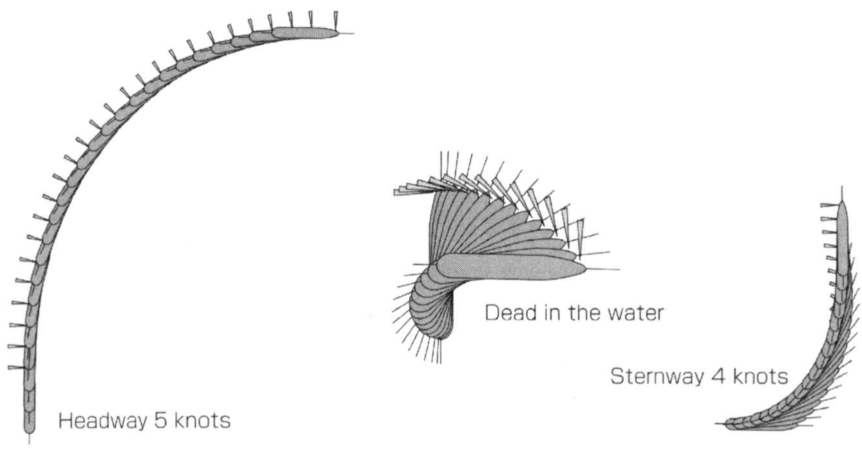

Fig. 2.3.4 Turning effect by bow thruster
(Headway, Dead in the Water and Sternway, h/d = ∞)

Comparing time required and rate-of-turn while turning by the bow thruster until the ship's heading turns to 90° when headway is 5 knots and sternway 4 knots, in the former case it took 15 minutes until 8°/min of rate-of-turn was gained while in the latter case it took 11 minutes to gain 10°/min of rate-of-turn when the ship's heading turned to 90°. Significant difference could not be found between the two cases above.

However, in the case which the bow thruster was started when the ship was dead in the water, rate-of-turn reached 18°/min within 7 minutes. This demonstrates the fact that when a ship has 4 to 5 knots headway or sternway, the effect of turning by bow thruster will be significantly reduced.

3. Reduction of Bow Thruster Effect under Increasing Headway

The effect of bow thrusters can be maximized when the ship is lying dead in the water. When the ship has headway, the force of discharge current from the bow thruster is weakened by the flow around the hull. The above fact that the effect of bow thrusters will be decreased as a ship's headway increases should be kept in mind when you use bow thrusters in ship handling.

The relation between the reduction of the effect of bow thrusters and ship speed can be expressed by the following equation (2.3.3).

$$T = T_0 \cdot e^{(-0.3V)} \quad \text{(2.3.3)}$$

where

T : effective thrust at "V" knots
T_0 : effective thrust when ship is dead in the water
V : ship speed (knot)

According to the equation (2.3.3), the effect of a bow thruster is reduced by half when the ship's speed is 2 to 2.5 knots, and its effect will be reduced further to less than 20% when headway is 6 knots. It would be better to think that 6 knots of headway is the limit of the effective use of a bow thruster. In addition, caution is necessary considering the fact that when the ship is in light draft or in trim by the stern so that the bow thruster may be placed close to water surface its ability will be decreased due to aeration or loss of discharge current pressure disturbed by surface waves.

4. Maneuvering by Bow Thruster

4.1 Turning

When proper headway is maintained, turning by the rudder is possible. However, when ship speed decreases to 5 to 6 knots, particularly if the ship is going only by inertia force, to maintain steerability of the ship by the rudder becomes increasingly difficult. In such cases, the use of a bow thruster is effective.

When headway is small, the ship can be made to turn around a certain point at the stern solely by the bow thruster. However, it should be kept in mind that when the bow thruster is used, the ship will simultaneously be moved laterally as time elapses to the right in starboard turning and to the left in port turning.

4.2 Course Keeping

When a ship is proceeding at low speed under wind, it is effective to use the bow thruster to make the ship's heading turn leeward to prevent its heading from turning into the wind and to keep the ship's course. However, it should be noted that in such cases sometimes unexpected large leeward drift may be created by the multiplied effect of leeway by the wind and leeward lateral shift by bow thruster.

4.3 Astern Straight

Although a ship is obliged to turn its rudder starboard while it is going astern to prevent clockwise turning of the heading without a bow thruster, a bow thruster is useful making a ship go straight astern canceling clockwise turning of the heading due to the lateral force created by impingement due to the propeller discharge current at the stern and the sidewise pressure. However, as the rudder effect sometimes becomes poor while the ship is going astern, remarkably, a bow thruster can be relied upon to adjust the deflection of ship's heading in such cases.

4.4 Anchoring

When a ship is riding on a single anchor under strong wind, she moves violently drawing a figure-eight swinging motion. A bow thruster is useful to restrain such swinging motion and minimize the risk of dragging the anchor.

4.5 Berthing and Un-berthing

When a ship is approaching the berth on her starboard and its engine is turned to astern in front of the berth, its bow tends to turn closer to the berth because of the effect of lateral force created by impingement due to propeller discharge current at the stern. As a ship in front of berth is almost dead in the water, a bow thruster is useful to adjust the ship's heading regardless of the alongside board and whether it is berthing or unberthing. However, caution should be paid to lateral shift by its force whenever a bow thruster is used.

Chapter 2 Maneuvering by Tugs

1. Maneuvering Assisted by Tugs

1.1 Necessity of Tugs

In the past when ships had a gross tonnage exceeding 10,000 they were considered as large ships; ship handling inside ports and harbors used to be done relying on a ship's own equipment and devices such as main engine, rudder, anchors, and mooring ropes.

Today, ports and harbors have become too narrow for hypertrophied ships to move by their own facilities within such restricted water areas without the assistance of tugs. At present, for such large ships, their own equipments such as the main engine and rudder have reduced to auxiliary means in ship handling and tugs have become leading and indispensable means for ship handling for berthing and un-berthing.

1.2 Effective Use of Tugs

One of the important skills required of a person engaged in ship handling inside a harbor is how to handle tugs effectively. In ship handling under the assistance of tugs, it is of great importance to make the tug masters employed in berthing or un-berthing operations to understand what the person in command of ship handling intends or envisions about the process and procedures of that ship's handling, including means of communication beforehand.

To perform safe ship handling the person engaded in ship handling and the tug masters must work cooperatively, relying on each other. In building reliability, the first step is for the person conning the ship to be conversant to the characteristics, performance and operation of tugs, and acquire skills on how to control a large ship's inertia using the ability of tugs.

2. Performance Characteristics of Tugs

2.1 Types of Tugs

Today, the numbers of tugs that are devoid of a rudder have been increasing instead of the conventional type of tugs fitted with both a propeller and rudder. The types of tugs without rudders can be sorted into two types, Voith Schneider Propeller and Z-Drive Propeller.

As shown in Fig. 2.3.5, a Voith Schneider Propeller comprised of a rotating horizontal disc and 4 to 6 vertical blades fitted to it, with which thrust can be adjusted by changing their pitch. Generally superior maneuvering performance can be gained by fitting two units per tug. However, the short coming of this propeller is that each propeller blade is obliged to be placed idle at every half rotation of the horizontal disc. In this respect, it is said that the ***Voith Schneider Propeller*** is inferior to ***Z-Drive Propellers*** in pulling and towing capacity.

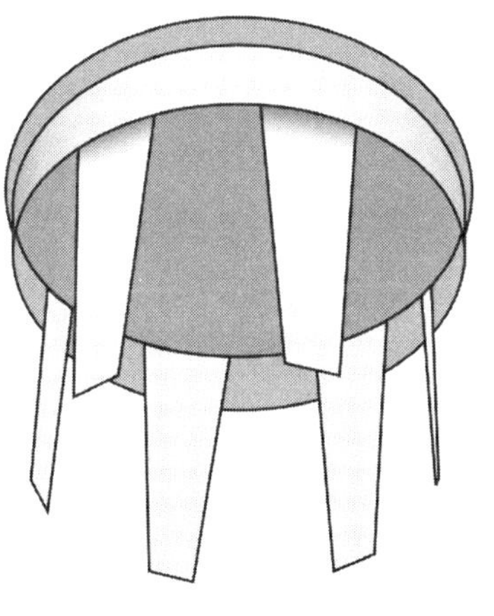

Fig. 2.3.5 Voith Schneider Propeller

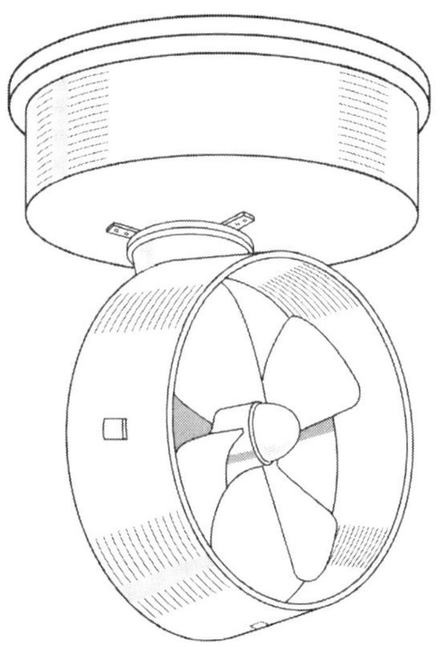

Fig. 2.3.6 Z-Drive Propeller

As shown in Fig. 2.3.6, a Z-Drive Propeller can be rotated 360°on the horizontal plane. If two units of this type are fitted, ahead, astern, turning, oblique and lateral movements, sudden stop and turn, and turning round on the spot will be possible by a combination of the directions of the two propellers.

Today, Z-Drive Propellers combined with Kolt Nozzles which are able to gain larger thrust are superior in towing ability and maneuverability compared with Voith Schneider Propeller and have been widely used according to their recognized performance.

2.2 Towing Ability of Tugs

The thrust value of a tug measured by tying one end of tow rope to a fixed shore bollard is called ***bollard Pull.*** The ability of a tug can be compared based on bollard pull. Generally, 100 PS bollard pull of a Voith Schneider Propeller can be converted to 1 ton in thrust power while that of a Z-Drive Propeller can be converted to 1.5 tons in thrust power. When astern, its thrust power is roughly 80% that of ahead.

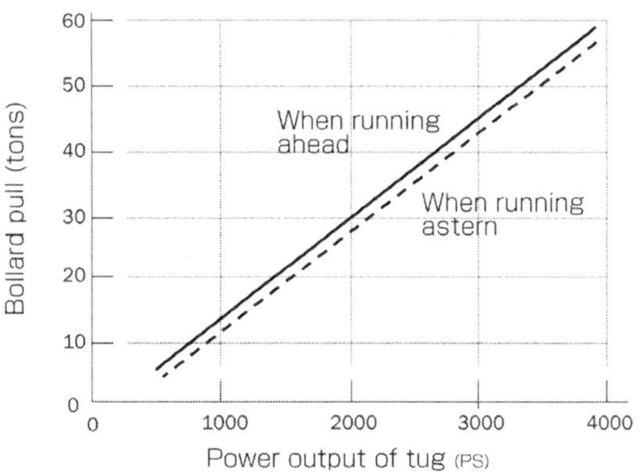

Fig. 2.3.7 Bollard pull of Z-Drive tug

Fig. 2.3.7 shows the relation between horse power of a tug and bollard pull of Z-Drive Propeller. From the graph, you will see examples of equivalent ahead and astern thrust power for each output, i.e. 34 tons in ahead and 32 tons in astern for 2,400 PS, and 52 tons in ahead and 48 tons in astern for 3,600 PS.

2.3 Limit of Headway for Tug Operation

When a ship keeps headway beyond a certain speed, pull and push of tugs will be ineffective because the tugs are dragged by the ship. Moreover, such conditions involve the risk of overturning tugs.

For instance, when tugs are used to push or tow laterally under the condition that ship speed is beyond 3 knots, special attention is necessary. When ship speed reaches 4 to 5 knots, safety of tugs becomes quite precarious. And it should be considered that when ship speed reaches 6 knots, the limit of a tug's assistance is overstepped. It is generally said that the limit of ship speed that a tug's assistance can be expected is 4 knots in the case of pulling and 5 knots in the case of pushing in view of safety and effectiveness. Even if a tug is taken at the stern like a dangling pendant to be used as a brake, its operational limit would be 6 knots although the above mentioned risk may be somewhat relaxed.

2.4 Utilization Method of Tugs

There are four fundamental patterns for tug arrangement as shown in Fig. 2.3.8, although the number of tugs, arrangement and each role may vary case by case. Persons engaged in ship handling should use tugs by understanding the appropriate role of each tug taking into account the number of tugs and their arrangement.

(1) Pulling

To assist the headway of a ship, a tow line can be attached from the tug's hook to the bow of the ship to be towed. This tug can also be used to turn the ship. If this type of tug is attached at the stern, it can be used as a brake. If two tugs are taken fore and aft on the same board, lateral or oblique movement of the ship is possible. By this mean of lateral or oblique movement, pulling a ship out from a berth is possible. In these cases, a tow line twice the length of the tug is necessary so that resistance due to the tug's astern discharge current towards the ship being towed will not impede the tug's pulling force.

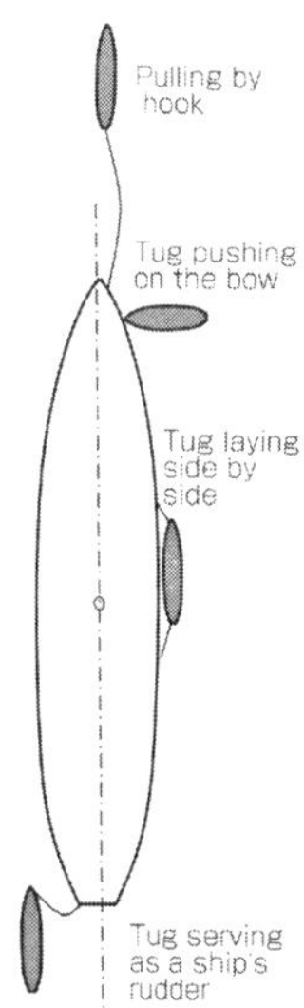

Fig. 2.3.8 Four basic patterns in using tugs

(2) Pushing

A tug is deployed to push a ship taking it at fore or aft, close to the bow or stern. A tug's thrust can be used to create lateral movement of a ship and to turn its heading. If a tug's lines are taken, they can be used for pulling as well.

For ships with a large flare at the bow and stern, if the position of tugs taken are too close to the flared parts, caution is necessary because of the increased risk of damage to the outer shell of the ship's superstructure or the tug's.

(3) Alongside

When a tug is taken alongside parallel to the ship with the tug's bow and stern lines, its fore and aft thrust can be used instead of the ship's main engine. This style cannot be used when the effect of waves is strong as a tug's free motion is quite restricted.

(4) Heading Control

One tug is taken at the stern with a long tow line to control headway of the ship by pulling. The tug can be used for pulling or pushing by appropriately adjusting the length of its tow line. When its line is not used, the tug is used exclusively to turn the ship like the ship's rudder.

3. Maneuvering under Tug Assistance

3.1 Shift of Pivoting Point under Tug Assistance

(1) Effect of Tug Force to Ship Lying Dead in the Water

The initial location of the pivoting point of a ship lying dead in the water emerges depending on a point where the tug's thrust is given. And it also moves along the longitudinal centerline of the ship to be pushed or pulled corresponding to the development or change of the rate-of-turn and lateral drifting speed of the ship.

Fig. 2.3.9 and Fig. 2.3.10 show the simulation results of the change of the pivoting point and the locus of a 2,800 TEU container ship for which specifications are given in Table 2.1.3 depending on a point where the tug's thrust is given.

Where, the condition of the container ship is under full load, water depth "h" to ship's draft "d" ratio is $h/d=1.3$, and the tug's thrust power is 22.5 tons, the line of action of the tug's thrust is always perpendicular to the longitudinal center line of the ship, the points of action of the tug's thrust are the following 6 points, (A) bow's end, (B) $0.2L$ after bow's end, (C) $0.3L$ after bow's end, (D) $0.7L$ after

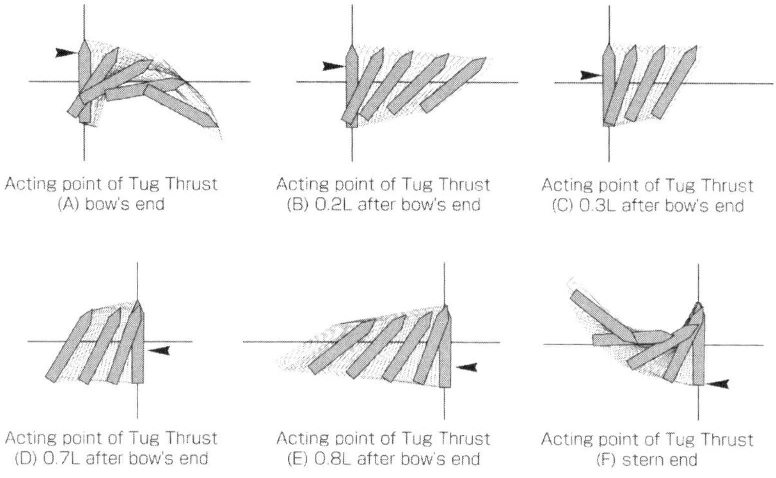

Fig. 2.3.9 Tug assistance when a ship is dead in the water and ship motion (2,800 TEU container ship, full load, h/d = 1.3, thrust power 22.5 tons)

bow's end, (E) 0.8L after bow's end and (F) stern end.

The initial location of the pivoting point at the instance when the tug's thrust is given emerges solely dependent on the point where its thrust was given regardless of the scale of its thrust force. When thrust is given to the end of the bow, the pivoting point at first emerges in the vicinity of 0.1L after the center of gravity. Thereafter, it moves aft as turning develops until turning angle reaches 60°. When the turning angle is beyond 60°, the pivoting point moves forward towards the midship as the headway motion prevails over turning motion. When

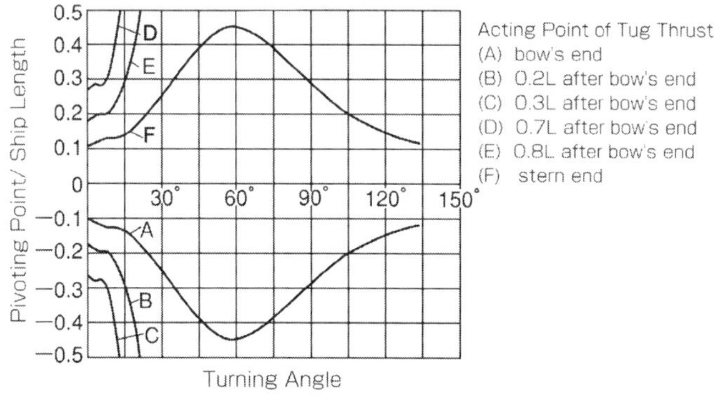

Fig. 2.3.10 Acting point of tug force to ship dead in water and pivoting point (2,800 TEU container ship, full load, h/d = 1.3, thrust power 22.5 tons)

thrust is given to the end of stern instead, the shift of the pivoting point will be reversed.

When a tug's thrust is given to a point close to the ship's center of gravity, the ship is moved laterally, turning slightly. In such a case, the pivoting point exists remote from the ship's bow end or stern end. Thus, you will understand that the pivoting point does not always exist onboard the ship but emerges on some remote point of the extended longitudinal centerline of the ship depending on the ship's motion.

(2) Effect of Tug Force in Large Angle Course-Alteration

Fig. 2.3.11 shows the change of the pivoting point of a 2,800 TEU container ship for which specifications are given in Table 2.1.3, derived from simulation calculation when it is turned to 90° by a tug. In which, letting the conditions of the ship be as follows, full load, initial speed 4 knots, angle of rudder 35° and water depth "h" to ship's draft "d" ratio $h/d=\infty$.

The thrust of each tug is 7.5 tons. The tug's thrust is given to the tips of bow and stern. Acting direction of the tug's thrust is always perpendicular to the longitudinal centerline of the ship. The arrangement of tugs is as follows; three patterns, (A) two tugs at the bow for simultaneous pushing and pulling in accordance with the ship's rudder actions, (B) two tugs at the stern for simultaneous pushing and pulling in accordance with the ship's rudder actions, (C) one tug at the bow and the other at the stern for simultaneous pulling at the bow and pushing at the stern in accordance with the ship's rudder actions.

In the three patterns above, the same scale of turning moment around the ship's center of gravity can be created in each case, but lateral motions differ. In the case of (A) where two tugs at the bow engage pushing and pulling simultaneously, at first the pivoting point emerges close to the center of gravity of the ship as the effects of the tugs and the ship's rudder offset each other, thereafter it moves forwards and finally rests at a position $0.2L$ forward of the center of gravity as turning motion develops. This is because the resultant lateral drifting speed becomes smaller as the effect of lateral force by the tug's thrust

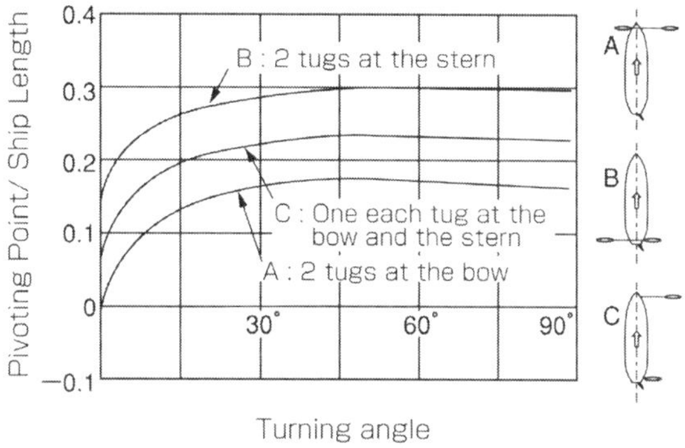

Fig. 2.3.11 Effect of tug force in large angle course-alteration and pivoting point (2,800 TEU container ship, full load, initial speed 4 knots, angle of rudder 35°, h/d = ∞, thrust power 7.5 tons per each)

and the effect of the ship's rudder force are offset each other. As a result, stern kick can be reduced to 90% of that when the ship is turned only by its own rudder.

In the case of (B) in which two tugs at the stern engage pushing and pulling simultaneously, the pivoting point is located on the farthest forward position among the three patterns above and exists closer to the bow at not only the initial stage but also during the process of turning, because both the forces of the tugs and the ship's rudder act on the stern. And finally it rests $0.3L$ forward from the center of gravity. This is because the resultant lateral drifting speed is multiplied by the lateral force of the tugs and the ship's rudder. As a result, stern kick is increased about 25% compared with that of when the ship is turned only by its own rudder. However, in view of turning efficiency, this can be the best method as its pivoting point is located on the farthest forward position among the three patterns.

In the case of (C) in which there are two tugs, one at the bow and the other at the stern, the location of the pivoting point exists between (A) and (B). However, the reduction of a ship's speed during turning is the least among the three patterns. Comparing the time and space required for turning of each of the three patterns in which tugs are used with or that of without tug assistance, they can be reduced to 70% in time and 50 to 60% in space.

(3) Effective Arrangement of a Tug in Large Angle Course-Alteration

When a ship is going to make a large angle course-alteration within the smallest turning circle under the assistance of a tug, an arrangement of one at the stern for pushing or one at the stern for pulling seems the most effective, even though stern kick increases, this is because the pivoting point is always located at the farthest forward position.

Comparing the efficiency of pushing with the efficiency of pulling, pushing is more effective. Therefore, when tug assistance is used in turning while a ship is going ahead, pushing is most effective and pulling follows. The same can be said in cases in which a ship is going astern.

3.2 Dynamic Analysis of Tug Assistance

(1) Effects of Resultant Force and Moment

It is an important ability for persons engaged in ship handling to estimate how the ship will react when multiple forces of tug thrust and other external disturbances act on their ship simultaneously. It is too complicated to make a precise estimation of a ship's motion after forces act on the hull taking the ship's inertia and resistance into account. However, rough estimation will be possible if the scale and direction of the resultant force to the ship's center of gravity can be known.

Here, I will show the analytical method to solve the resultant force and moment when multiple forces act on a ship's hull. For example, when as shown in Fig. 2.3.12, the force of two tugs, one at the starboard bow for pulling to an oblique forward direction (thrust power : F_1) and the other at the starboard stern for pushing to an oblique forward direction (thrust power : F_2), in addition to a ship's own bow thruster (thrust power : F_3), main engine (ahead thrust power : F_4) and rudder force (F_5), fundamental procedures to solve how the resultant force and moment act on the ship's center of gravity are as shown below.

① Define a perpendicular coordinate (G - XY) centered on the ship's center of gravity as the origin, the ship's longitudinal center line is the Y-axis and

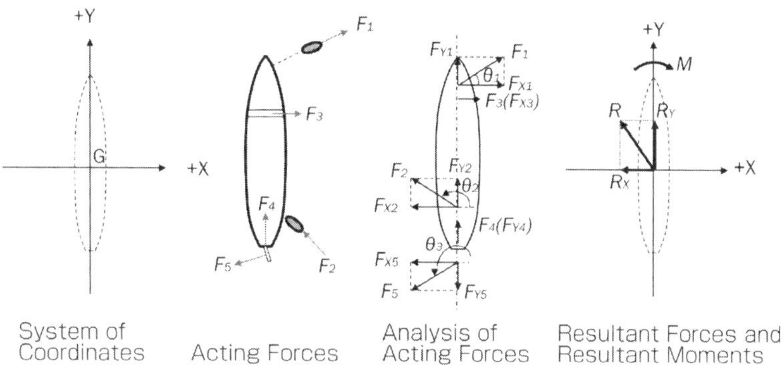

Fig. 2.3.12 Dynamic analysis of acting forces and moments

transverse line passing the ship's center of gravity is the X-axis. As shown in Fig. 2.3.12, a positive sign is given for forward and starboard side.

② Decompose acting forces (F_i, $i = 1 - 5$) on the coordinates.

③ Sum up X and Y-axis components of acting forces according to equation (2.3.4) and find the resultant transverse (X-axis) force "R_X" and the resultant longitudinal (Y-axis) force "R_Y". In this instance, angle "θ" must be measured counterclockwise from the positive X-axis.

$$\left.\begin{aligned} R_X &= \Sigma(F_{X_i} \cdot \cos\theta_i) \\ R_Y &= \Sigma(F_{Y_i} \cdot \sin\theta_i) \end{aligned}\right\} \quad \cdots\cdots (2.3.4)$$

④ Based on the characteristics of vectors and explanation of ③ as shown in Fig. 2.3.2, when "R_X" and "R_Y" are deemed to act on the center of gravity, a resultant force integrated "R" will be,

$$\left.\begin{aligned} R &= \sqrt{R_X^{\,2} + R_Y^{\,2}} \\ \theta &= \tan^{-1}\left(\frac{R_Y}{R_X}\right) \end{aligned}\right\} \quad \cdots\cdots (2.3.5)$$

⑤ The Y-axis component "R_Y" and X-axis component "R_X" of a resultant force "R" of all acting forces act respectively to the ship's center of gravity to move the ship longitudinally and laterally. As a result the ship will be moved to an oblique forward direction by the resultant force "R".

⑥ Then integrate turning moment of the ship around its center of gravity (F_i, $i=1-5$) according to the following equation (2.3.6) giving positive sign to a counterclockwise moment, in which "r" means moment lever.

$$M=\Sigma(F_i \times r_i) \quad \cdots\cdots (2.3.6)$$

⑦ According to the scale and sign of the resultant moment "M" derived from the above equation, the ship will turn.

⑧ In this example, taking the resultant force ⑤ and moment ⑦ into account, the ship will move obliquely forward toward the left and turn its bow to the right.

(2) Balance of Resultant Force and Moment

What all affecting forces are in balance in the X-Y perpendicular coordinates means that the resultant force is zero. Therefore, when the following equations (2.3.7) are held in respect to the resultant forces of the X and Y-axis, all forces acting are in balance and a ship would not move in any direction.

$$\left.\begin{aligned} R_X &= \Sigma(F_i \cdot \cos\theta_i)=0 \\ R_Y &= \Sigma(F_i \cdot \sin\theta_i)=0 \end{aligned}\right\} \quad \cdots\cdots (2.3.7)$$

And when the following equation (2.3.8) is held, all acting turning moments are in balance and the ship would not be turned in either direction.

$$M= \Sigma(F_i \times r_i)=0 \quad \cdots\cdots (2.3.8)$$

3.3 Utilization of Tugs and Consequential Ship's Motion

The Figs. 2.3.13 to 2.3.21 show the motion of an LNG carrier for which specifications are given in Table 2.2.1 in various maneuverings under tug assistance in deep water systematically together with the results of its responsive motions obtained from simulation calculations. For ease of reference, in these figures, solid arrows are used to indicate the directions of acting tug force to the ship's center of gravity and moments around it, and dotted arrows for the moments by hydraulic resistance against the oblique movement of the ship. These tests were carried out assuming that each tug thrust is 39 tons, its acting direction is always perpendicular to the ship's longitudinal center line or 45° if oblique force is applied.

(1) Effect of Tugs on Ship Dead in the Water

• Turning by one Tug

Fig. 2.3.13 (a) shows turning by the push of one tug taken at the bow while Fig. 2.3.13 (b) shows turning by the push of one tug taken at the stern. In the former case the ship turns centered on its stern and in the latter case the ship turns centered on its bow.

In either case, there is no significant difference in time required to turn the ship's heading to 90° and it took 6 to 7 minutes. In the former case the ship gains headway while in the latter case gains sternway during turning although both speed of headway and sternway is not more than one knot. However, caution should be paid in practice that in either case, to turn the ship, the water area of about one ship length for the turning side is necessary, because the lateral force acting on the ship's center of gravity pushes out the ship's hull toward the opposite side of the tug force acts upon.

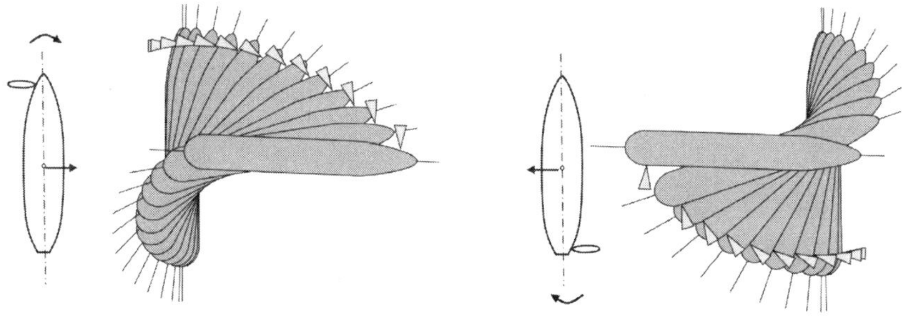

(a) Pushing by one tug at the bow　　(b) Pushing by one tug at the stern

Fig. 2.3.13 Turning by one tug (ship dead in water)

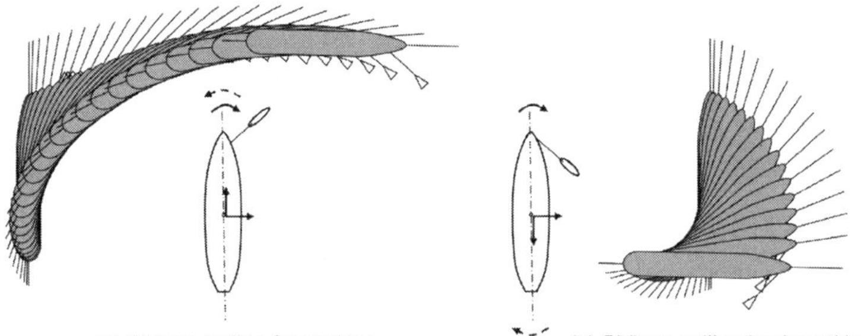

(a) Oblique pulling forward by one tug on the bow

(b) Oblique pulling backward by one tug on the bow

Fig. 2.3.14 Oblique pulling one tug at bow (ship dead in water)

• Oblique Pulling by One tug at Bow

Fig. 2.3.14 (a) shows the case of oblique pulling forward by one tug taken at the bow. To a ship moving obliquely forward, a counterclockwise moment by water resistance affects it. As a result, the development of rate-of-turn is impeded and turning of the bow becomes sluggish as clockwise moment by the tug and counterclockwise moment by water resistance are offset by each other. In this pattern, the headway of the ship is created easily and it grows larger because the resistance of the water is comparatively smaller in its direction. And the ship's headway reaches about 3 knots accelerated until the ship turns 90° as the development of rate-of-turn is impeded in case of oblique pulling forward and it will take a longer time to turn the ship.

Fig. 2.3.14 (b) shows the case of oblique pulling backward by one tug taken at the bow. In this pattern, a clockwise moment by current coming from starboard quarter affects the ship as it moves obliquely backward to the right. As a result, the ship can turn more quickly as a clockwise moment created by the tug and that of current are multiplied. Although sternway is also created at this instance, significant sternway will not be created until the ship's heading is turned 90° because resistance of water is larger due to the form of the stern part of the hull and time required is shorter compared with the preceding case.

It takes about 13 minutes to turn the ship to 90° in the case of oblique pulling forward and about 9 minutes in the case of oblique pulling backward. From the above facts you can understand that the latter pattern is more effective in turning.

In addition, the water area required to turn the ship by pulling forward is about a half ship length to the front and 2.5 ship lengths to the turning side, but in pulling backward the ship can turn centered on her stern almost on the spot.

• Oblique Pulling by One Tug at Stern

Fig. 2.3.15 (a) shows the case of oblique pulling backward by one tug taken at the stern. In this pattern, a counterclockwise moment by water resistance affects the ship as it moves obliquely backward to the left. As a result, the development of rate-of-turn is impeded and turning of the bow becomes sluggish as clockwise moment by the tug and counterclockwise moment by water resistance are offset by each other. And it takes about 12 minutes for the ship to turn 90° and about 2 knots sternway is created.

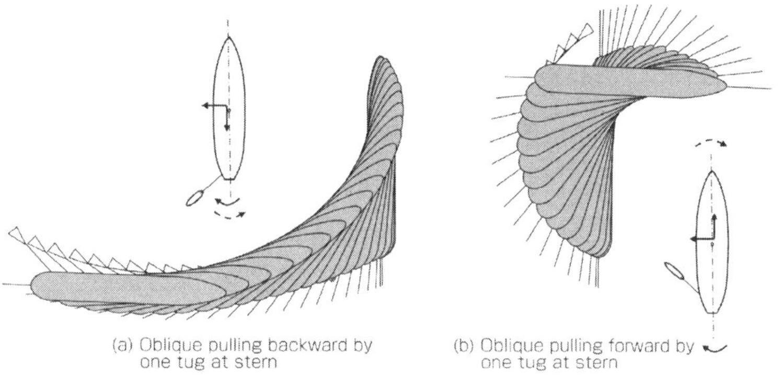

Fig. 2.3.15 Oblique pulling by one tug at stern (ship dead in the water)

Fig. 2.3.15 (b) shows the case of oblique pulling forward by one tug taken at the stern. In this pattern, on the ship a clockwise moment due the current coming from port bow affects it. As a result, the ship can turn more quickly as a clockwise moment created by the tug and that of current are multiplied. Although headway is created for this period, it does not grow significantly until the ship turns to 90° as the ship can turn quickly corresponding to rapidly developing rate-of-turn.

It takes about 12 minutes in the case of oblique pulling backward while it takes about 8 minutes in the case of oblique pulling forward. The time required in the latter case is comparatively shorter than that in the former case. In addition, the water area required to turn the ship by oblique pulling backwards is about a half ship length to the rear and 2.5 ship lengths to the turning side, but in oblique pulling forward the ship can turn centered on her bow almost on the spot.

From the above, it can be said that in the case of oblique pulling at bow or stern, pulling backward in the case of oblique pulling at the bow and pulling forward in the case of oblique pulling at the stern can achieve faster and more effective turning within a smaller water area to maneuver.

• Turning by Two Tugs

Fig. 2.3.16 shows turning centered on one point on the ship's longitudinal line by pushing simultaneously with two tugs, one at the bow and the other at the stern on the opposite board. In this pattern, the ship can turn effectively around one point by multiplied moment of the two tugs as lateral forces acting on the ship's center of gravity by the two tugs are offset by each other. Rate-of-turn grows, accelerated to 20°/min when its turning angle reaches 30°, 30°/min when turning angle reaches 45°, and 40°/min when turning angle reaches 90°. In this case, it is important to adjust and control rate-of-turn appropriately.

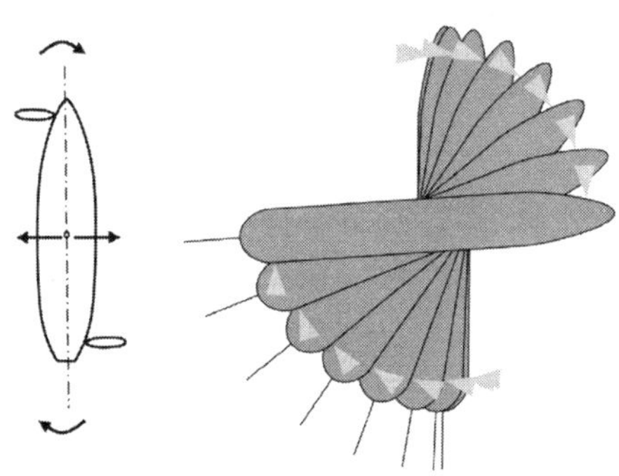

Fig. 2.3.16 Turning by two tugs, one at the bow and the other at stern (ship dead in the water)

• Oblique Pulling by Two Tugs: one at Bow and the other at Stern

Fig. 2.3.17 shows the case of oblique pulling 45° forward by two tugs, one at the bow and the other at the stern on the same board. In this pattern, the ship can be moved obliquely forward in about 3 minutes from the start of pulling.

In this instance, although turning moment by two tugs at the bow and stern are offset by each other, a counterclockwise moment due to water resistance is created on to the ship as she moves obliquely to starboard forward.

As a result, more than one knot headway and at the same time 5°/min rate-of-turn after 5 minutes will be created.

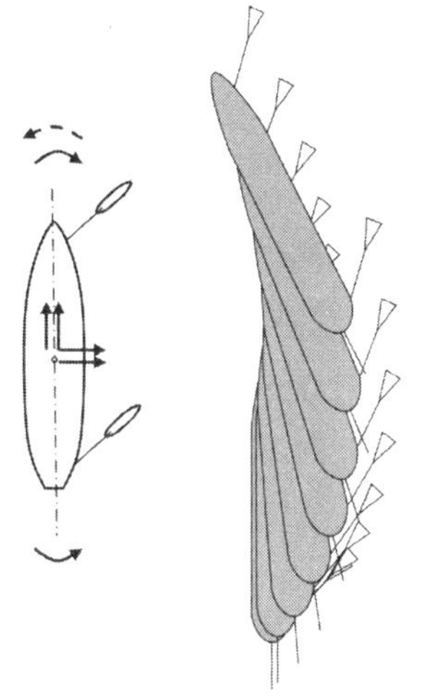

Fig. 2.3.17 Oblique pulling forward by two tugs at bow and stern (ship dead in the water)

If pulling continues further, the ship will slide obliquely, swinging its stern and increasing headway which finally reaches around 5 knots. Caution is necessary when the ship departs from the berth where any other ship is berthed close ahead of her. In such a case, pulling should be stopped at once when the ship has been pulled off about a distance of the ship's breadth from the berth face.

Fig. 2.3.18 shows the case of oblique pulling 45°backward by two tugs, one at the bow and the other at the stern on the same board.

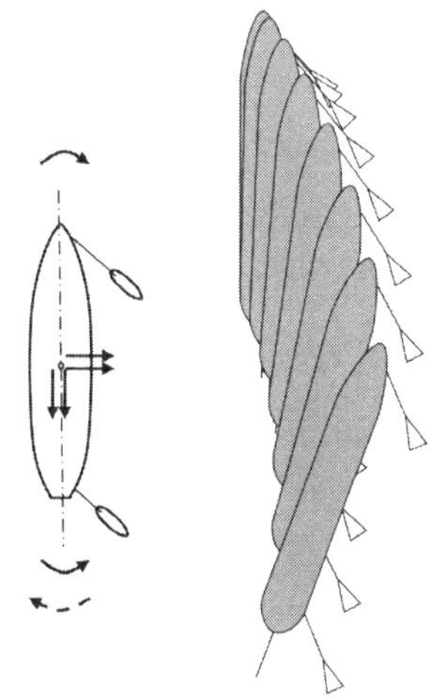

Fig. 2.3.18 Oblique pulling backward by two tugs at bow and stern (ship dead in the water)

In this case, turning moments by two tugs at the bow and stern are offset by each other as in the preceding case. In this pattern, more than one knot sternway and at the same time about 5°/min rate-of-turn will be created after 5 minutes as a clockwise moment due to water resistance affects on the ship while she is being pulled obliquely backward. Therefore, in cases of pulling the ship away from the berth obliquely backwards, pulling should be stopped at once before sternway and rate-of-turn develop to adjust and control the effects of tugs.

To deduce the difference between the pulling direction of tugs at "θ" and the actual direction "α" in which the ship moves with the usage of tugs as can be seen in Figs. 2.3.17 and 2.3.18, a simplified formulas are showen in the equation (2.3.9) and (2.3.10), where "K_{mx}" is a longitudinal added mass coefficient and "K_{my}" is a lateral added mass coefficient.

$$\tan\alpha = \left(\frac{1+K_{mx}}{1+K_{my}}\right)\tan\theta \quad \cdots\cdots (2.3.9)$$

In general, as "K_{mx}" and "K_{my}" respectively can be considered about 7% and about 75% of the ship's mass, the following equation holds

$$\tan\alpha = 0.6\tan\theta \quad \cdots\cdots (2.3.10)$$

From the above equation, the actual direction of movement will be 30° when the ship is pulled in the direction of 45° by tugs.

• Lateral Pulling by Two Tugs at Bow and Stern

Fig. 2.3.19 shows the case of simultaneous lateral pulling by two tugs, one at the bow and the other at the stern on the same board. In this pattern, it is possible to pull the ship out almost parallel to its original position as there is not much difference between the water resistance affecting fore and aft part of the hull. However, caution is necessary in the following case; the larger the trim, the larger the disarray of the ship's posture occurs.

After 5 minutes from the time tugs start to pull, the speed of lateral shift reaches 50 cm/sec until the ship is pulled off to about twice of the ship's breadth from her original resting position. Thereafter the same posture will be kept if pulling is lateral by two tugs.

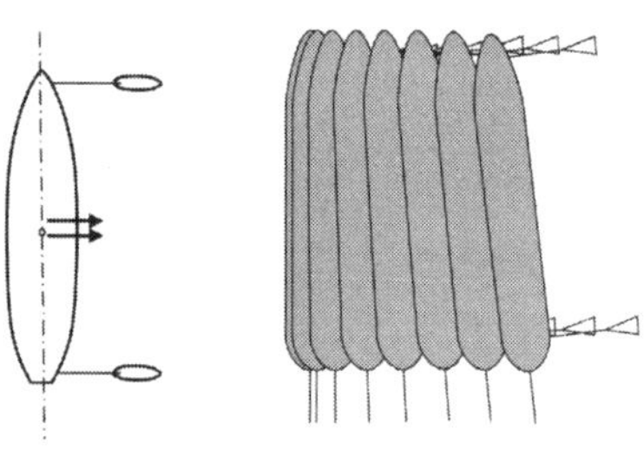

Fig. 2.3.19 Lateral pulling by two tugs at bow and stern (ship dead in the water)

(2) Effect of Tugs on Ship with Headway

• Turning by One Tug

Figs. 2.3.20 (a) and (b) show cases of turning by pushing by one tug at the bow and one at the stern when the ship is going ahead at 5 knots. In these two cases, the scale of stern kick differs as each direction of lateral force acting on a ship's center of gravity is opposite. In the case of pushing at the bow, significant stern kick is not created but large stern kick emerges in the case of pushing at the stern. In this respect, the two methods are distinctly different from each other.

Time required turning the ship 90° is about 10 minutes in the case of pushing at the bow and 9 minutes in the case of pushing at the stern. There is no significant difference of time required and the development of rate-of-turn between the two methods. The water area required to turn the ship by pushing at the stern is somewhat smaller than the ship's breadth to the forward direction and one ship length to the lateral direction than by pushing at the bow. However, it should be noted that in actual cases, the effect of turning may be decreased as it is considerably more difficult for a tug to keep its pushing posture against the ship running at 5 knots.

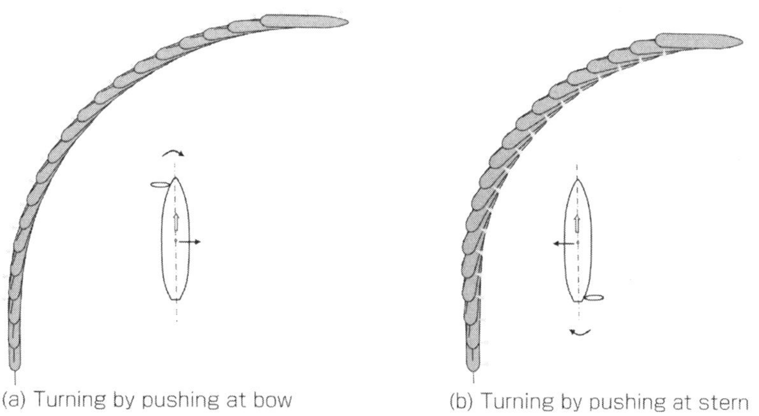

(a) Turning by pushing at bow (b) Turning by pushing at stern

Fig. 2.3.20 Turning by one tug (headway 5 knots)

• Turning by Two Tugs

Fig. 2.3.21 shows the case of turning with pushing by two tugs, one at the bow and the other at the stern, each on a different board when the ship is running ahead at 5 knots. Turning motion can be accelerated effectively by the thrust of two tugs while lateral forces acting on the center of gravity of the ship in the opposite direction by two tugs are offset by each other. However, water area required for 90° turning will extend corresponding to the headway. The water area of about 2.5 times the ship's length to the forward direction and 1.5 times the ship's length to the lateral direction is required.

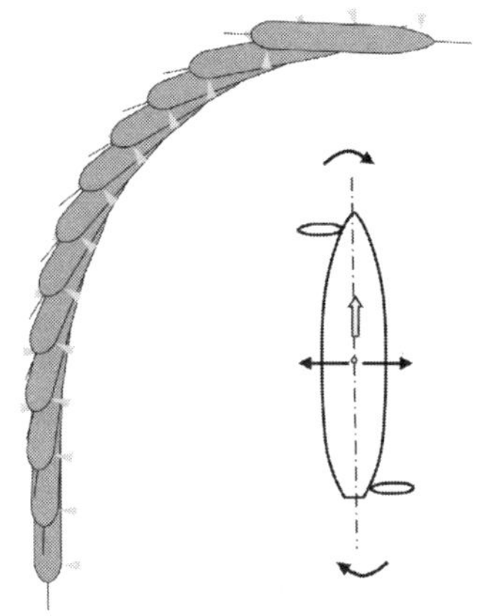

Fig. 2.3.21 Turning by two tugs (headway 5 knots)

However, it should be noted that in actual cases, the effect of turning may be decreased as it is considerably difficult for a tug to keep its pushing posture against a ship running at 5 knots.

3.4 Ship's Motion under Constant Pulling Direction

The simulations of maneuvering as shown in the Figs. 2.3.13 to 2.3.21 are made on the premise that the tug's thrust works at a constant angle to the ship regardless of its movement. Apart from this premise, the ship is towed in a direction so that the moment created by the tug and by the resistance of water affecting the hull may be balanced.

When the tug's line is taken and the ship is towed in the direction as shown in Fig. 2.3.22 (a) without taking the resistance of water into consideration, the

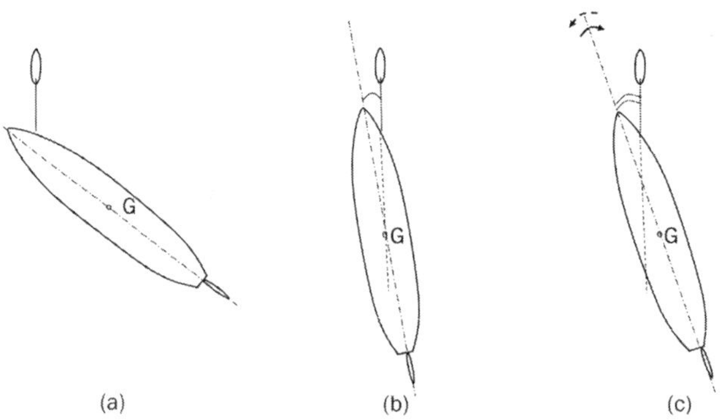

Fig. 2.3.22 Ship's motion under constant pulling direction

direction of tow and the tow line coincides with the line passing the position where the tug line is belayed and the ship's center of gravity as shown in Fig. 2.3.22 (b). However, in actuality, moment affects the ship because the ship moves obliquely to the relative current. As a result, as shown in Fig. 2.3.22 (c), the ship moves obliquely under the condition that the clockwise moment by the tug and the counterclockwise moment by the resistance of water are in balance.

Consequently, the actual angle of towing direction to the longitudinal centerline of the ship may become larger than that shown in Fig. 2.3.22 (b).

3.5 Turning in Bow Current

When a ship is turned in a water area where current exists, caution is necessary; because it may drift toward downstream even under the assistance of a tug. In a stream, if a tug is used for pushing or pulling at the stern as shown in Fig. 2.3.23 (a), the drift of the ship can be restrained to some extent as the lateral force acting on the ship's center of gravity by the tug works opposite to the fluid pressure of the stream, although it depends on the strength of flow.

On the other hand, if a tug is used for pushing or pulling on the bow as shown in Fig. 2.3.23 (b), the lateral force acting on the ship's center of gravity by the tug and the fluid pressure of the stream are multiplied and the risk to drift downstream will be increased. Therefore, it would be better to avoid this method when there is not enough water area downstream.

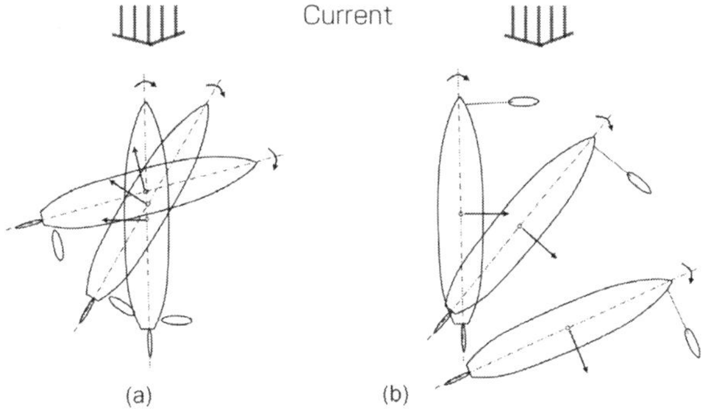

Fig. 2.3.23 Turning in bow current

3.6 Necessary Thrust of Tugs

For a ship maneuvered under the assistance of a tug, the time tug thrust is needed most is during the maneuvering phase when the ship is stopped parallel to the berth face and is readied to move laterally for laying the ship alongside to the berth. Speed of the lateral shifting at the final contact with the berth should be not more than 15 cm/sec.

Fig. 2.3.24 shows the necessary thrust of a tug under different water depths "h" to a ship's draft "d" ratio h/d when the ship approaches the berth at 15 cm/sec of lateral shifting speed, under 10 m/sec of offshore wind and 0.1 m/sec of counter current. From the graphs for this figure you can deduce the necessary tug thrust depending on the ship's weight.

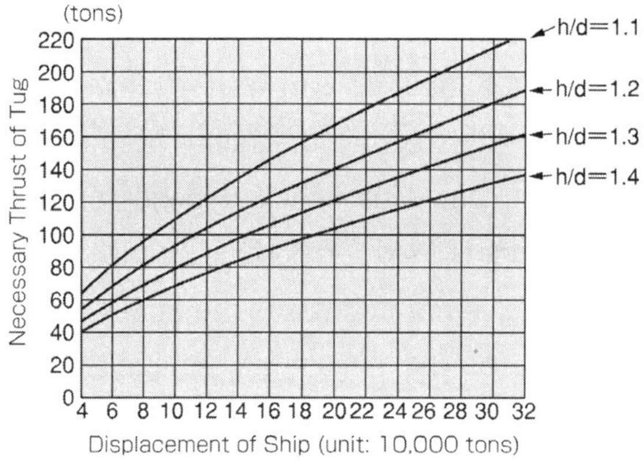

Fig. 2.3.24 Necessary thrust of tug when a ship is laid alongside a berth

3.7 Number of Tugs Necessary

Although there is no established standard in deciding the number of tugs to be used in the assistance of a ship handling, considering the following, it is generally understood that environmental conditions, including sea and weather conditions at the time at a particular location, topography, depth of berth, plans for berthing or un-berthing, including maneuvering, and the ship's own physical characteristics such as the ship's weight, draft and trim or other restrictions due to the ship's particular design, e.g., the type and power of engine(s), and if thrusters are available are of importance.

In practice (in Japanese ports), 2 tugs for ships for which the dead weight is up to about 50,000 tons, 3 tugs for ships up to 100,000 tons, 3 to 4 tugs for ships above 100,000 tons, and 5 to 6 tugs for super large ships like VLCCs.

3.8 Caution when Using Tugs

(1) Limit of Availability of Tug Assistance

In Japanese ports, in view of safe and effective ship handling, the limits of weather and sea conditions for availability of tug assistance were once considered as 8 m/sec in wind speed and 0.5 m in wave height. Recently however, these limits seem to have been relaxed to some extent owing to the advent of larger powered and larger sized tugs; it would be safe to think that the operational limits are now 15 m/sec in wind speed and 1 to 1.5 m in wave height.

(2) Matters Requiring Attention in Operation

- Limiting the ship's speed to not more than 6 knots, at which a tug is capable to assist.
- Even when the ship's speed is lower than 6 knots, caution should be exerted as to prevent the risk of overturning the assisting tugs by dragging.
- Handling in which tugs used are dragged laterally by the ship's headway must be avoided.
- Empirically it is said that at any one time, the maximum number of tugs that can be handled by one person engaged in ship handling is six.
- When multiple tugs are handled, constant confirmation on what order and to which tug orders are given must be made carefully.
- For the person in charge of ship handling, it is important to make intentions and methods of maneuvers clear to the tug masters beforehand.
- When tugs are used for proper maneuvering, for the safety of the tugs and also for the most effective use of them, it is necessary to appropriately lower the speed of the ship.

Ⅳ Maneuvering under External Disturbances

Chapter 1 Effect of Wind

1. Estimation of Wind Pressure

1.1 Hughes's Formula

Fig. 2.4.1 shows the relation between relative wind and wind pressure when a ship is running at a constant speed under the effects of wind. Let the resultant wind pressure affecting various parts of the ship's hull be "F" and the point affected by the resultant wind pressure be the ***center of wind pressure*** on the hull.

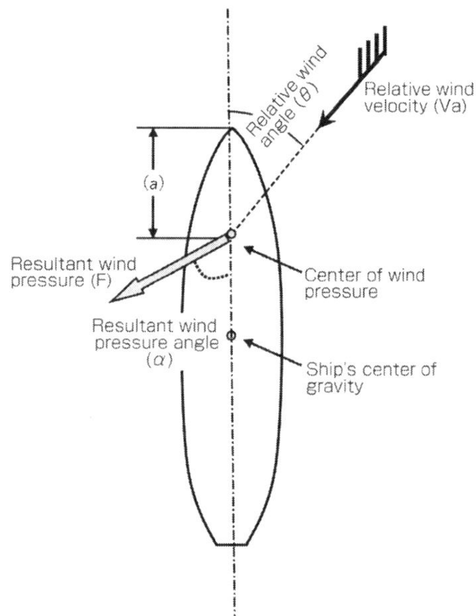

Fig. 2.4.1 Relation between relative wind and wind pressure

In this case, the distance from the bow to the center of wind pressure is "a" and a resultant wind pressure angle is "α" which is an angle made by the action line of resultant wind pressure "F" and the longitudinal centerline of the ship.

To deduce the resultant wind pressure, Hughes's formula as shown in the equation (2.4.1) is often used.

$$F=\frac{1}{2}\rho_a\, C_a(A\cdot\cos^2\theta+B\cdot\sin^2\theta)V_a^2 \quad \cdots\cdots (2.4.1)$$

where

F : resultant wind pressure

ρ_a : density of air (0.125 kg · sec^2/m^4)

C_a : resultant wind pressure coefficient

A : front projected area above water line

B : side projected area above water line

V_a : relative wind velocity

θ : relative wind angle

1.2 Resultant Wind Pressure Coefficient

The scale of resultant wind pressure may vary depending on relative wind velocity "V_a", its angle "θ" and projected area "A" and "B". On the other hand, the effect of the ship's shape is reflected in resultant wind pressure coefficient "C_a" by the type of ship.

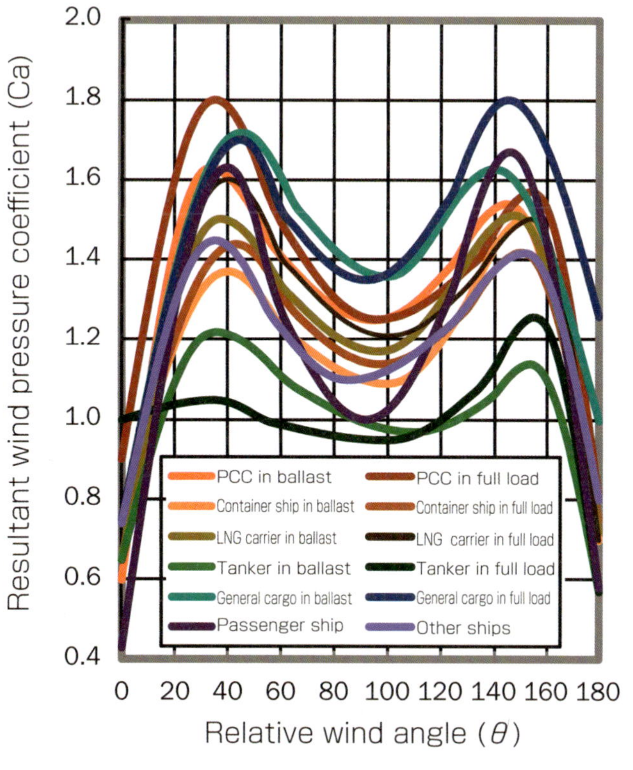

Fig. 2.4.2 Resultant wind pressure coefficients of typical ship types

Although a precise resultant wind pressure coefficient of an actual ship must be decided through the ***wind tunnel model test***, an approximate value obtained from Fig. 2.4.2 showing the resultant wind pressure coefficients of typical ship types may be applied to a ship of the same type.

1.3 Angle of Resultant Wind Pressure

Fig. 2.4.3 shows the relation between relative wind angle "θ" and resultant wind pressure angle "α" of typical ship types. In the curves among the chosen types of ships other than VLCCs in full load conditions, there is little difference by ship type and conditions including VLCCs in ballast conditions.

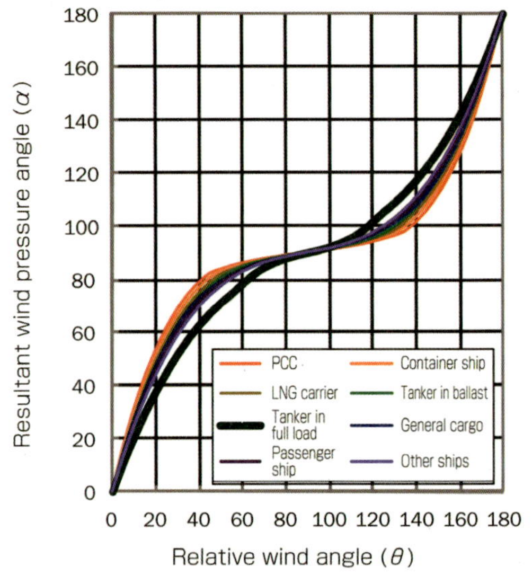

Fig. 2.4.3 Relationship between relative wind angle and resultant wind pressure

1.4 Center of Wind Pressure

The center of wind pressure moves as the relative wind angle "θ" changes.

Fig. 2.4.4 shows the change of the center of wind pressure corresponding to relative wind angle "θ". In this case, an approximate trend of all types of ships can be understood except in the case of a VLCC in full load conditions.

The center of wind pressure is located forward close to the bow as far as wind comes from forward and relative wind angle is small. However, it moves backward as the relative wind angle grows larger. And it is located almost in the middle of a ship to beam

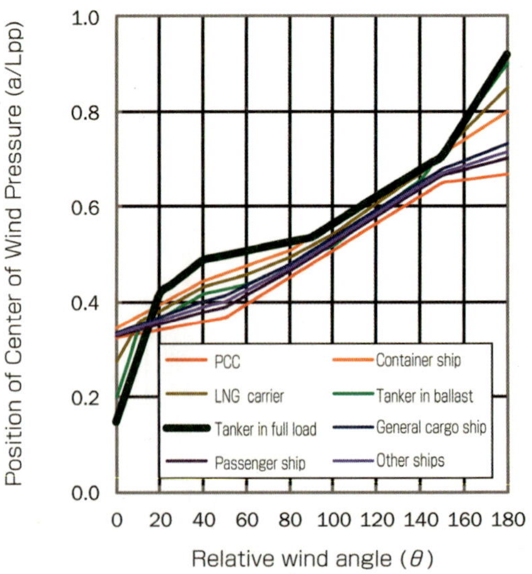

Fig. 2.4.4 Relationship between relative wind angle and position of center of wind pressure

wind and moves further backward as wind veers to the stern.

1.5 Approximate Value of the Lateral Component of Wind Pressure

Even if wind velocity does not change, the effect of wind increases noticeably when a ship's speed decreases because the ratio of wind velocity to ship speed increases. This effect will become particularly significant in ship handling inside port for berthing or un-berthing where ship speed is obliged to be reduced, and particular caution against lateral drift is necessary. It is important to decide allocation of tug(s) taking lateral drift of the ship by the effect of wind into consideration. Such lateral component of wind pressure "Y" can be derived from the equation (2.4.2) below based on resultant wind pressure "F" and resultant wind pressure angle "α".

$$Y = F \cdot \sin\alpha \qquad (2.4.2)$$

For wind velocity, to calculate resultant wind pressure according to the equation (2.4.1), 1.25 times of an assumed average wind velocity is usually used because actual wind is not stable. This figure is called gust factor. If stronger wind is assumed, 1.5 instead of 1.25 should be used.

Figures 2.4.5 to 2.4.8 show the estimated lateral component of wind pressures of a 300,000 DWT VLCC, a 6,000 unit PCC, a 6,000 TEU container ship, and a 135,000m^3 LNG carrier. In the graphs, wind speed 18.75 m/sec and 12.5 m/sec

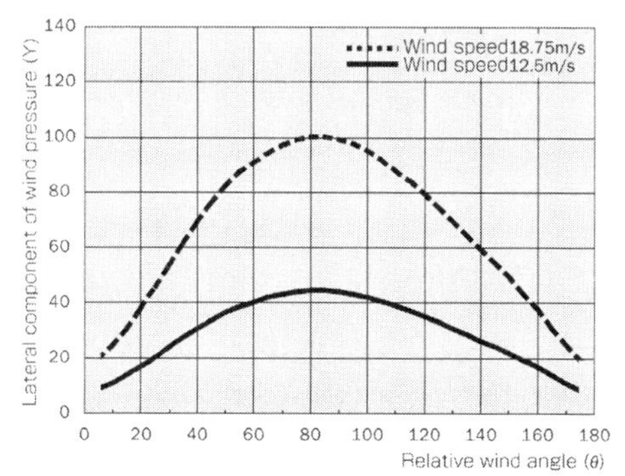

Fig. 2.4.5 Deduced lateral component of wind pressure (300,000 DWT VLCC in full load)

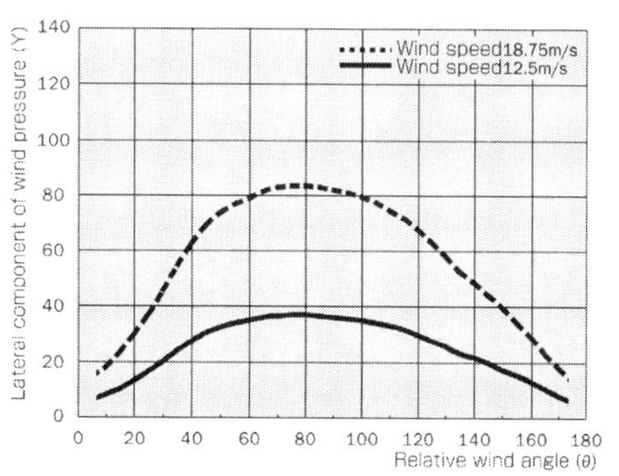

Fig. 2.4.6 Deduced lateral component of wind pressure (6,000 unit PCC in full load)

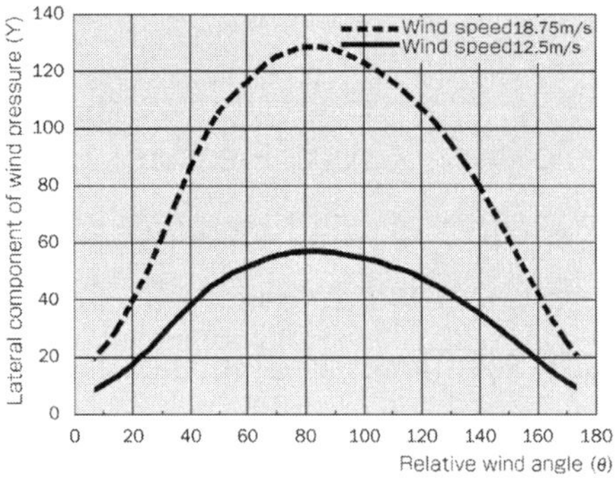

Fig. 2.4.7 Deduced lateral component of wind pressure (6,000 TEU container ship with containers stuck 4 high on the deck in full load)

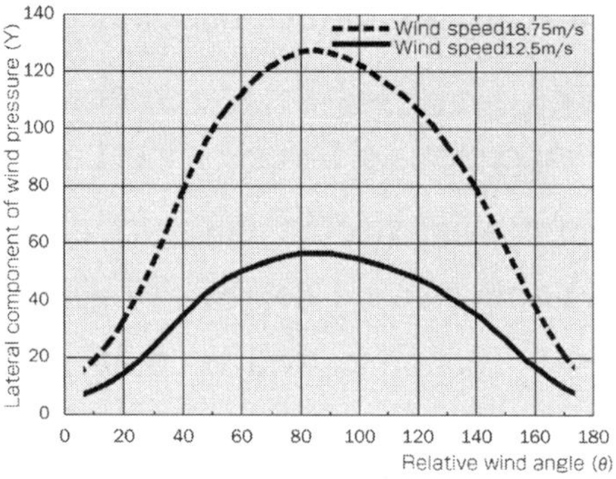

Fig. 2.4.8 Deduced transverse wind pressure (135,000 m^3 LNG carrier in full load)

correspond to the average wind velocity of 15 m/sec and 10 m/sec respectively because 1.25 of the gust factor were multiplied in calculations.

1.6 Wind Pressure Moment

The lateral component "Y" of resultant wind pressure "F" derived from the equation (2.4.2) creates a turning moment around the center of the gravity to turn the ship's bow. That moment can be derived from equation (2.4.3). Assuming the center of the wind pressure from the bow is "a" and the ship's center of gravity is located $1/2L$ (L: ship length), moment lever is $(1/2L-a)$. Then,

$$M = Y \times \left(\frac{1}{2}L - a\right)$$

$$= F \cdot \sin\alpha \times \left(\frac{1}{2}L - a\right) \quad \cdots\cdots (2.4.3)$$

As shown in Fig. 2.4.3, wind pressure moment is not created when relative wind angles "θ" are 0° and 180° as resultant wind pressure angles "α" are 0° and 180°. For other relative wind angles, when it comes from oblique forward, moment to turn the bow to the lee side will be created and when it comes from oblique backward, moment to turn the stern to the lee side will be created, as you can see in the graphs of Fig. 2.4.4 on the position of the center of wind pressure obtained from the test results.

Following that the approximation to derive the experimental values of the position of the center of wind pressure was made by the equation (2.4.4), and assuming that the lateral component of wind pressure as shown in the figures 2.4.5 to 2.4.8 are approximated by the equation (2.4.5), the wind pressure moment can be deduced by the equation (2.4.6).

$$a = (0.291 + 0.0023\,\theta)L \quad \cdots\cdots (2.4.4)$$

$$Y = Y_{max} \cdot \sin\theta \quad \cdots\cdots (2.4.5)$$

$$M = Y_{max} \cdot \sin\theta \times \{0.5L - (0.291 + 0.0023\,\theta)L\}$$

$$= Y_{max} \cdot L \times (0.209 + 0.0023\theta)\sin\theta \quad \cdots\cdots (2.4.6)$$

Fig. 2.4.9 shows the result of approximation in which wind pressure moment to each relative wind angle was derived by equation (2.4.6). According to Fig. 2.4.9, that moment reaches its maximum when relative wind angles are around 40° and 140°. From this fact, caution is necessary when the relative wind comes from 40° forward or sternward.

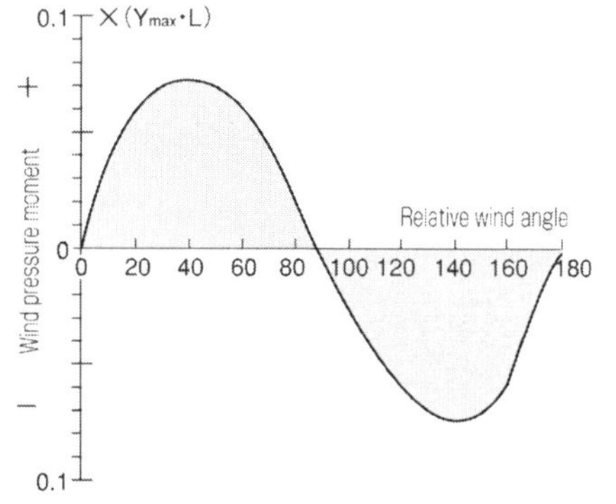

Fig. 2.4.9 Turning moment by wind pressure derived from equation (2.4.6)

2. Disposition of Bow and Stern to Turn toward or away from Wind

2.1 Turning of Bow by Wind

When a ship is running under a head wind, its speed decreases due to wind pressure while running with a tailwind its speed increases. When a ship receives oblique wind, consequential ship track becomes oblique in the water due to the drift by the lateral component of wind pressure. As shown in Fig. 2.4.10, when a ship moves obliquely in water, the water resistance affecting the lee side of the ship's bow under the water line creates turning moment to turn its bow toward the wind. This effect is called ***head pusher*** and always works to turn a ship's bow toward the wind.

Thus, the turning moment by head pusher due to a consequential oblique track in the water and that by wind simultaneously affect a ship running under the effect of wind, and its heading is turned toward windward or leeward depending on the resultant effect of these two moments.

2.2 Disposition of Bow to Turn toward Wind when Ship is Running Ahead

When a ship is running ahead along an oblique track under the effect of wind consequentially, the affected point of the head pusher is situated closer to its bow while relative wind angle remains small. And the center of wind pressure also is situated closer to the bow. Therefore, the ship will be turned to either direction of the moment by head pusher or of that by wind pressure, whichever is superior. In most cases, ship heading turns toward the wind because the resistance of water is by far larger than that of air.

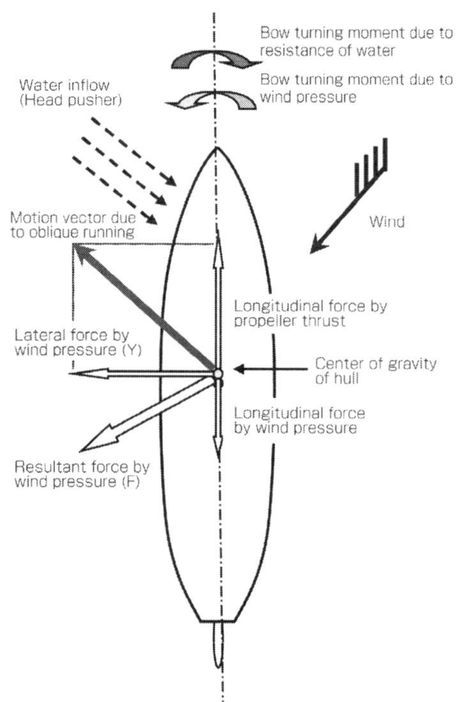

Fig. 2.4.10 Forces and moments affecting the ship while running ahead

As relative wind angle grows larger and changes close to abeam, turning of the bow toward the wind will be enhanced because the center of wind pressure which comes closer to the ship's center of gravity and the moment by the wind to turn the bow leeward decreases. When relative wind angle becomes so large that the ship is obliged to run with a tailwind, turning of the bow will be further enhanced as the center of wind pressure moves sternward from the ship's center of gravity and its moment acts to turn ship's stern leeward in addition to the effect of head pusher to turn bow windward. The trend of bow turning toward the wind while a ship is running under the effect of wind is called ***disposition of bow to turn toward the wind.***

Fig. 2.4.11 shows how a ship's bow turns toward the wind in the course of time based on the simulation results of a 3,500 TEU container ship when the ship is running at 8 knots under 20 m/sec of oblique wind keeping its rudder midship. The figure demonstrates that the disposition of bow turning toward the wind becomes quite distinctive to all relative wind angles, and the ship finally runs straight keeping its turned bow to an extreme angle in equilibrium.

Fig. 2.4.12 shows the closeup tracking of a ship for 3 minutes after the simulation was started. While relative wind angles are from ahead to abeam, turning of the bow is quite slow because the wind pressure moment resists toward the head pusher moment.

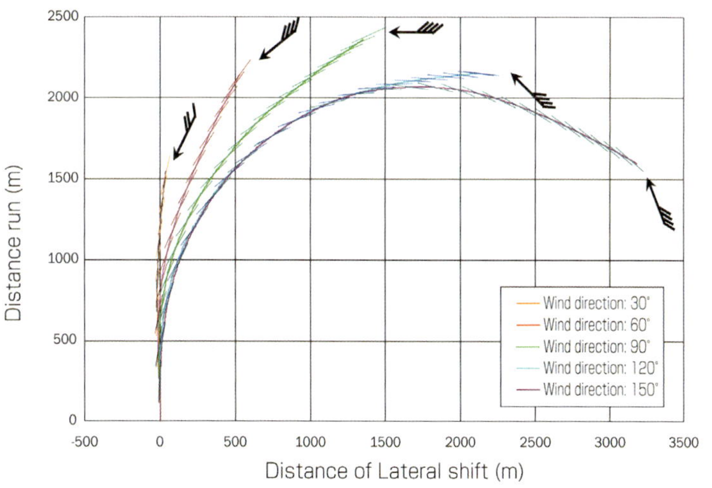

Fig. 2.4.11 Tracks to an extreme turning angle in equilibrium (3,500 TEU container ship in full load, initial speed of 8 knots, helm midship, wind speed 20 m/sec, h/d = 1.3)

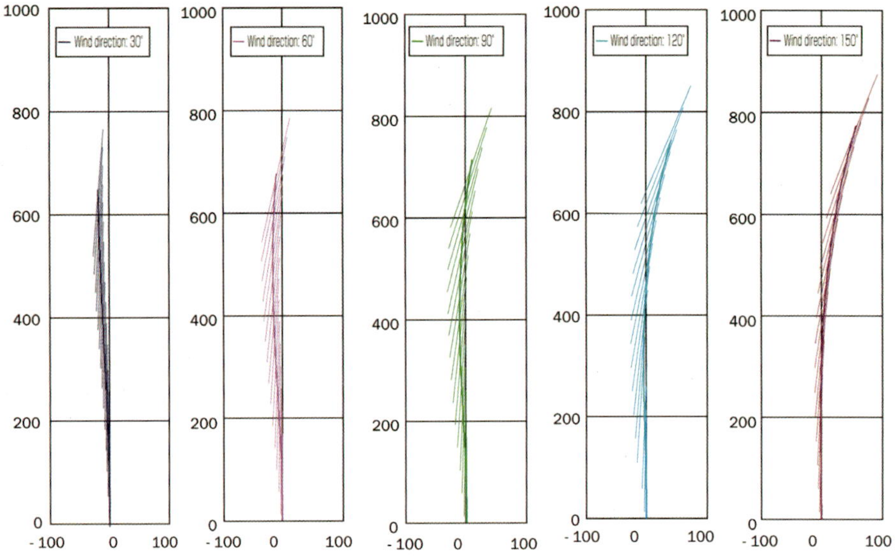

Fig. 2.4.12 Tracks in 3 minutes (3,500 TEU container ship in full load, initial speed of 8 knots, helm midship, wind speed 20 m/sec, h/d = 1.3)

However, for this period, the ship is drifting largely leeward. On the other hand, when a ship runs with a tailwind, disposition of the bow to turn toward the wind will be enhanced because the moment by wind pressure and that by head pusher act in the same direction.

2.3 Disposition of Stern to Turn toward Wind when Ship is Running Astern

When a ship is running astern under the effect of oblique wind blowing from the stern, its stern always turns windward because of strong head pusher due to an oblique track of the ship in water acting on it as shown in Fig. 2.4.13.

While relative wind angle from the stern is small on the ship under the effect of wind blowing from the stern, the center of wind pressure is situated closer to the stern and consequentially the moment acts to turn the stern leeward.

Therefore, the stern will be turned in either direction of the moment by head pusher or that by wind pressure, whichever is superior. In most cases, the stern turns toward the wind because the resistance of water is by far larger than that of air.

As relative wind angle grows larger and changes close to abeam, turning of stern toward the wind will be enhanced because the center of wind pressure comes closer to the ship's center of gravity and the moment by the wind to turn the stern leeward decreases. When relative wind angle becomes so large that the ship receives wind from the bow, turning of stern will be further enhanced as the center of wind pressure moves forward from the ship's center of gravity and turning moment created by the wind acts to turn the ship's bow leeward, in addition to the effect of head pusher to turn stern windward. The trend of stern turning toward the wind while a ship is running astern under the effect of wind is called ***disposition of stern to turn toward the wind.***

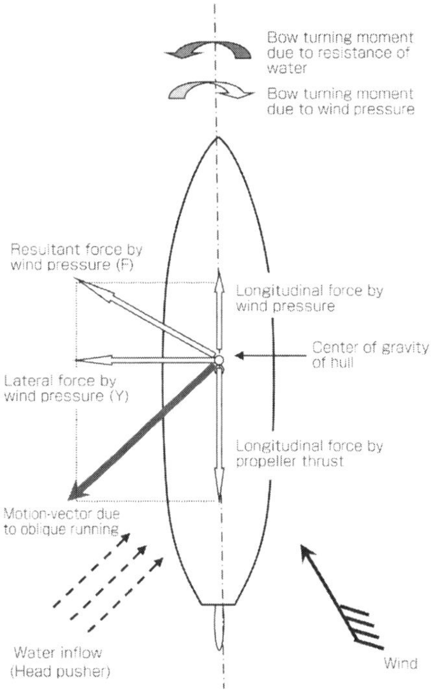

Fig. 2.4.13 Forces and moments acting on ship's hull when running astern

It should be noted that when a ship is running astern, it is difficult to control turning of the stern toward the wind only by its rudder as enough steerability cannot be expected compared with that of when a ship is running ahead. And it is advisable that to control this trend auxiliary means such as anchor, bow thruster or tug should be used particularly in the case of strong wind.

2.4 Disposition of Bow and Stern to Turn Away from Wind

When a ship is running at slow speed and wind velocity is relativery larger compared with ship's speed, the effect of the moment by the wind pressure appears distinctly. As shown in Fig. 2.4.9, the effect of wind pressure moment to turn the bow or stern leeward becomes strong when a ship is running ahead receiving wind from close to its bow or running astern receiving wind from close to its stern at a smaller relative wind angle. In such case, sometimes the bow or stern may be turned to leeward overwhelming the disposition of bow or stern to turn toward the wind. This tendency is called ***disposition of bow and stern to turn away from wind.***

3. Effect of Wind in Maneuvering

3.1 Limit of Steerability under Wind Pressure

(1) Minimum Speed to Keep Heading

To offset the disposition of the bow to turn toward the wind and to keep a ship's heading, keeping heading by steering is a basic means. However, it becomes difficult as steerability decreases as the ship speed decreases. The same can be said when the ratio of wind velocity to ship speed grows larger even if the wind velocity is unchanged. Therefore, it is important to know how to keep minimum speed for steerability which is capable to offset the disposition of the bow to turn toward the wind and to keep ship's heading.

To enable control of the ship's heading by the rudder at a certain ship speed under the effect of wind pressure, the moments by wind pressure and water resistance due to the ship's oblique track in the water need to be offset by the turning moment by the rudder as shown in equation (2.4.7).

$$M_a + M_w + M_r = 0 \quad \cdots\cdots (2.4.7)$$

where

M_a : moment by wind pressure

M_w : moment by water resistance due to ship's oblique track in water

M_r : moment by rudder at a certain speed

Fig. 2.4.14 shows the estimated minimum speed necessary for 2,300 units PCC, one of the most wind-sensitive types of ship, to keep its heading with the helm angle of 15°; the estimated minimum speed was derived using the static equilibrium condition to meet the equation (2.4.7).

The graphs indicate that the ship is capable of keeping its heading of at least 6 knots under 15 m/sec of head wind of small relative wind angles, but when relative wind angle changes from abeam to stern, a speed of 8 knots or higher is necessary.

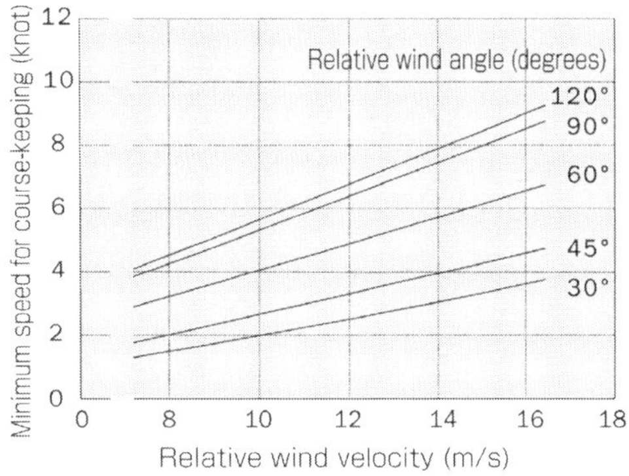

Fig. 2.4.14 Minimum speed to keep heading (2,300 unit PCC)

(2) Limiting Conditions to Keep Heading or Course Line

In view of safe ship handling, it would be ideal to offset the effect of wind pressure to keep a ship's heading within a normal helm angle of 15°.

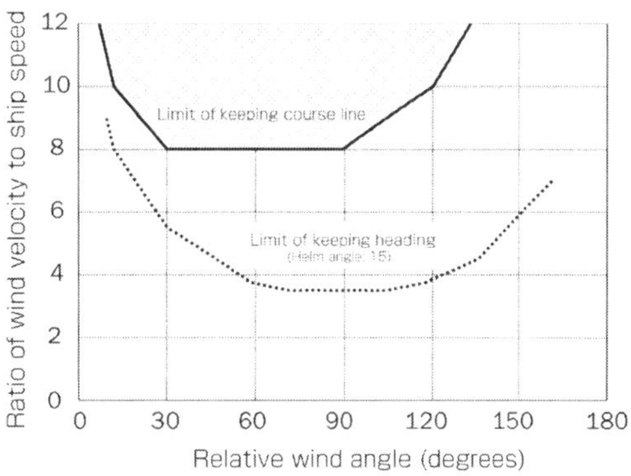

Fig. 2.4.15 Limits of keeping heading and course line (4,500 unit PCC, h/d = 1.3)

However, under strong winds, drift of the ship to leeward often becomes larger even if the ship's heading can barely be kept. In other words, even if ship's heading could be kept, its original course line could not be maintained.

In such cases, it is necessary to make the ship run on the original course, offsetting drift angle. The solid line in Fig. 2.4.15 shows the simulation result of a 4,500 unit PCC shown in Table 2.2.6, capable of keeping the original course by a maximum counter helm angle to offset drift angle. By contrast, the dotted line shows the limit of the same ship capable of keeping its heading by a helm angle of 15° counter rudder.

From the graphs, when the helm angle is 15° or lower, if the ratio of wind velocity to the ship's speed exceeds 4 against an oblique wind from fore to abeam, it should be deemed as out of the steerability limit, meaning the ship cannot keep its heading overpowered by the effect of wind pressure.

By contrast, against a head wind, maintenance of heading by the same helm angle will be possible up to 6 to 8 in the ratio of wind velocity to a ship's speed. For instance, if ship speeds are assumed 6 knots and 8 knots at the ratio of wind velocity to a ship's speed 4 actual corresponding wind velocities will be 12 m/sec and 16 m/sec respectively. In addition, when the helm angle is 35°, if the ratio of wind velocity to a ship's speed exceeds 8 against an oblique wind from fore to abeam, it should be deemed as the out of the steerability limit, meaning the ship cannot keep its course line overpowered by the wind pressure pushing the ship toward leeward.

3.2 Limit of Maneuverability under Wind Pressure

(1) Limiting Conditions for Altering Course

Fig. 2.4.16 shows the limit of maneuverability for the capability of 4,500 unit PCC, for which specifications are as shown in Table 2.2.6, to alter the course to 90° with a helm angle of 15° under the effect of wind.

In this case, if a rate-of-turn has been lost before its heading reaches 90°, it should be deemed that maneuvering is impossible. In the figure, the horizontal axis represents relative wind angle, the vertical axis represents the ratio of wind velocity to a ship's speed and the shaded part represents the area where 90° alteration of course is impossible.

When the ship is running to turn starboard (windward) against the wind from its starboard side, a 90° alteration of course becomes impossible if the ratio of wind velocity to the ship speed exceeds 8. While when the ship is turning starboard (leeward) following the wind from its port side, 90° alteration of course becomes difficult if the ratio of wind velocity to ship speed becomes 4. To explain why 90° alteration of course to starboard side becomes difficult sooner when the ship receives wind on its port side compared with the wind on its starboard side, we will see typical examples when the ship is running to turn under the wind from starboard ahead and when the ship is running to turn before the wind from the port quarter in comparison.

At the initial stage of turning, as in both cases previously mentioned, the turning moment by wind pressure acts opposite to that of the rudder. However, the moment by water resistance due to an oblique track of the ship (head pusher) acts in the same direction as that by the rudder when wind comes from starboard ahead but it (head pusher) reverses when wind comes from the port quarter. This is the reason why starboard turning under the wind coming from its port side becomes more difficult.

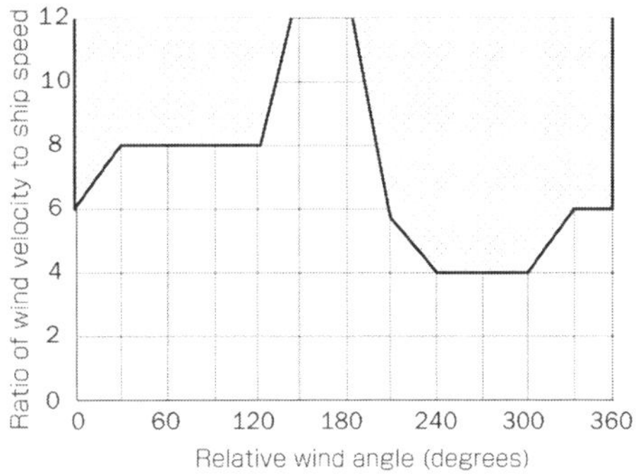

Fig. 2.4.16 Limits of altering course (4,500 unit PCC, h/d = 1.3)

In addition, whether rate-of-turn will be maintained or lost before completing a 90° turn depends on the relationship among the moments of wind pressure, water resistance and rudder in the process of turning.

(2) Wind Pressure Effect in Approaching to Berth

☞ Refer to Vol. 4 Ship Handling Characteristics Data Base

When a ship is approaching a berth, its engine is stopped leaving some distance from the berth so that the ship may approach to the berth front at few knots under its inertia with its propeller rotating freely. At this stage, the effect of wind grows relatively larger as the ratio of wind velocity to ship speed increases.

As the result, while the ship is approaching a berth with its hull being pushed leeward, ship heading will deviate from its original course. In such case, persons engaged in ship handling are requested to take optimum countermeasures estimating the extent of the heading deflection and consequential drift due to the effect of wind, keeping in mind that it is difficult to maintain a ship's steerageway by its own rudder.

Fig. 2.4.17 presents the indices to estimate a deflection angle of course and deflection angle of heading due to the effect of wind for persons engaged in ship handling. A deflection angle of course is the same as leeway of a ship running in normal condition under the effect of wind and current and is equal to the angle between a ship's track and its original approaching course while the ship runs a certain distance. While a deflection angle of heading is equal to an angle between the ship's heading and its original approaching course while the ship runs a certain distance.

Fig. 2.4.18 shows the simulation result of a deflection angle of course and deflection angle of heading of a 4,500 unit PCC, for which specifications are given in Table 2.2.6, while it runs a half ship length and one ship length, with its helm midship by inertia after its engine is stopped. The results of the effect of wind while the ship runs a certain distance can be applied to the speed range of any ship approaching a berth and will be useful information to enable persons engaged in ship handling to estimate the effect of wind beforehand.

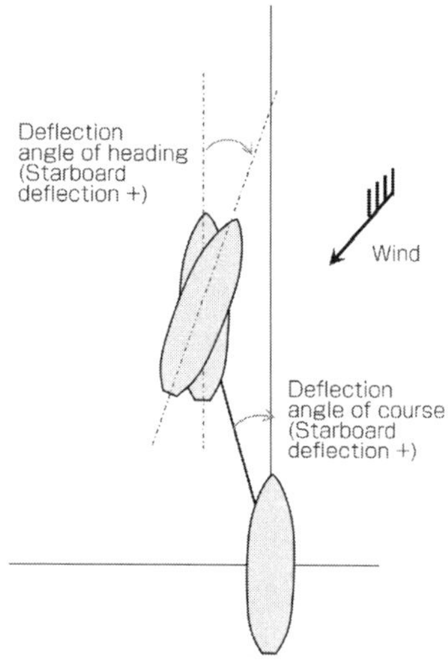

Fig. 2.4.17 Index for wind effect

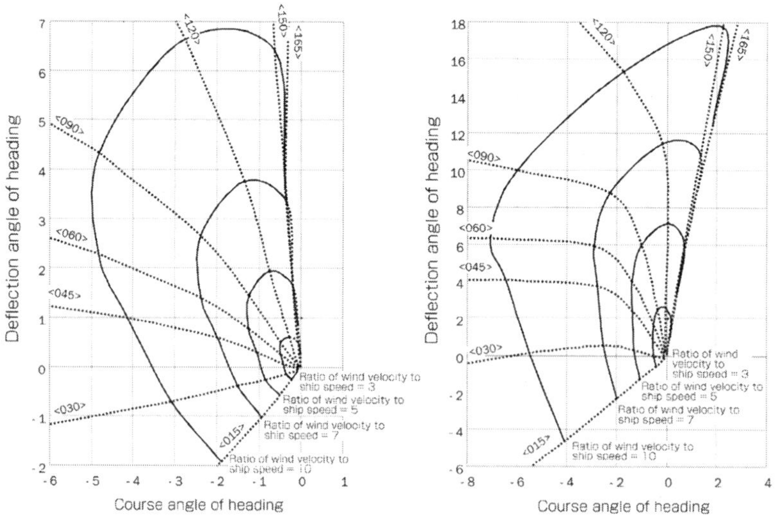

(a) Wind effect while the ship runs 1/2 a ship's length

(b) Wind effect while the ship runs one ship's length

Fig. 2.4.18 Wind effect while the ship runs a certain distance (4,500 unit PCC, h/d = 1.3)

In addition, you can convert the deflection angle of course of the horizontal axis to the drifted distance based on the ship's length. For instance, 1/2 of the ship's length × sin (deflection angle of course) and one ship length × sin (deflection angle of course) equals to respective drifted distance while the ship runs half a ship's length or one ship's length. Thus persons engaged in ship handling can approximate the effect of wind from the scale of drift by using either angle or distance.

(3) Lateral Drift of Ship Dead in the Water

Persons engaged in ship handling should be most careful of the ship's lateral drift when the ship receives abeam inshore wind when the ship is lying dead in the water in front of the berth. In such conditions the ship will be carried laterally toward a berth under beam wind.

Drifting speed of ships can be calculated according to the following equation (2.4.8) assuming that the ship is drifting at a constant rate on the condition that wind pressure acting on the hull above water line and water resistance acting on the hull below water line are in equilibrium.

$$\frac{1}{2}\rho_a C_a B_a V_a^2 = \frac{1}{2}\rho_w C_w B_w V_w^2 \qquad (2.4.8)$$

where

ρ_a : density of air (air mass per unit volume, 0.125 kg · sec^2/m^4)

ρ_w : density of water (water mass per unit volume, 104.5 kg · sec^2/m^4)

C_a : coefficient of the lateral component of wind pressure above water

C_w : coefficient of the lateral component of water resistance below water

B_a : side projected area above water

B_w : side projected area below water

V_a : relative wind velocity

V_w : relative velocity of a ship's hull against water

From the above,

$$V_w = \sqrt{\frac{\rho_a}{\rho_w} \cdot \frac{C_a}{C_w} \cdot \frac{B_a}{B_w}} \cdot V_a \quad \cdots\cdots (2.4.9)$$

In the above, although both C_a/C_w and B_a/B_w vary ship by ship, assuming that the approximate value of C_a/C_w is 1.3 and that of B_a/B_w is 3.0 referring to PCCs, then, the equation (2.4.9) can be approximated by the following equation.

$$V_w = 0.068\, V_a \quad \cdots\cdots (2.4.10)$$

The value estimated by the equation (2.4.10) represents that the ship is drifting laterally at a steady rate under the effect of beam wind. In actuality, the drifting speed of the ship increases gradually from the onset of the effect of wind, then reaches a steady rate 2 to 3 minutes later. Fig. 2.4.19 shows the simulation results of lateral drifting speed at 120 seconds after the onset of the effect of wind of a 4,500 unit PCC in various combinations of the ratio of wind velocity to ship speed and relative wind angle.

For instance, according to Fig. 2.4.19, when the PCC reduces its headway to

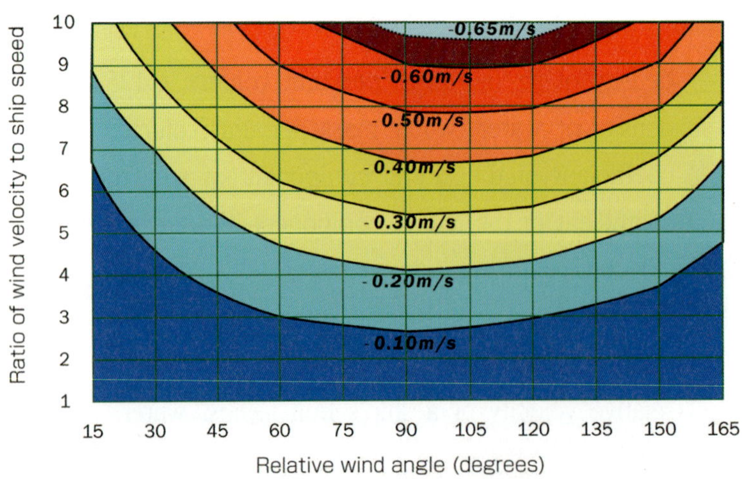

Fig. 2.4.19 Lateral drifting speed at 120 seconds after the onset of the effect of wind (4,500 unit PCC, h/d = 1.3)

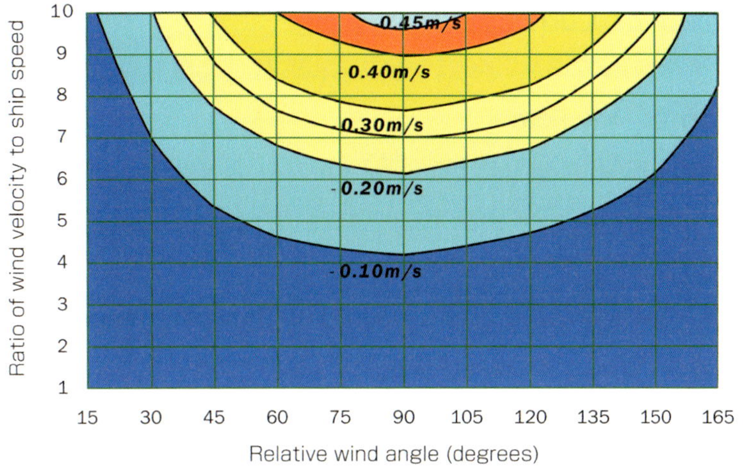

Fig. 2.4.20 Lateral drifting speed at 30 seconds after the onset of the effect of wind (4,500 unit PCC, h/d = 1.3)

2 knots under 10 m/sec beam wind in the vicinity of the berth, the ratio of wind velocity to ship speed will be 10. And the estimated lateral drifting speed of the ship after 120 seconds will be 0.65 m/sec. This estimated value almost coincides to that derived from the equation (2.4.10). Fig. 2.4.20 shows lateral drifting speed 30 seconds after the onset of the effect of wind.

Chapter 2 Effect of Current

1. Deduction of Drift Force

To a ship running under the effect of current is significant since the density of water is 836 times greater than that of air. When a ship runs in static water, water resistance affects the ship. Water resistance gives longitudinal and lateral forces to the ship's center of gravity and turning moment around the ship's center of gravity and acts as a resistance to the ship's motion. By contrast, when a ship runs in a steady current the effect of water resistance can be handled in relation to the direction and velocity of the current just like wind.

Fore and aft components of water resistance affect ship speed. When a ship is running against head current its speed over the ground is reduced and in following current it is increased. When a ship receives water resistance obliquely due to drift, fore and aft components of water resistance are small and most of their resistance takes affect as lateral force to make a ship drift leeward and gives consequential turning moment. The scale of lateral component of water resistance (drift force) varies depending on the ship's side area projected under water and relative current velocity as shown in the following equation (2.4.11). In the equation, the difference of ship shape under water is reflected in the coefficient of lateral component of water resistance "C_w".

$$R_w = \frac{1}{2}\rho_w C_w L d V_w^2 \qquad \cdots\cdots (2.4.11)$$

where

R_w : lateral component of water (drift force) resistance
ρ_w : density of water (water mass per unit volume, 104.5 kg·sec^2/m^4)
C_w : coefficient of the lateral component of water resistance
L : ship length
D : ship draft
V_w : relative velocity of a ship's hull against water (relative current velocity)

Although the coefficient of the lateral component of water resistance varies ship by ship and needs to be defined through model tests, the equation (2.4.12) can give an approximate value for a VLCC. Assuming that $C_w \fallingdotseq 1$ at 90° of relative current angle "β" for enough water depth "h" to the ship's draft "d" ratio h/d (the larger h/d, the bigger C_w grows), an approximate value of "C_w" will be almost 1. The values of "C_w" at 90° of the relative current angle "β" for LNG carriers, container ships and PCCs are about 90%, 85% and 75% that of VLCC respectively.

$$C_w = \left\{ \frac{0.750}{\frac{h}{d} - 0.900} + 1 \right\} \sin\beta \quad \cdots\cdots (2.4.12)$$

And turning moment can be derived according to the following equation (2.4.13).

$$M_w = \frac{1}{2} \rho_w C_{mw} L^2 d V_w^2 \quad \cdots\cdots (2.4.13)$$

The approximate value of the coefficient of water resistance moment "C_{mw}" at relative current angle "β" for VLCC can be derived from the following equation (2.4.14), if the water depth is sufficient.

$$C_{mw} = 0.1 \sin(2\beta) \quad \cdots\cdots (2.4.14)$$

2. Effect of Current on Maneuvering

☞ Refer to Vol. 4 Ship Handling Characteristics Data Base

Let's see how a ship behaves when it is turned in a steady current. In turning, a ship's heading changes gradually with the direction of current.

The water resistance acting on the hull pushes the ship to run along or against the current and the locus of the ship changes subject to the velocity and direction of current.

Figs. 2.4.21 and 2.4.22 show the simulation results of an LNG carrier when it is turned to make U-turn at an initial speed 12.3 knots by a helm angle of 15° in a steady current where the ratio of water depth to the ship's draft is 1.3 and current velocity is 4 knots. Fig. 2.4.21 shows the loci for comparison when the directions of current are 0° and 180° and Fig. 2.4.22 shows those when the directions of current are 90° and 270°.

As you can see in Fig. 2.4.21, running ahead distance is extended when the ship is turned following the direction of current at 0° while it is shortened when the ship is turned against the direction of current at 180°. However, in development of rate-of-turn in the two cases, there is little difference. After the completion of a U-turn, if the relative current angle is 0°, the movement of the ship is impeded by the adverse current but if relative current angle is 180°, the ship is carried downstream largely by the following current.

Fig. 2.4.22 shows cases where a ship is turned following 90° beam current and where a ship is turned opposing 270° beam current. In the former case, the ship is diverted from its initial course quickly and drifted largely downstream. While in the latter case, although the ship's head turns, as the ship is carried downstream against the current, it takes a long time to turn and consequently its running ahead distance is extended.

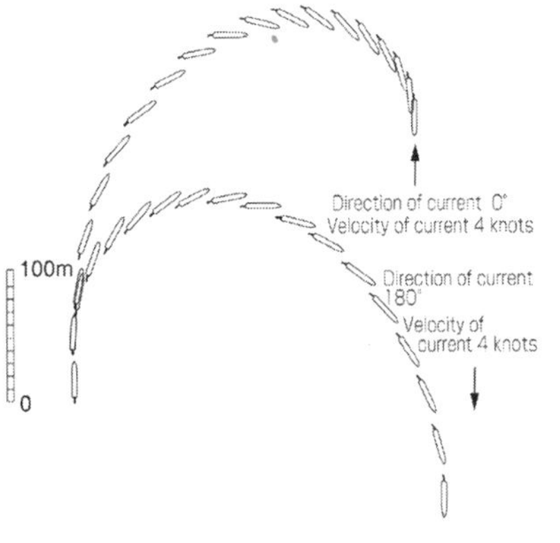

Fig. 2.4.21 Locus of U-turn in current (135,000 m^3 LNG carrier in full load, initial speed 12.3 kt, h/d : 1.3, relative current angle : 0°/180°, helm angle 15°)

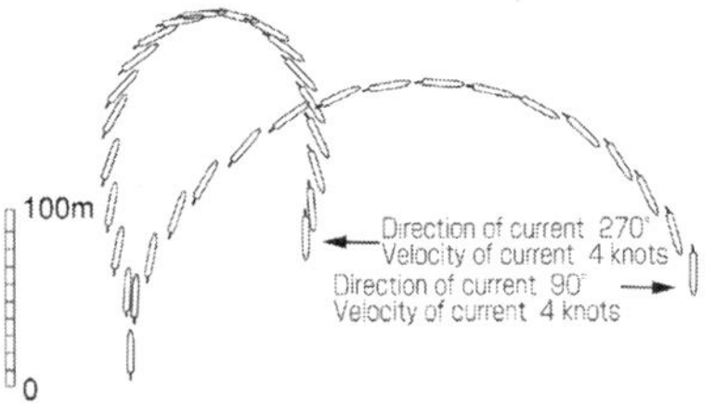

Fig. 2.4.22 Locus of U-turn in current (135,000 m^3 LNG carrier in full load, initial speed 12.3 kt, h/d : 1.3, relative current angle : 90°/270°, helm angle 15°)

After the completion of the U-turn, in the former case, the ship is largely carried downstream, while in the latter case, the ship is pushed back toward the original course by the current. In both cases there is little difference in the development of rate-of-turn.

3. Matters to be Kept in Mind while Maneuvering in Current

The followings are characteristics of maneuvering in current.

- During a U-turn in fair current, the ship's locus is drawn further downstream as the ship drifts in the current while turning.
- During a U-turn in adverse current, the ship's locus shrinks as the ship is pushed back while turning.
- There is little difference in the development of rate-of-turn between the two cases.
- The same characteristics can be seen during a U-turn in an oblique current.
- Caution is necessary with the fact that the effect of the direction of current is noticeable particularly when ship speed is low.
- Maneuvering to avoid any obstruction, fishing ship(s) engaged in fishing or other ship(s) should be made keeping the above characteristics in mind.

Chapter 3 Effects in Restricted Channel

1. Shallow Water Effect

☞ Refer to Vol. 4 Ship Handling Characteristics Data Base

1.1 Cause of Shallow Water Effect

When a ship moves in water, water around it is displaced and the displaced mass moves backward along the hull. However, as depth decreases, a part of the displaced mass of water which could normally pass through the space between the ship's bottom and the sea bed is interrupted and forced to flow to the sides of the ship.

Such interruption of flow around the ship's hull will give rise to added mass, added turning moment of inertia, increased hull resistance, and increased resistance of turning moment. In addition, due to change of flow around the hull, velocity of the water flowing around the side of the ship is accelerated and inversely pressure distribution on the hull changes. Such phenomenon of fluid dynamics caused by restricted water depth is called ***shallow water effect*** and it affects ship handling in the following aspects.

- Reduction of ship speed
- In maneuverability, although responsiveness to helm rises, turning ability becomes poor.
- Sinkage of hull
- Change of trim

1.2 Effect on Ship Maneuverability

Between the maneuverability indices of "T" (Index of responsiveness to the helm) and "K" (Index of turning ability) derived from the theory of a ship's maneuvering motion, the following relations in contrast to physical phenomenon hold true.

$$T=\left(\frac{I}{b}\right) \cdots\cdots (2.4.15)$$

$$K=\left(\frac{a}{b}\right) \cdots\cdots (2.4.16)$$

Where, "I" means turning moment of inertia, "a" and "b" represent gain coefficients. As explained in the previous subsection, added turning moment of inertia and resistance of turning moment increase as water depth decreases. Rudder force also increases slightly as propeller discharge current is enhanced by increased slip due to the reduction of ship speed.

Therefore all the values of "I", "a" and "b" increase when the water depth decreases. However the value of "T" and "K" of the equation (2.4.15) and (2.4.16) turn out to be smaller because the increment in "b", which is the denominator of the equations, is bigger than those in "I" and "a". This means that although responsiveness to the helm rises, turning ability becomes poor.

Fig. 2.4.23 shows the results of turning tests of an LNG carrier, for which specifications are shown in Table 2.2.1, in $h/d=\infty$ and $h/d=1.3$ in comparison. You can see the clear difference between the two turning circles due to the shallow water effect. The final diameters of turning circles in shallow water where $h/d=1.3$ are almost two times larger than those in deep water. However, comparing the increasing rate of advance with that of tactical diameter due to the shallow water effect, the increasing rate of the latter becomes larger than that of advance and consequently turning circles become bulging to the lateral direction.

In Fig. 2.4.23, in respect to the difference of the coefficient of responsiveness to helm "T" due to the shallow water effect, it was not possible to identify in the difference of head reach,

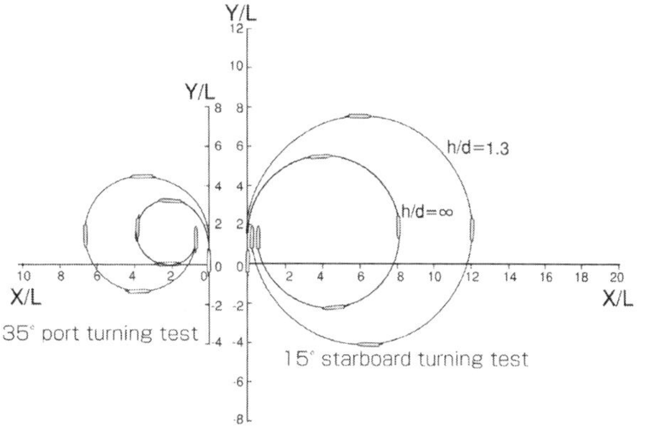

Fig. 2.4.23 Shallow water effect on locus ($135,000m^3$ LNG carrier in full load, helm angle : 15° starboard and 35° port, h/d = ∞ and h/d = 1.3)

because in the simulation the initial speeds in both cases, 12.33 knots in $h/d=\infty$ and 11.9 knots in $h/d=1.3$, were different. The difference in the initial speed was caused by the reduction of propeller revolution due to shallow water effect when the ship entered into the shallow water area under the same condition of S/B full ahead. Although such reduction of speed in shallow water area varies depending on ship type, it generally ranges around a few percent to 10% under the condition of $h/d=1.3$.

1.3 Sinkage of Hull and Change of Trim

When a ship runs in shallow water areas, the part of interrupted flow to pass through the space between the ship bottom and the sea bed is forced to seek its exit to the sides of the ship. Accordingly, flow around the hull transforms from three-dimensional to two-dimensional and velocity of the water flowing around the side of the ship is accelerated and the pressure at the midship part of the ship decreases. The distribution of pressure becomes high at the bow and stern and low at midship. As a result, the hull sinks and its trim changes as the ship is forced to float keeping its balance.

In general, this phenomenon brings about the change of trim by the head at low speed and trim by the stern at high speed. In addition, the shallower the water depth, the lower the speed range of the ship in which the change of trim by the stern becomes. However, within the speed range of merchant ships, ships sink and become trim by the head in shallow water area. Fig. 2.4.24 shows the results of the average hull sinkage and change of trim due to the shallow water effect by model tests. Where, a ship's speed is substituted by Froude number "F_n"($F_n = V/\sqrt{L \cdot g}$). For instance, when a tanker which has a length of 300m is running ahead at 11 knots, its Froude number will be 0.1

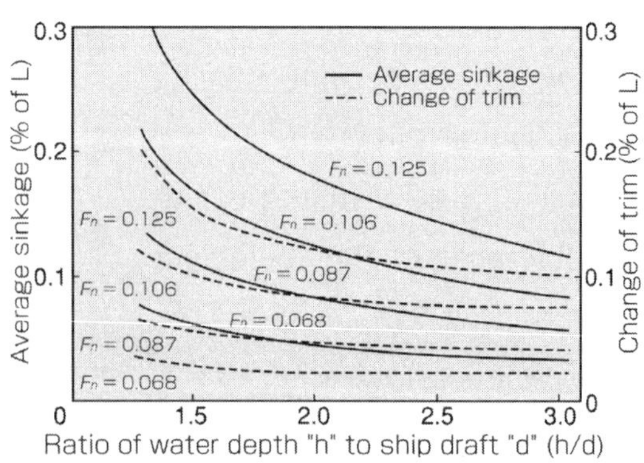

Fig. 2.4.24 Average sinkage and change of trim

from $F_n = 5.66\text{m/sec}\sqrt{300\text{m} \times 9.8\text{m/sec}^2}$. Therefore, in this case, in shallow water area with $h/d = 1.3$, sinkage of the hull and change of trim will be 0.2% and 0.1% of ship length respectively.

Equation (2.4.17) can be used to obtain an approximate value of sinkage at the bow when a ship is running in shallow water. As you see in the equation (2.4.17), the larger the "C_b", the smaller the L/B. This means that for broad beam and hypertrophied ship, the shallower the water depth and higher the ship's speed, the larger the sinkage at the bow becomes.

A typical example of this type is a VLCC which has a comparatively larger draft considering its water depth. For these types of ships, special caution is necessary for the squat phenomenon at the bow in shallow water in deciding under keel clearance for the safe maneuvering.

$$\text{Sinkage at the bow} = 1.5\left(\frac{d}{h}\right) C_b \left(\frac{L}{B}\right) F_n^2 L \cdots\cdots (2.4.17)$$

For your information, the effect of shallow water appears when the under keel clearance becomes less than 50% of the ship's draft. To prevent such undesirable effects, reduction of ship speed is most effective.

1.4 Change of Pivoting Point in Shallow Water Area

Fig. 2.4.25 illustrates the change of the pivoting point of a 2,800 TEU container ship for which specification is given in Table 2.1.3 while it is turned by hard over rudder (35°) at initial speed of 10.62 knots in various depths. In shallow water, the pivoting point changes characteristically as the development of rate-of-turn and lateral drifting velocity are deterred by increased resistance due to shallow water effect.

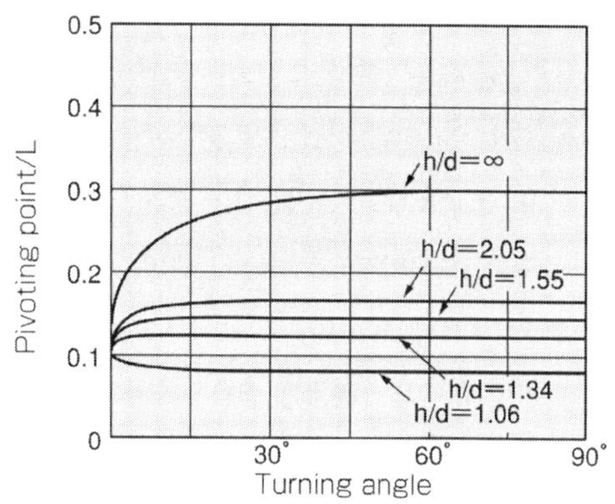

Fig. 2.4.25 Change of location of the pivoting point in shallow water area (2,800 TEU container ship in full load at initial speed 10.62 kt, helm angle 35°)

In shallow water, the pivoting point finally rests on the position about $0.15L$ forward from or closer to the ship's center of gravity, while it rests about $0.3L$ forward from the ship's center of gravity in deep water. When depth becomes shallower than $h/d=1.1$, the pivoting point will be situated a little backward from the ship's center of gravity.

1.5 Safe Maneuvering under Keel Clearance in Port

The space between a ship's bottom and the sea bed is called ***under keel clearance***. For maneuvering in port where water depths are restricted, special care is necessary to secure safe under keel clearance. For maneuvering in port, the following matters need to be taken into consideration.

- Sinkage of hull and change of trim of ship under way : When a ship is running at high speed in shallow water, sinkage of the hull becomes larger and generally its trim changes to the trim by the head.
- Sinkage due to the difference of specific gravity between salt water and fresh water : Sinkage of the hull may be caused in bays or harbors where the river discharges fresh water and specific gravity of sea water is diluted by the inflow of fresh water and lightened as a result.
- Sinkage due to the ship motion : the hull may hit or touch the sea bed instantaneously due to the ship's motions, such as heaving, pitching and rolling under the effect of waves.
- The size of the anchor dropped : when anchor is used in maneuvering, their size, such as the thickness of anchor head and tripping palm must be taken into account lest the ship's bottom may touch them while ship handling. To drop anchor for a larger ship, about one meter clearance is necessary.
- Cooling water inlet for main engine : to guard the cooling water inlet for the main engine from inflow of mud or sand from the seabed, extra under keel clearance of 1.5 to 2 times its diameter is needed.
- Reliability of chart datum and tide tables : For the margins of error of chart datum and tide table, about 10% for chart datum and 0.3m for the data of tide table, should be expected.

- Effects of meteorological, hydrographic and geographic conditions: sea level decrease by one centimeter per 1hPa (Hectopascal) atmospheric pressure increase.
- In addition, tidal heights, bottom quality and shape and sunken obstructions also must be taken into account in deciding safe under keel clearance.

2. Bank Effect

When a ship is running along the wall of a canal or close to a wharf, the velocity of flow between the ship's side and wall is quickened and the pressure between the ship's side and the wall decreases as discussed previously. As the asymmetric pressures affect to press the ship to the wall while a turning moment created by repulsive force between the wall due to built up pressure at the bow acts to turn the ship's head away from the wall.

When a ship runs along a restricted waterway close to its wall, ***cushion moment*** at the bow and ***bank suction*** to the ship's center of gravity simultaneously act as illustrated in Fig. 2.4.26. This phenomenon is called ***bank effect***. The closer the distance from the wall, the faster the ship speed and the shallower the water depth, the larger the effect becomes.

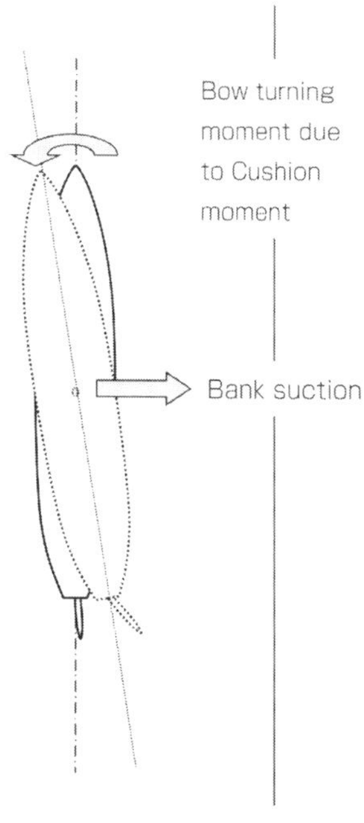

Fig. 2.4.26 Bank effect in restricted water

Turning of the bow created by bank effect makes the ship lie oblique to the relative current in the water. While lateral force due to oblique running acts to offset the suction to the wall, the turning moment due to oblique running acts on the same direction as the cushion moment at the bow. To offset this multiplied effect, the rudder needs to be turned toward the wall. When a ship is running along restricted water close to its wall, steering to keep balance among lateral forces and turning moment is required.

In such a case, it is recommended to keep an appropriate distance between the wall so that the ship may be controlled within normal helm angle 15°, or 5° as much as possible. If it is difficult to keep an appropriate distance, it would be better to reduce ship speed. An appropriate distance free from bank effect is said to be one ship length.

It should be noted that bank effect may also affect when a ship is running along the slope of the sea bottom. Even when the gradient of slope is gentle, a ship's bow sometimes turns to the deeper water repulsed by the slope due to high pressure build up at the bow depending on under keel clearance at that time. To avoid this phenomenon, its rudder needs to be turned toward shallow water. In this instance, caution is necessary to avoid grounding as the shallower the water depth, the closer the slope and the faster the ship's speed, the larger the helm angle becomes.

3. Interaction between Ships

3.1 Interaction between Two Ships

The distribution of pressure around a ship's hull running ahead is high at the bow and stern and low at the midship. For this reason, when two ships are running parallel closely in the same direction as shown in Fig. 2.4.27, at the part where high pressure coincides, cushion force is created, while at the part where low pressure coincides, suction force is created. This phenomenon created by two ships running parallel in the same direction is called ***interaction between two ships***.

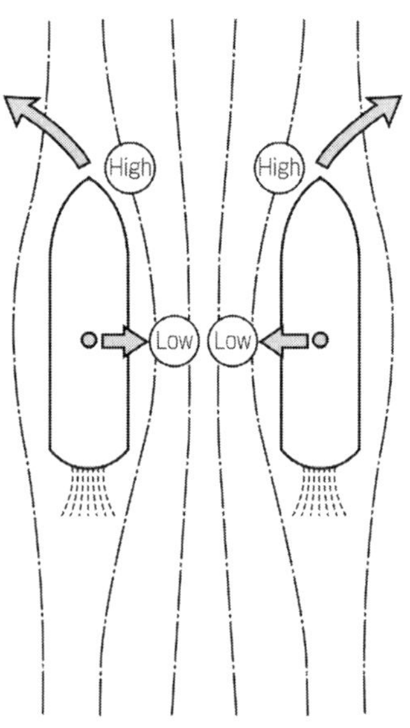

Fig. 2.4.27 Interaction between two ships

The effect of interaction between two ships meeting reciprocally may be transitory as they may pass clear of each other before the effect evolves.

In contrast, in the case of overtaking, caution is necessary as the same situation continues longer with the smaller relative speed and its effects of cushion and suction can be stronger and consequentially those ships could become too close to each other or the ship's bow could turn unexpectedly.

In addition, these effects will be enhanced in shallow water and affect stronger to the smaller ship. To reduce these effects, it is advisable to keep enough distance between the two ships and reduce speed. Safe distance free from these effects is about one ship length.

Fig. 2.4.28 illustrates typical interactions between two ships meeting reciprocally port to port to each other. In the figure, arrow marks show the direction of the effect.

① When the bows of two ships coincide, repulsive force to the hulls and turning moment to turn each ship's bow outward are affected.

② When two ships come abeam each other, suction at the part of low pressure between two ships and turning moment to turn the bow of each ship inward are affected.

③ When the sterns of two ships coincide, repulsive force to the hull and turning moment to turn each ship's bow outward are affected.

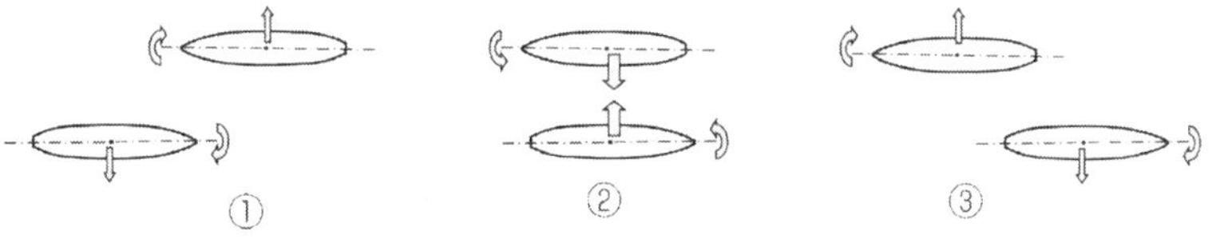

Fig. 2.4.28 Interaction between two ships meeting on reciprocal courses

3.2 Interaction between Ships in Overtaking

Fig. 2.4.29 illustrates the interactions of a typical overtaking case where ship A overtakes ship B on its starboard side. In the figure, arrow marks show the direction of the effect. Focusing the overtaking ship A and its consequential moment to be created,

① When A is approaching B from behind its starboard, weak turning moment to turn the bow of A toward B is created.

② When the bow of A comes almost abreast of the midship of B, turning moment to turn the bow of A toward B is enhanced.

③ When the two ships become parallel, strong turning moment to turn the bows outward from each other is affected.
④ When the stern of A comes abreast of the midship of B, turning moment to turn the bow of A outward weakens but remains.
⑤ When A almost clears B, although it is not so strong, the direction of turning moment changes to turn the bow of A toward B.

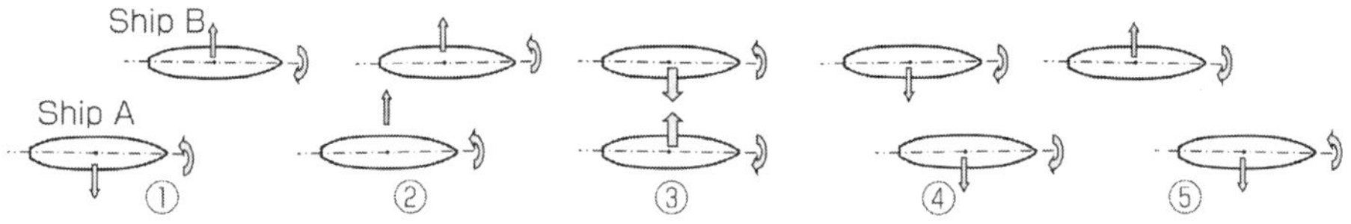

Fig. 2.4.29 Interaction between two ships in an overtaking situation

In respect to suction and cushion forces to be created on A change as follows,
① When A is approaching B from behind, weak cushion force is created,
② then, it changes to suction force which pulls A to B,
③ very strong suction force is created affecting both ships when the two ships come parallel,
④ suction force acting on A changes to cushion force when the stern of A comes almost abreast of the midship of B,
⑤ cushion force remains weakened until A clears B.

If the sizes of two ships of overtaking and overtaken are almost the same, the effects to B also may be deduced from the direction of arrow marks in Fig. 2.4.29. If either of the two is smaller, the interaction effect is stronger on the smaller ship. Special caution is necessary to prevent abrupt parallel approaching at ③ and possible contact of the bow or stern to the other ship's side at ② and ④.

The interactions between the two ships are also enhanced in shallow water in overtaking, especially when it is made at high speed. To reduce these effects, it is necessary to keep enough distance between the two ships and reduce speed. Safe distance free from these effects is about one ship length.

3.3 Effect on a Moored Ship by a Passing Ship

The effects of interaction between the two ships can be recognized when a ship is passing by another ships moored to the berth. In this case, a moored ship may be affected more strongly by the effects. When a passing ship approaches a moored ship from behind, the stern of the moored ship is at first pushed toward the wharf and slightly ahead by cushion force due to high pressure at the bow of the passing ship. Conversely, the moored ship is pulled away from its berth by suction force due to low pressure at the midship of the passing ship when the two ships are parallel. When the stern of the passing ship comes abreast of the bow of the moored ship, the bow of the moored ship is turned toward the berth being pushed backward as a whole by cushion force due to high pressure at the stern of the passing ship.

Thus, a ship running close to a moored ship is capable to inflict damage to mooring lines of the moored ship and fenders of the wharf by the effects of interaction between the two ships. Even after the passing ship has gone away from the moored ship, the moored ship may still be damaged by surging, swaying, heaving, rolling pitching and yawing due to divergent waves built up by the passing ship or the reflected waves from the wharf. These aftershocks also may cause not only damage to the ship's mooring ropes and fenders of the wharf but also interrupt the cargo works of the moored ship.

Such effects may be worsened when a ship runs close to a moored ship at high speed in shallow water. Special care is necessary when a ship is passing by any other moored smaller ships. To reduce these effects and prevent consequential damage to moored ships, particularly for those smaller ships, speed reduction is the most effective means.

Chapter 4 Effect of Waves on Maneuvering

1. Dangerous Phenomena Caused by Waves in Maneuvering

The longer strong wind lasts over the sea surface, the higher the waves grow. Strong wind and high waves are mortal enemies of ships. For instance, sea water gives high impact force to ships as the weight of water is more than one ton per cubic meter, and brings heavy shocks to ship structures, which are beyond a non-seafarer's imagination.

Persons engaged in ship handling need to make every effort to avert and reduce such shocks to prevent consequential damage by reducing speed or changing course or other effective means available.

The effect of waves increases as ship size decreases. If the hull is damaged, the ship immediately faces the risk of capsizing due to, at worst, taking on water. In stormy seas, a ship's motions, such as rolling, pitching, heaving and yawing, are magnified. Violent pitching causes the racing of the propeller which is detrimental to the main engine. And it also brings about shocks at the bow and panting of the bottom accompanied by intense heaving. Pooping down waves may invite flooding or damage the rudder system. Abeam waves magnify a ship's roll which may lead to capsizing or structural damage due to a scooped mass of sea water on deck. The role of a person engaged in ship handling is how to mitigate such effect of waves and the ship's motion.

2. Particularly Dangerous Phenomena in Rough Seas

2.1 Panting

Panting is the phenomenon that shocks the bow plates, given by waves when a ship runs against them. Smaller ships are sometimes thrown back by such shocks.

2.2 Slamming

Slamming is the phenomenon when the ship's bottom plates at the bow hit violently on the surface of the sea when the ship is running through swells against waves accompanied by large pitching. This may sometimes cause severe structural damage due to tremendous hydraulic pressure.

2.3 Broaching

Broaching is the phenomenon when the ship's heading suddenly becomes unstable and turns to make the ship lay abeam against a swell due to yawing when the ship is running down the slope of a swell following its propagating direction. This sometimes happens when there is little difference between ship speed and wave velocity. In this phenomenon, the risk of capsizing increases due to beam waves in addition to roll.

2.4 Poop down

Poop down means the onrush of a mass of sea water on the poop deck when ship is running in following seas. Due to the fundamental design concept of ship sterns, the rudder or structures at the stern are liable to be damaged by this phenomenon.

2.5 Pitch poling

Pitch poling is quite a dangerous phenomenon in which a ship turns a somersault thrusting its bow into the rear face of the facing wave when small and light ships are running through steep waves at high speed, and are going down the slope of a wave.

3. Avoidance of Synchronous Ship Motions

When a ship is running in rough seas, utmost caution must be paid to prevent the magnification of ship motion by synchronous motion to the waves. Among the ship's motions due to waves, rolling and pitching are most prominent. Lurch which may lead to capsizing will be caused when a ship's roll is synchronized to the cycle of waves.

When pitching is synchronized to the cycle of waves, slamming happens. Either of these phenomena is extremely dangerous to a ship.

3.1 Courses Likely to Cause Synchronous Rolling

Rolling will be enhanced by beam and quartering waves. When a ship runs in the following wave as shown in Fig. 2.4.30, its encountering cycle "T_E" can be given by the following equation (2.4.18).

$$T_E = \frac{\lambda}{V_W - V\cos\theta} \quad \cdots\cdots (2.4.18)$$

where

λ : wave length (m) $\quad \lambda = 1.56T_W^2$ (T_W: wave period)

V_W : wave velocity (m/sec) $\quad V_W = 1.25\sqrt{\lambda}$

V : ship speed (m/sec)

θ : encountering angle to wave (degree)

Let the ship's natural rolling period be "T_R", a synchronous course can be derived from the equation (2.4.19), as $T_E = T_R$ stands in this case.

$$\cos\theta = \frac{\frac{\lambda}{T_R} + V_W}{V} \quad \cdots\cdots (2.4.19)$$

An encountering angle when a ship's roll and wave become synchronous can be derived replacing "T_R" from the relation, $T_R = 0.8B/\sqrt{GM}$. And the ship's speed enabling it to avoid synchronous rolling can be deduced by the following equation (2.4.20).

$$V = \frac{\frac{\lambda}{T_R} + V_W}{\cos\theta} \quad \cdots\cdots (2.4.20)$$

In the wavy seas, ships need to run avoiding at such course and speed.

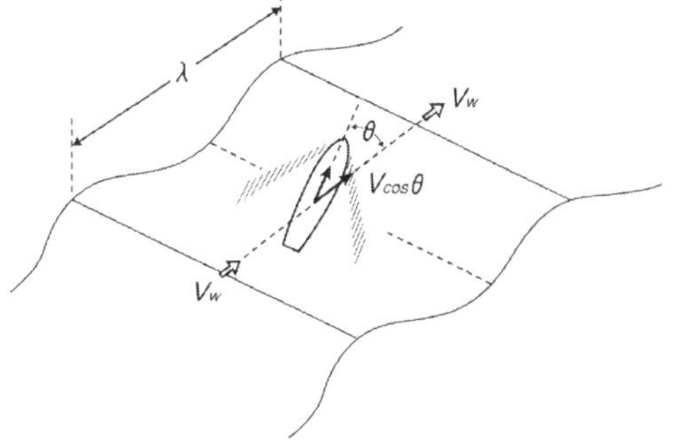

Fig. 2.4.30 Encounter with waves (quartering seas)

3.2 Speeds Likely to Cause Synchronous Pitching

Pitching will become intense in head seas. The wave encountering cycle "T_E" when a ship runs in bow waves can be derived by the equation (2.4.21).

$$T_E = \frac{\lambda}{V_W + V} \quad \cdots\cdots (2.4.21)$$

Let the ship's natural pitching period be "T_P", the synchronous speed can be derived from the equation (2.4.22), as the condition of the period of the ship's pitching and waves become synchronous is $T_E = T_P$.

$$V = \frac{\lambda}{T_P} - V_W \quad \cdots\cdots (2.4.22)$$

Substituting $T_P = 0.5\sqrt{L}$ (L: ship length) in the equation (2.4.22), a speed synchronous for pitching can be derived. When a ship runs in head seas at the speed obtained from the above equation, synchronous pitching can be caused. In such conditions, slamming may also be aggravated. Such synchronous speed must be avoided in the wavy sea.

Volume 3

Practical Ship Handling

Ⅰ. Navigation, Berthing and Un-berthing

Ⅱ. Mooring and Anchoring

Ⅰ
Navigation, Berthing and Un-berthing

Chapter 1 Course Keeping Maneuvers

1. Key to Course Keeping Maneuvers

The fundamental operation of navigation to run a ship along its preplanned course is to keep and alter the course regardless of the water areas where the ship navigates whether it is the ocean, coastal seas, inland seas or narrow channels. Ship handling to have a ship run along the preplanned routes by keeping and altering course are respectively called ***course keeping maneuvers*** and ***course changing maneuvers***.

To keep navigating along the preplanned course, once the direction of the heading is set, the ship should be held on the intended course line by using either autopilot or manual steering as appropriate depending on the navigational situation to control the ship's heading to offset deviation from the preplanned course due to external disturbances.

2. Leeway

Without any external disturbances, a ship can sail on its course lines as preplanned. Actually, for a ship navigating at sea, various lateral forces have effect, causing it to deviate from its original course due to external disturbances such as wind, sea and tidal currents.

As a result, the ship moves being drifted in the direction of the resultant vector of the effects of wind and currents and ahead force acting on the ship's center of gravity as shown in Fig. 3.1.1. And the ship's track will be oblique to its original course under the conditions that external disturbances such as wind, currents, rudder force, and the resistance of water are in equilibrium. The angle between the original course and the actual track is called ***leeway***.

From the figure on the right, you can understand that if the ahead force is decreased by reducing ship speed, leeway will increase as the effect of lateral force by wind and currents becomes relatively larger and fluid pressure decreases simultaneously. Therefore it should be noted that even under the same level of external disturbances, with lower speed, a ship might drift largely in an unexpected scale.

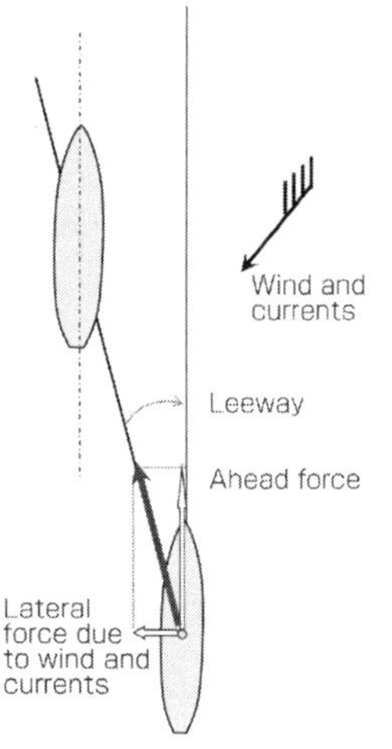

Fig. 3.1.1 Leeway

3. Scale of Leeway

Leeway is an angle that is formed under the condition that rudder force, the lateral force due to wind and currents and fluid resistance due to a ship's oblique track are in equilibrium, in which the resultant lateral force affecting the ship's center of gravity and resultant moment around the ship's center of gravity become zero as shown in the equations (3.1.1)

$$\left.\begin{aligned} Y &= Y(e) + Y(\beta) + Y(\delta) = 0 \\ M &= M(e) + M(\beta) + M(\delta) = 0 \end{aligned}\right\} \qquad (3.1.1)$$

where

Y : Resultant lateral force
- $Y(e)$: lateral force due to wind and current
- $Y(\beta)$: lateral component of fluid resistance
- $Y(\delta)$: lateral component of rudder force

M : Resultant moment around the ship's center of gravity
- $M(e)$: moment by wind and currents
- $M(\beta)$: moment by fluid resistance
- $M(\delta)$: moment by rudder

When the lateral forces and turning moments by wind and currents, fluid resistance and rudder are given to equation (3.1.1) in which a ship is running obliquely in equilibrium at a fixed angle under the effect of wind and currents, leeway "β" can be derived from equation (3.1.2) as follows.

$$\beta = k \cdot \frac{B_a}{Ld}\left(\frac{V_a}{V_s}\right)^2 \qquad \cdots\cdots (3.1.2)$$

where

B_a : side projected area above water (m2)
L : ship length (Lpp : m)
d : ship draft
V_a : relative wind speed (m/sec)
V_s : ship speed (m/sec)
k : a ship's natural parameter

Fig. 3.1.2 illustrates deduced leeway of a 6,000 unit PCC running under the effect of wind.

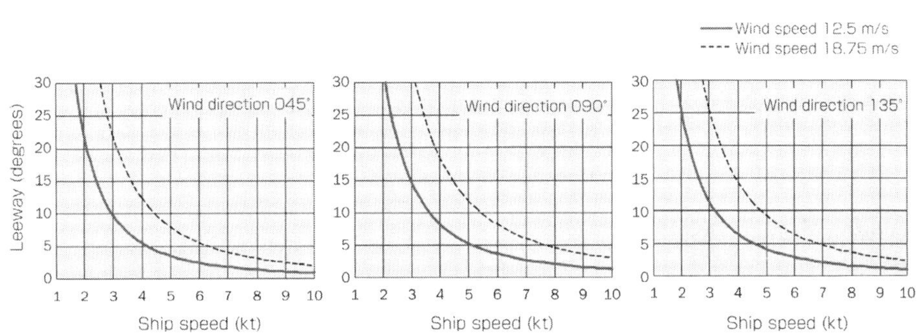

Fig. 3.1.2 Deduced leeway of 6,000 unit PCC in full load, h/d = 1.3

Chapter 2 Course Changing Maneuvers

1. Key to Course Changing Maneuvers

Usually, a person engaged in ship handling changes their ship's course by taking a bearing of an available navigational aid or a land mark like conspicuous mountain peak or the tip of a promontory that is abeam selected beforehand. In such cases, to make the ship run on the new course line as preplanned, they need to steer her applying easing or checking the helm as appropriate to control the ship's turning inertia following the initial helm angle to turn to settle the ship on the new course.

Therefore when actually at sea, according to the turning angle for a new course, appropriate judgments should be made based on sufficient knowledge about previously mentioned characteristics of turning inertia of one's own ship.

To work out such steering well, it is necessary for a person engaged in ship handling to be conversant to the characteristics of responsiveness to the helm of their own ship in respect to timing and/or the extent of easing helm, checking rudder and over shoot angle, including development of rate-of-turn, stern kick and speed reduction in turning as necessary, although the process, procedures and extent of steering may vary depending on the situation.

2. New Course Distance

One of important things to be kept in mind at the alteration of course is the timing of when the order should be made to turn the ship. For instance, if the new course is drawn bearing on any navigational mark abeam and the ship is turned at that point, the ship will veer from its course outward. In addition, if the ship is turned too early, its track will be off the right course inward. This timing is particularly important for safe maneuvering especially when a new course line is set to avoid obstructions inside or outside of the course line.

To make the ship run right on course, the helm needs to be turned before the turning point on the course line taking the time lag due to the responsiveness of the rudder into account. Such distance between the turning point of the course line and the point to turn the helm is called ***distance to new course***. When turning angle and helm angle are respectively "ψ" and "δ", the distance to the new course corresponds to the segment AC in Fig. 3.1.3. AC is the sum of AB and BC where AB corresponds to reach shown in the equation (2.1.26).

From the above,

$$\begin{aligned} AC = AB + BC &= V\left(T + \frac{\tau}{2}\right) + R \cdot \tan\frac{\psi}{2} \\ &= V\left(T + \frac{\tau}{2}\right) + \frac{V}{K\delta}\tan\left(\frac{\psi}{2}\right) \quad \cdots\cdots\cdots (3.1.3) \end{aligned}$$

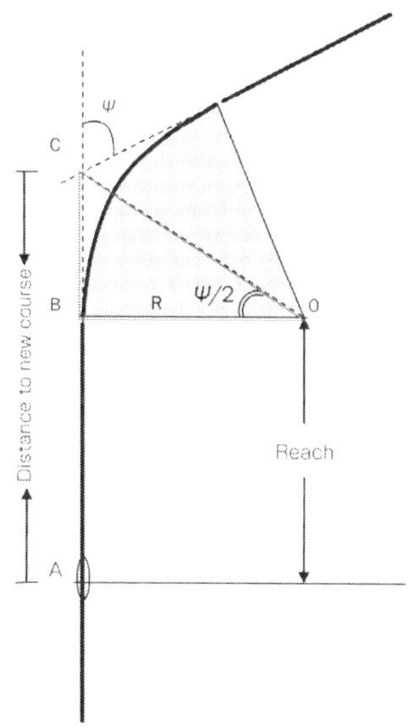

Fig. 3.1.3 Distance to new course

Distance to a new course can be derived from the equation (3.1.3). As you see in the equation, the distance to a new course is closely related to the maneuverability indices, "T" and "K". For a ship which index of responsiveness to the helm is large and index of turning ability is small, that is to say, responsiveness to the helm is slow and difficult to make sharp turns, it is necessary to turn its helm sooner, anticipating that the ship may run a longer distance to the new course. In addition, it can be said the same for a ship which speed "V" is high or angle of turn "ψ" is large, and/or time required for steering "τ" is long.

3. Test for New Course Distance

It is indispensable for a person engaged in ship handling to be conversant their own ship's characteristic of distance to new course corresponding to angle to be turned and timing of when to turn the helm thereof. For this purpose, a ***course change test*** is carried out. In this test, the development of rate-of-turn corresponding to a helm angle applied and over shoot angle followed by easing helm and checking rudder in addition to distance to new course are measured.

For instance, when a ship is turned by $\delta°$ of helm angle at first, then, when it turns to $\psi°$, the rudder is reversed $\delta°$ to the opposite board, and the overshoot angle at that moment is measured. Also, the overshoot angle after a ship has been turned to an intended direction and its rudder is returned to midship is measured. If the data in various combinations of change of heading angle and helm angle are measured in this way, the scale of distance to a new course, necessary helm angle, timing of easing and checking the rudder corresponding to the change of heading angle can be deduced beforehand. It is very useful to organize and apply such data in prior to giving actual orders for changing course by deciding the correct angles and timings.

Chapter 3 Collision Avoidance Maneuvers

1. Key to Collision Avoidance Maneuvers

If a risk of collision exists between two ships underway, immediate action to avoid such a risk must be taken complying with the appropriate international or domestic rules at sea. When two ships are meeting in a head-on situation, both ships must take appropriate collision avoidance action respectively. And if the two ships are in a crossing situation with each other, the ship which has the other ship on her own starboard side must give way to the other ship. Such action to avoid the risk of collision is called ***collision avoidance maneuver.***

Collision avoidance maneuvers by steering are an unexpected transient course changing maneuver when a ship faces a dangerous encounter with another ship, which is different from ordinary course changing maneuvers. To avoid the risk, the person engaged in ship handling is required to take a series of actions promptly after finding another ship which has possibility to interfere the safe maneuvering of their ship in any way. Such actions should include the assessment of the risk for collision, and if there is any, decisions when to initiate proper actions, how to react and to what extent.

In such instances, the decision of timing to initiate collision avoidance maneuvers, course to be headed on and helm angle to be ordered including checking rudder thereof should be made taking their own ship's turning characteristics into account.

2. Distance to Initiate Collision Avoidance Action

The distance between the two ships that collision avoidance actions are initiated should be decided taking a ship's turning circle and allowance of it into account. In the case of a ship which responsiveness to the helm and turning ability are poor, collision avoidance actions need to be initiated well in advance.

Therefore, the distance that collision avoidance actions should be initiated may vary depending not only on a ship's own maneuvering abilities but also the environmental conditions at that time such as the relative position of two ships involved, ship size, ship speed, available water area, water depth and the skill of a person engaged in ship handling. At present, the distance to initiate collision avoidance actions are entrusted to the individual decision of each person engaged in ship handling. For your reference, the equation (3.1.4), which is derived from the statistics of marine traffic surveys, is presented to deduce distance to initiate collision avoidance actions.

$$D = (3.3R_v + 6)\sqrt{(L_o^2 + L_t^2)/2} \quad \cdots\cdots (3.1.4)$$

where

R_v : relative ship speed (m/sec)
L_o : length of subject ship (m)
L_t : length of the other ship (m)

3. Safe Distance to be Maintained between Two Ships

3.1 Safe Distance to be Maintained in Broad Water Area

In addition to the distance to initiate collision avoidance action, distance to be maintained for safety reasons in a broader water area is also entrusted to each person engaged in ship handling. Thus, it will depend on each of the person's discretion engaged in ship handling, but its scale tends to be larger in broader water areas than in narrow water areas. This seems to reflect the fact that relative ship speed is higher in broader water areas than in narrow water areas.

Fig. 3.1.4 shows the analytical result of marine traffic surveys on safe distance to be maintained in broad water areas. According to the results, it ranges from 8 ship lengths fore and aft and 3.2 ship lengths port and starboard centered on one's own ship.

However, for ocean going ships, the distance seems to be estimated double to triple that due to the recent tendency of increased size and speed of ships, bringing about a wider, safer range of approach.

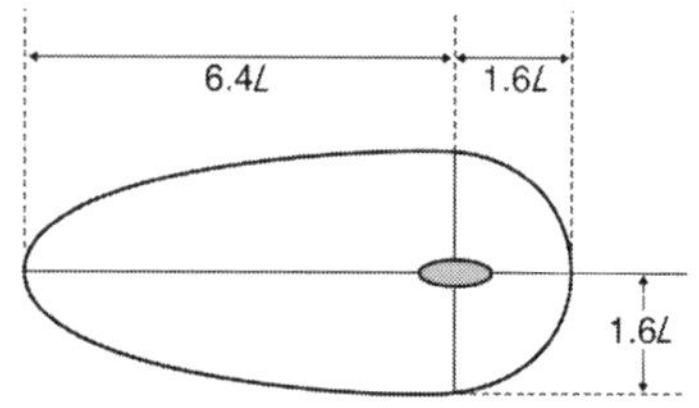

Fig. 3.1.4 Safe distance to be maintained in broad water area

3.2 Safe Distance to be Maintained in Narrow Water Area

The closest distance between two ships as a result of collision avoidance actions is called ***passing distance***. There is also no established standard for safe range of passing distance to be kept between ships and it is entrusted to each person's discretion engaged in ship handling.

Fig. 3.1.5 shows the analytical result of the consciousness of persons engaged in ship handling about the passing distance from a questionnaire. Equation (3.1.5) presents the permissible minimum passing distance and equation (3.1.6) presents safe passing distance. The segment FA and SP in Fig. 3.1.5 respectively represent safe passing distance fore and aft and that for port and starboard centered on one's own ship. In the same figure, "L_o" and "L_t" represent the length of one's own ship and that of the other ship.

Minimum permissible passing distance (in port)

$$\left.\begin{aligned} FA &= (0.015L_t + 2.076)L_o \\ SP &= (0.008L_t + 0.667)L_o \end{aligned}\right\} \cdots\cdots (3.1.5)$$

Safe enough passing distance (in port)

$$\left.\begin{aligned} FA &= (0.025L_t + 3.125)L_o \\ SP &= (0.012L_t + 1.096)L_o \end{aligned}\right\} \cdots\cdots (3.1.6)$$

Ships moving in port are obliged to clear many other obstructions in addition to other ships under way, including ships lying at anchor.

The followings is the data on the minimum permissible passing distance between one's own ship and obstructions or ships lying at anchor obtained from questionnaires given to persons engaged in ship handling. Where, "L_o" represent the length of one's own ship.

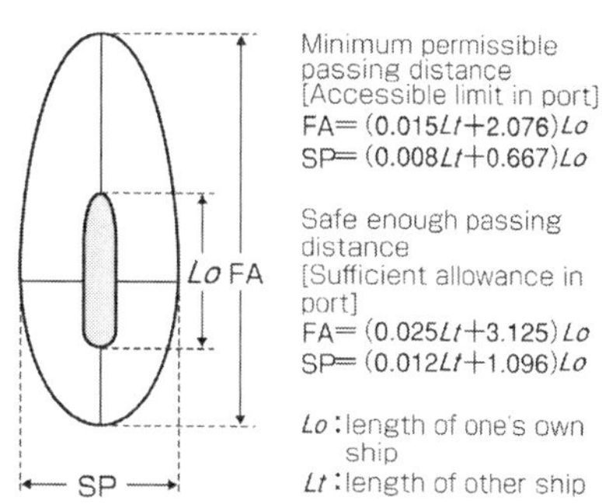

Fig. 3.1.5 Safe distance to be maintained in narrow water areas

Clearance between obstacles:

Buoy	: 0.33 L_o
Breakwater	: 0.54 L_o
Wharf	: 0.68 L_o
Ship lying at anchor	: 0.89 L_o

4. Minimum Distance Possible to Avoid Collision by Steering

Minimum distance possible to avoid collision with another ship will be defined by the size of turning circle and particularly maximum advance by extreme helm angle. As the scale of a turning circle has almost no change by the ship's speed, the knowledge of the size of the maximum advance can be useful to estimate minimum distance to avoid collision in any cases of ship speed. However, caution is necessary because the size of the turning circle increases considerably where the depth is shallow.

Chapter 4 Berthing and Un-berthing Maneuvers

1. Utilization of a Ship's Ropes in Berthing and Un-berthing Maneuvers

1.1 Effects of a Ship's Ropes in Berthing and Un-berthing Maneuvers

In many cases the spring line is the line to be sent out first at berthing and let go last when leaving from the berth. It is certain that the spring line is the line which can be used as auxiliary means in berthing and un-berthing maneuvers. Particularly, the back spring has important roles and can be used as follows.

(1) Control of the Ship's Headway

At berthing, if a back spring is first sent out, its tensile strength can be used to check the ship's headway. Its role will be particularly important when astern engines are not available.

(2) Pulling the Ship Closer to the Berth

When the ahead engine is ordered to operate in the state that a back spring is taken to a bit, the hull is pulled to berth by the reaction force of the rope. It is effective for ships without lateral thrust.

(3) Swinging out the Ship's Stern while Un-berthing

When ships are leaving from the berth, their sterns are swung away from the wharf first so that their propellers, rudders, and stern structures are not damaged. In such cases, if the ahead engine is used keeping the back spring belayed, the ship's stern will be swung out off the berth; with helm angle on the inshore side, the swing out of the ship's stern will be larger.

1.2 Dynamic Analysis of the Effect of Utilization of Ropes

To understand a ship's motion under multiple forces such as fore and aft thrust, rudder force, reaction force of mooring ropes and other external disturbances by wind and currents more easily and clearly, analysis using a polygon of vectors of acting forces is useful.

For example, assuming only two affecting forces, fore and aft thrust and reaction force of ropes, I will illustrate the polygonal analysis hereafter. The same concept can be applied to other cases of berthing or un-berthing maneuvers under the effects of external disturbances.

(1) Effect of a Rope when a Ship has Headway

As shown in Fig. 3.1.6, assuming that a ship approaching a berth, sends out a rope from A to the bitt B on the berth. Now the ship is under two forces, ahead thrust to the ship's center of gravity G and the tension on the rope acting on A on the ship, let's look at how to analyze these forces. Figures 3.1.6 (a), 3.1.6 (b) and 3.1.6 (c) respectively show the conditions of a rope sent out from the bow, from the stern and from the midship.

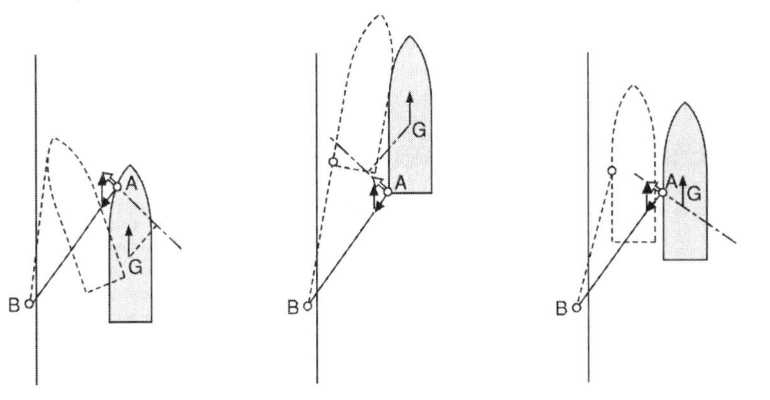

(a) Rope sent out from Bow (b) Rope sent out from Stern (c) Rope sent out from Midship

Fig. 3.1.6 Effect of a rope when a ship has headway

Let's look at how the resultant forces of these two forces act on A. From the starting point A, draw a vector of the tension on the rope and draw a vector of ahead thrust from the tip of the vector of the tension. Form a polygon consisting of vectors by connecting each tip of all the vectors. Then find the vector directing from the starting point towards the end point of the polygon, which is the resultant force acting on A.

Due to the resultant vector, when the tensile force starts affecting the rope,

A on the ship which is moving ahead will be moved to the direction of the resultant vector as if it was pulled by the rope toward the berth.

When the acting line of the resultant force does not pass through the ship's center of gravity as shown in the Figs. 3.1.6 (a) and (b), the ship will be turned by the moment around its center of gravity.

Turning moment in Fig. 3.1.6 (a) will act to turn the bow toward the berth while in Fig. 3.1.6 (b) it will act to turn the stern toward the berth. In contrast, as in the case of Fig. 3.1.6 (c), as the acting line of the resultant force passes through the ship's center of gravity, turning moment will not be created.

From the above, you can understand that the ship will be pulled to the berth turning its bow or stern inshore when its headway is checked by a back spring or stern line, and pulled parallel to the berth without turning when its headway is checked by a rope taken closer to the midship.

(2) Effect of a Rope when a Ship has Sternway

Fig. 3.1.7 illustrates the analysis of the effect of rope to a ship with sternway while berthing. The Fig. 3.1.7 (a), Fig. 3.1.7 (b) and Fig. 3.1.7 (c) respectively illustrate when a line is taken from the bow, stern, and midship.

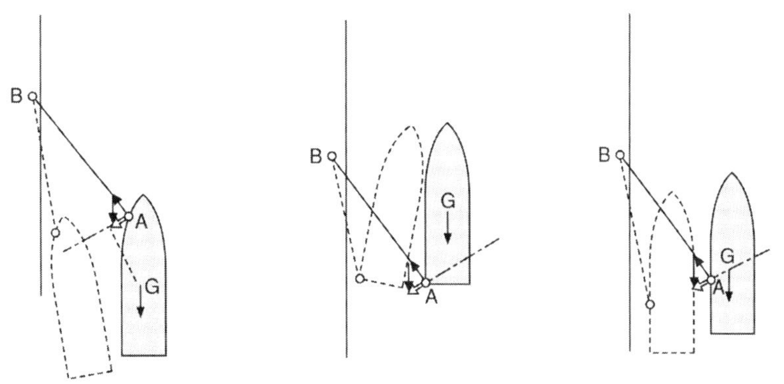

(a) Rope sent out from Bow

(b) Rope sent out from Stern

(c) Rope sent out from Midship

Fig. 3.1.7 Effect of a rope when a ship has sternway

As you can see from Fig. 3.1.7 the ship will be pulled to the berth turning its stern inshore when a rope is taken closer to its stern. While when a rope is taken closer to the bow, the ship is pulled to the berth turning its bow inshore. And when a rope is taken closer to the midship, the ship is pulled parallel to the berth without turning.

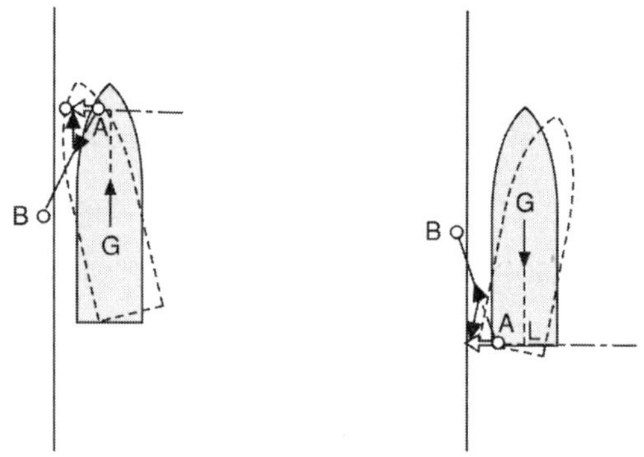

Fig. 3.1.8 Effective use of a rope at un-berthing

(3) Effect of a Rope when a Ship Leaves from Berth

Fig. 3.1.8 illustrates the analysis of the effect of a rope to swing out the bow or the stern when a ship is leaving from the berth. As shown in Fig. 3.1.8 (a), when the ahead engine is used keeping the back spring belayed, its stern will be swung out off the berth using the support of the rope. If the rudder is turned inshore at this moment, swing out of the stern will be enhanced. By contrast if the astern engine is used keeping the aft spring belayed as shown in Fig. 3.1.8 (b), its bow will be swung out off the berth. If the rudder is turned inshore at this moment as well, swing out of the bow will be enhanced.

2. Use of Anchor in Berthing and Un-berthing Maneuvers

2.1 Utilization of Drag Force of Anchor

Although the main purpose of the anchor is to keep the ship's position riding to its anchor, it is often used as an auxiliary means in berthing or un-berthing maneuvers. In particular, in restricted water area it can often be used effectively to control a ship's motion or posture. To utilize the drag force of an anchor is one of the various usages.

In this method, the anchor and its chain, usually 1.5 to 2 times of the depth of water, is actually dragged over the sea bed without being held firmly in the sea bed. To create right amount of dragging force, the length of the chain needs to be adjusted properly depending on the ship's speed and should not be too long. The drag force of the anchor in such conditions will be about 1.5 times of the weight of anchor. In addition, anchor can be used in the following cases.

(1) Control of Headway

It can be used to control residual headway in maneuvers for approaching the berth keeping steerageway and in an emergency as a last resort when the astern engine fails in approaching the berth.

(2) Prevention of Turning of Bow under Astern Engine

It is useful to restrain turning of the bow under astern engine and to keep a steady heading in approaching the berth.

(3) Auxiliary Use in Turning

When going along side at 180° turning before the berth, or in a narrow water area where space for turning is restricted or the effect of wind and currents are strong, a ship can make a small turn centered on the dredging anchor in addition to engine and rudder.

(4) Straight Astern

It can be used to prevent the turning of the bow when a ship is obliged to go astern, straight in a narrow channel.

(5) Prevention of Drift Due to External Forces

It can be used to prevent drift by wind and currents in berthing and un-berthing and is particularly useful to control the speed of lateral shift to the berth.

2.2 Utilization of Holding Power of Anchor

In this case, adequate length of anchor chain needs to be paid out according to ordinary anchoring procedures, using brake of windlass or walk back appropriatly, to assure the holding power of the anchor.

(1) Lateral Shift Clearing from Berth

If an outboard anchor is dropped before berthing, it can be available for clearing from the berth by heaving in that anchor.

(2) Braking in an Emergency

When a ship is obliged to stop in an emergency, both anchors can be, and in such a case rather should be used as a brake in addition to full astern engine.

2.3 Cautions for Handling of Anchor

- Before berthing, either anchor to be used should be prepared for letting go.
- Headway when an anchor is letting go should be not more than 2 to 3 knots regardless of its drag force or holding power that will be used.
- The rate of paying out anchor chain needs to be adjusted corresponding to the ship's speed so that anchor chain will not be piled up on the sea bed.
- Before the anchor is going to let go, sufficient caution should be paid if wires or anchor chains of other ships, fishing bank, rope handling boats and tugs are clear from the dropping anchor.
- Caution also should be taken if there is enough water area for swinging of the stern when the anchor is brought up.

3. Maneuvering Methods for Berthing and Un-berthing

3.1 Fundamentals of Berthing and Un-berthing Maneuvers

During berthing and un-berthing, the conditions affecting ship maneuvering, such as location of berth, type of berth, water depth, existence of other ships in the vicinity, size, type, draft and the maneuvering characteristics of one's own ship, whether or not there are side thrusters, sea and weather conditions and availability of tugs, needs to be taken into account. Therefore, flexible maneuvering depending on varying conditions and circumstances is required for both berthing and un-berthing.

However under certain environmental conditions, there are the fundamentals of ship maneuvering for persons engaged in ship handling to rely on. When actual situations at sea do not fit into such conditions, their role is to make the right decisions applying the fundamentals in concert with their personal skills built up by their long experience at sea.

In this section, we will discuss fundamental ship maneuvering for berthing and unberthing on the premise that;

- The shape of the berth is not L or comb type but straight.
- There are no ships moored to adjacent berths, lying or moving in the vicinity, and the ship to be berthed is not required to perform extraordinary ship handling, such as berthing ahead of a moored ship or between two moored ships.
- The propeller of the ship is single and right handed. This implies that the ship's stern will be swung to the portside while the bow will be swung to the starboard by the effect of impingement due to propeller discharge current at the stern and the sidewise pressure of the propeller when its engine is driven astern.
- There is no effect from wind and currents; thus maneuvering using current and wind is excluded.
- The ship will be berthed only by its engine and rudder without using thrusters and support of tugs.

3.2 When Approaching the Berth at a Small Angle

(1) Head-in Port side to (refer to Fig. 3.1.9)

① The ship proceeds to the berth with its heading at angle of 10° to 20° to the berth face at a low inertia speed aiming at the point of one ship length or 4 to 5 times the length of the ship's beam ahead from the appointed berth. The headway at this moment needs to be stopped

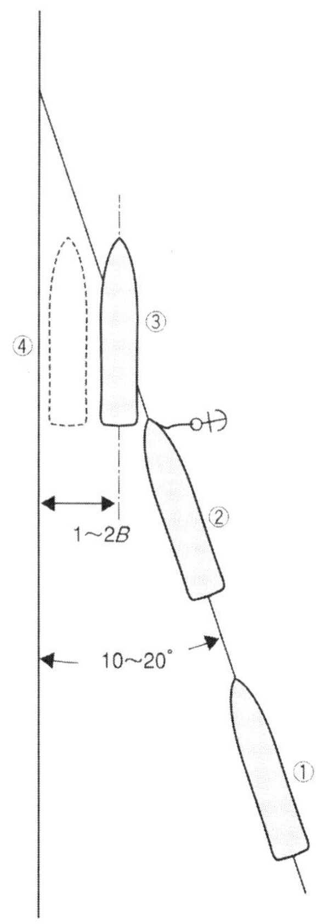

Fig. 3.1.9 Maneuvering: Head-in port side to

without a large swing of bow when the astern engine is used to stop.

② Send out the head line and back spring to the shore before one ship length before the appointed berth. When anchor is used, let go the starboard side anchor by the time the ship reaches the position of a half ship's length before the appointed berth and pay out anchor chain 1.5 to 2 times of water depth.

③ Putting engines astern a half ship's length before the appointed berth, the ship can be stopped parallel to the appointed berth by the effect of impingement due to propeller discharge current at the stern and the sidewise pressure of the propeller by the astern engine. The distance from the berth at this moment will be 1 to 2 times the ship's beam.

④ Send out stern line and heave it in together with head line to secure the ship to berth.

(2) Head-in Starboard Side to (refer to Fig. 3.1.10)

① In this case, maneuvering is more difficult because when an astern engine is used near the appointed berth, the ship's stern will be swung outward from the berth by the effect of impingement due to the propeller discharge current at the astern and sidewise pressure of propeller by the astern engine while the bow will be swung to the berth. Therefore, approach to the appointed berth should be in low inertia speed and made parallel to or at a small angle to the berth as much as possible keeping distance 1 to 2 times the ship's beam from the berth face.

② At one ship length before the appointed berth, send out head line or back spring to the shore. When anchor is used, drop the port side anchor by the time the ship reaches the position of a half ship's length before the appointed berth and pay out anchor chain 1.5 to 2 times the water depth.

③ Use the astern engine so that the ship may be stopped right

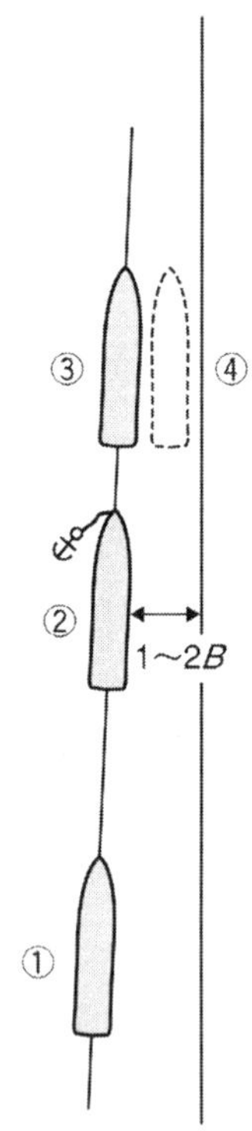

Fig. 3.1.10 Maneuvering: Head-in starboard side to

beside the appointed berth. Utilize anchor and its chain to restrain inshore swing of the bow.

④ Send out stern line and heave in it together with head line to secure the ship to berth.

3.3 Stern-in Mooring

(1) **Stern-in Port side to** (refer to Fig. 3.1.11)

① Approach parallel to the berth face keeping a 1/2 to 1 ship's length from the berth face under 2 to 3knots headway so that the ship can be stopped by the brake force of the anchor and chain paid out.

② Drop the anchor opposite to the alongside board about one ship's length before the appointed berth.

③~⑤ Keep going about one ship's length ahead paying out anchor chain. When the anchor chain is held on, the ship's stern begins to turn centered on the dredging anchor. Stop headway and make the rudder starboard hard over to turn the ship by boosting.

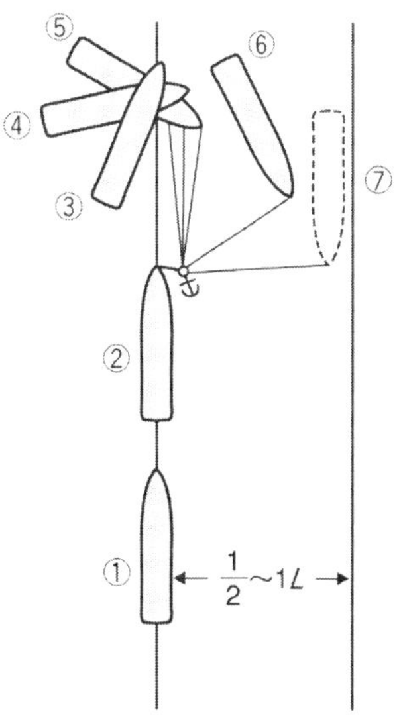

Fig. 3.1.11 Maneuvering : Stern-in port side to

⑥~⑦ Heave the anchor chain in using the starboard rudder and slow ahead engine as appropriate to bring the ship closer to the berth and send out the ship's lines to the shore to secure the ship to the berth.

(2) **Stern-in Starboard Side to** (refer to Fig. 3.1.12)

Although it is possible to bring the ship stern-in starboard side to in the same procedures as the previous section (1) in a mirror image, here, the berthing with the support of anchor by right turning in front of the berth making the use of the turning characteristic (easiness of right turning) of single right handed propeller is shown below.

① Approach parallel to the berth face keeping a 1/2 to 1 ship's length from it in low inertia speed. Drop the starboard anchor before the appointed berth taking the starboard rudder to prepare for a starboard turning with the support of dragging force of the anchor.

②~③ Make starboard turn using full astern engine and hard over port rudder by which ship's stern swings to port side by the effect of impingement due to propeller discharge current at the stern and the sidewise pressure of the propeller.

④~⑤ Approach closer to the berth using the starboard rudder and dead slow ahead. When the ship comes close to the appointed berth, send out the ship's lines to the shore and heave them in to secure the ship to the berth.

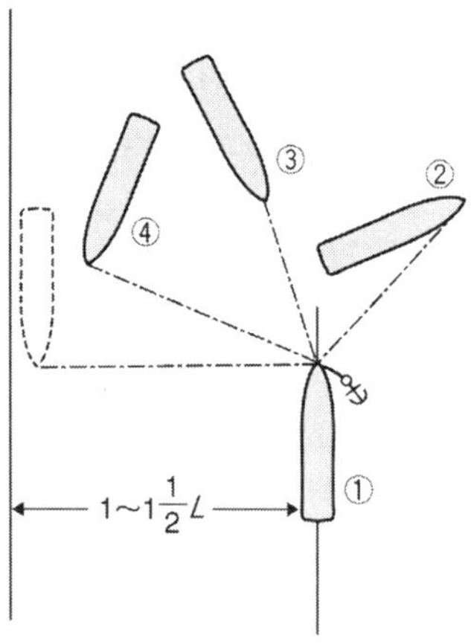

Fig. 3.1.12 Maneuvering : Stern-in starboard side to

3.4 When Approaching the Berth at a Larger Angle

(1) Head-in Mooring (refer to Fig. 3.1.13)

① Approach the berth at a low inertia speed. Drop the anchor opposite to the alongside board more than a 1/2 to 1 ship's length before the appointed berth.

② Approach the berth closer making use of the drag force of the anchor to reduce headway and turning of the stern centered on the anchor and the effect of boosting in combination use of rudder and engine.

③ When the ship comes closer to the appointed berth, send out the ship's lines to the shore and heave it in to secure the ship to the berth.

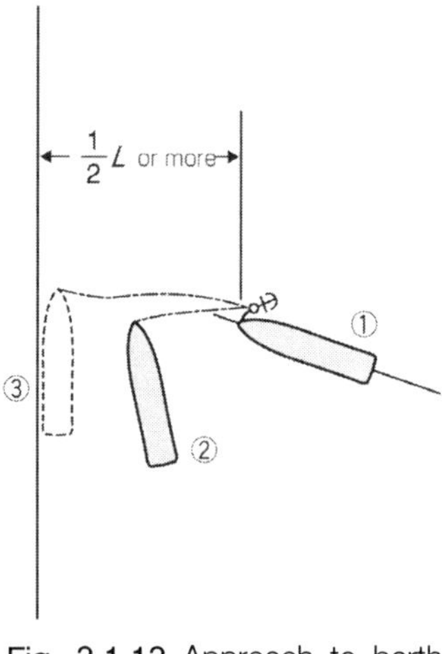

Fig. 3.1.13 Approach to berth at a larger angle, maneuvering for head-in mooring

(2) Stern-in Mooring (refer to Fig. 3.1.14)

① Approach the berth at a slow inertia speed. Drop the anchor opposite to the alongside board more than a 1/2 to 1 ship's length before the appointed berth.

② Approach closer to the berth making use of the drag force of the anchor to reduce headway and turning of the stern centered on the anchor and the effect of boosting in combination use of rudder and engine.

③ When the ship comes closer to the appointed berth, send out the ship's lines to the shore and heave it in to secure the ship to the berth.

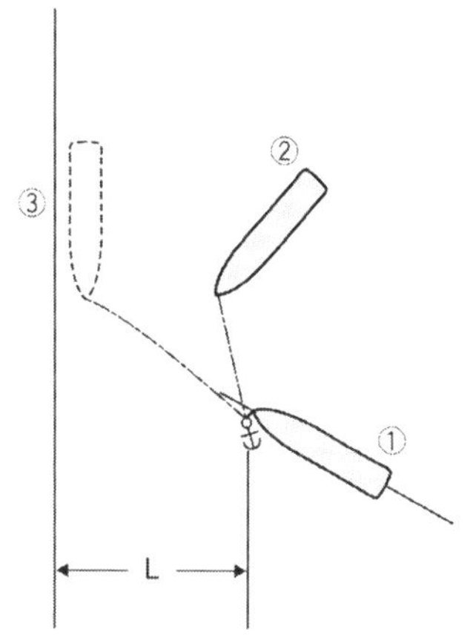

Fig. 3.1.14 Approach to berth at a larger angle, maneuvering for stern-in mooring

3.5 Stern Mooring (refer to Fig. 3.1.15)

This is mooring method uses a single anchor or two anchors at the bow and lines at the stern.

① Drop the anchor 2 ship's length before the berth.

② Approach closer to the berth making use of the drag force of the anchor to reduce headway and turning of the stern centered on the anchor and the effect of boosting in combination use of the rudder and engine.

③~④ When the ship is turned 180°, approach closer to the berth going astern making use of the drag force of the anchor.

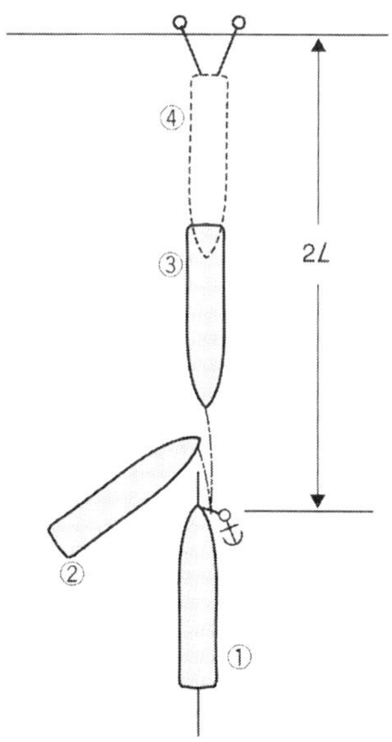

Fig. 3.1.15 Maneuver for stern mooring

3.6 Departure from Head-in Mooring (refer to Fig. 3.1.16)

The fundamental process of departing from the berth for a ship with a head-in mooring is to swing out its stern first, then go astern and turn regardless of alongside board.

① Swing out its stern by using the ahead engine and turning the rudder to berth side keeping a back spring. Otherwise by heaving in the head lines that had been taken from the outboard, the turning moment can be acquired to prepare swinging out its stern.

② After clearing its stern from the berth, go astern until the ship reaches the point where there is enough water area to turn by its own engine and rudder, heaving in its anchor and chain if they were used at berthing.

③ Make the necessary turn for leaving port.

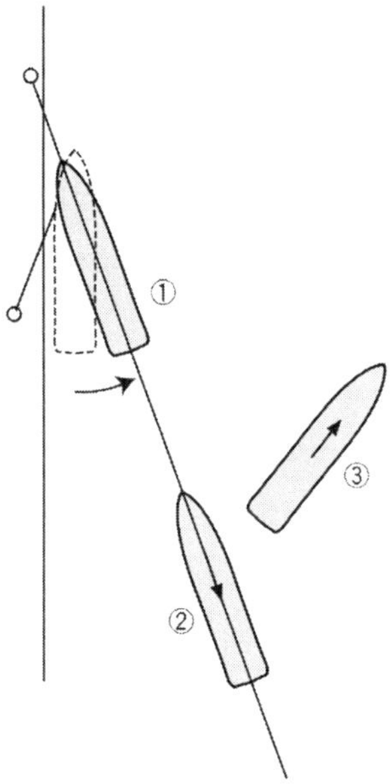

Fig. 3.1.16 Maneuver for clearing from berth in case of head-in mooring

3.7 Departure from Stern-in Mooring (refer to Fig. 3.1.17)

When there is enough water area behind the ship to swing out its stern, the fundamental process will be almost the same as in the previous case. If there is no space behind the ship available, the ship will be obliged to swing out its bow first.

① If its anchor was dropped at berthing, its bow can be swung out from the berth by heaving in the anchor. In the cases in which anchors are not available, if the ship is small, by controlling aft spring line while stern line that had been taken from the outboard is heaved in, its bow

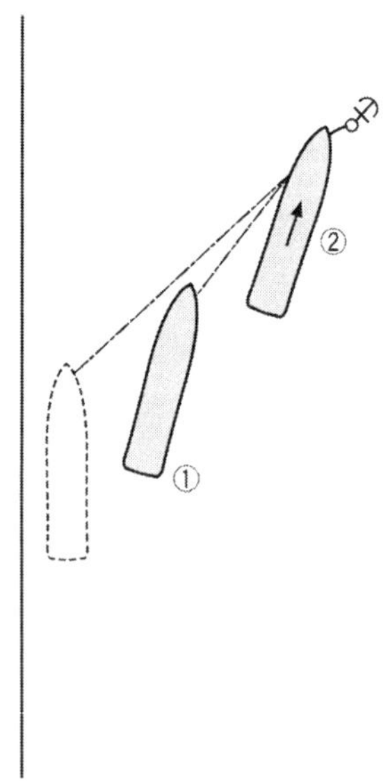

Fig. 3.1.17 Maneuver for clearing from berth in case of stern-in mooring

also can be swung out from the berth somewhat. However it would be better to drop anchor at berthing to prepare for departure.

② Make necessary turns for leaving port.

3.8 Departure from Stern Mooring (refer to Fig. 3.1.18)

① Going ahead by heaving in the anchor.

② Make necessary turns for leaving port.

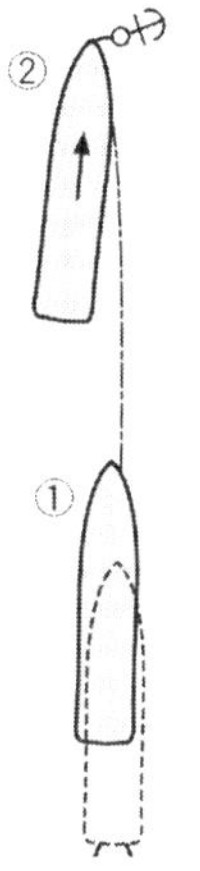

Fig. 3.1.18 Maneuver for clearing from berth in cases of stern mooring

3.9 Maneuvering Large Ships for Berthing and Un-berthing

It should be noted that there has been a change in maneuvering methods with the advent of large ships, e.g.; the methods illustrated in the previous sections using the ship's rope and anchor can not be applied to larger size ships of today for which utilization of thrusters or a tug's assistance are indispensable.

Such large ships going to berth need to be brought once to the position in front of the berth parallel to its face, about 2 ship's beams off, and then they can be moved to the berth by thruster or a tug's force. When they are clearing from the berth, thruster or a tug's pulling are also necessary.

However, it is believed that traditional berthing and unberthing methods still contain valuable lessons for maneuvering of larger ships. It is hoped that persons engaged in ship handling of larger ships make every effort to build up their skills on theoretical and empirical foundations to make the best use of modern maneuvering assistances such as thrusters and tugs based on the principals of ship maneuvers.

Chapter 5 Control of Headway and Lateral Shift Motion

1. Control of Headway in Approaching the Berth

1.1 Risk of Overrun

For persons engaged in ship handling, the phase of ship maneuvers approaching the berth is one of the most nervous and tense filled moments. In most cases, there are obstructions such as quay-walls and shoals and, in many cases other ships lying inside the port. In such circumstance, overrun passing the stopping point may lead to an accident such as contact with another ship, objects, or grounding. Persons engaged in ship handling are required to adjust the ship's speed corresponding to ship type, size, condition, inertia, maneuverability and external disturbances into account to prevent over-run.

1.2 Estimation of the Level of Risk of Overrun

(1) Concepts of Safety Margin

The extent of risk related to approaching speed can be evaluated by the concept of the safety margin. The safety margin in approaching the berth can be deduced by the difference between stopping distance in which the ship's headway can be stopped by its astern thrust and the distance to the berth. When the ship can be stopped well before the planned point that the ship is to be stopped, its safety margin is deemed as large, and when the ship needs to come closer to the planned point that the ship is to be stopped, it is said the residual distance is small, and its safety margin is deemed to be small.

(2) Definition of Safety Margin

As shown in Fig. 3.1.19, when a ship running at a speed of "V" at the position where distance from a predetermined stopping point is "D" and it can be stopped at the position where the residual distance to a predetermined stopping position is "A" by the ship's brake force, the safety margin in this case can be deduced by the equation (3.1.7).

$$\text{Safety margin } R = \frac{A}{D} \quad \cdots\cdots (3.1.7)$$

where

A : difference of distance between a predetermined stopping point and the point which the ship is actually stopped

D : distance between the point where the ship's brake is activated and a predetermined stopping point

The safety margin can be represented by numerals from 0 to 1. The safety margin of "1" means that the ship can be stopped promptly after the activation of the ship's brake force, thus it can be defined as the maximum safety margin. While the safety margin of "0" means that the ship can be stopped exactly at a predetermined point.

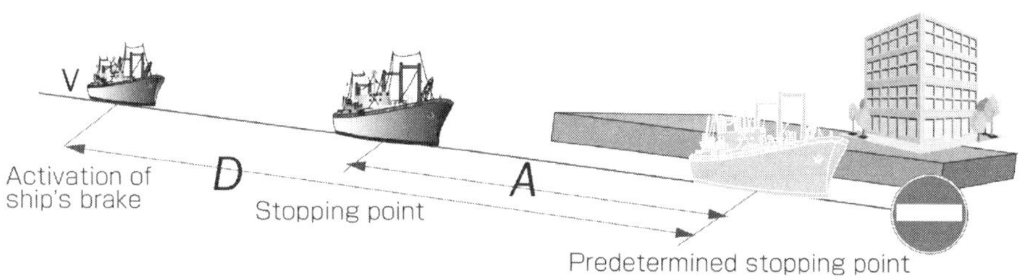

Fig. 3.1.19 Definition of safety margin (approaching maneuver)

(3) Deduction of Safety Margin

In deducing the safety margin, the ship's brake force (astern thrust) is a parameter. Although the tug's astern thrust can be used as a brake force, here, we will define it as a ship's own astern thrust alone in four steps, namely, Full, Half, Slow, and Dead Slow.

The graphs in Fig. 3.1.20 illustrate the result of calculations in the case of safety margin of zero taking parameters at each astern thrust of Full, Half, Slow, and Dead Slow into account. Respectively, μ and $\pm\sigma$ in the graphs show the average figure of the frequency of answers and the standard deviation deduced from the questionnaires for pilots for speed reduction plans of VLCCs, LNG carriers, PCCs

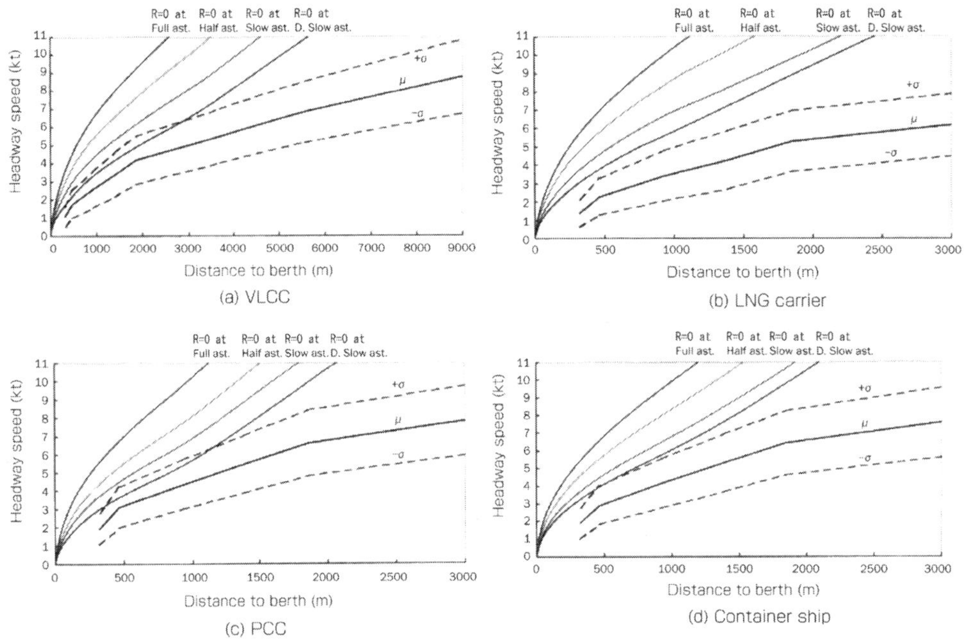

Fig. 3.1.20 Safety margin zero at every brake force (including pilots' views of safety margin)

and container ships. The area between $\pm\sigma$ corresponds to the area of safety margin from 0.3 to 0.6 using Dead Slow Astern thrust.

1.3 Guideline to Control Headway

Concept of safety margin for most pilots regardless of ship type whether it is a VLCC, LNG carrier, PCC or container ship falls in the area of safety margins from 0.3 to 0.6 by Dead Slow Astern thrust as a brake force.

This may be because of prudent consideration of pilots to restrain their ship's motion as low as possible when the astern engine is activated. Fig. 3.1.21 shows the practical guidelines for speed reduction in approaching a berth. To help make proper judgment in controlling headway, it takes into account prudent consideration of pilots as explained previously as well as the concept of the safety margin. For LNG carriers, PCCs and container ships, the same guidelines are presented as there is no difference among them. In addition, the guidelines are made based on the concepts and the results of the questionnaires as discussed hereafter.

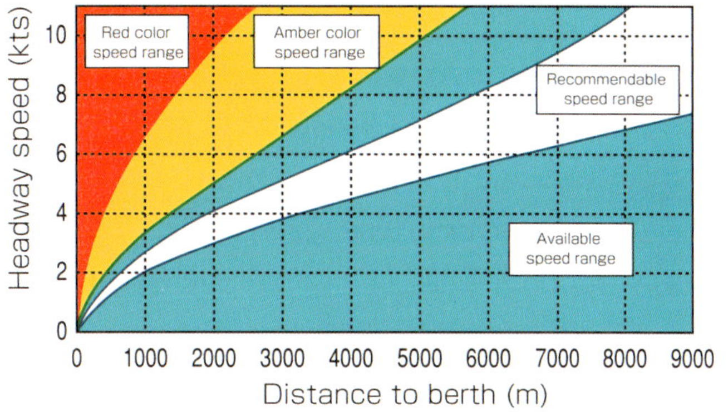

(a) VLCC

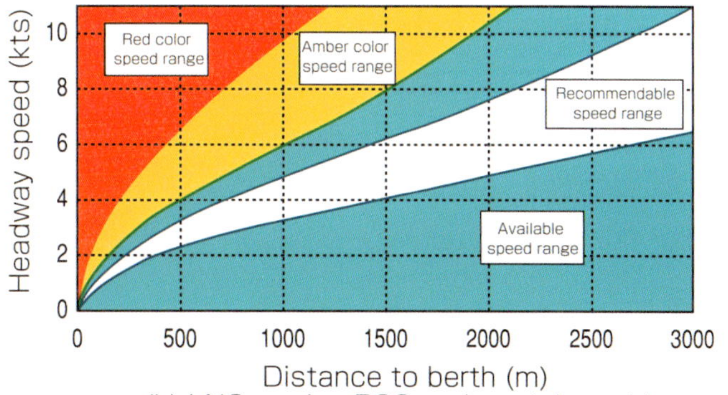

(b) LNG carrier, PCC and container ship

Fig. 3.1.21 Guidelines for speed reduction in approaching to berth

(1) Available Speed Range

All areas below the curve that indicate a safety margin of zero at Dead Slow Astern are defined as available speed range, which is where all responses from pilots in the survey lie. This range is decided based on the pilots' know-how obtained from adequate experience of berthing and trying to restrain their ship's motion to its minimum when the astern engine is activated.

(2) Recommendable Speed Range

From the results of the questionnaire, in light of their experience, two-thirds of the pilots answered thinking that a safety margin between 0.3 and 0.6 is reasonable. The fact is that since most of the answers belong to this category it suggests that it is the most practical speed range with the lowest risk of ship going out of control.

Considering this, the range of speed is defined as recommendable speed range.

However, it should be noted that the answers of a safety range 0 to 0.3 at Dead Slow Astern include 'audacious' type while a safety range below 0.6 include 'prudent' type. Therefore, both sides of the recommended area are not defined as the recommended area but just as available ranges.

(3) Red Color Speed Range

The curve to the utmost left in each graph shows the zero safety margin by Full Astern. If residual speed falls in the area to the left side of this curve, it means that even if Full Astern thrust is activated at such residual speeds, the ship could not be stopped within a predetermined stopping point and would overrun it.

Therefore, such excessive speeds must not be maintained within a predetermined stopping point. This area where utmost caution is necessary is defined as the red color speed range.

(4) Amber Color Speed Range

In the area between the red color and available range, theoretically ships can be stopped before a predetermined stopping point by Slow, Half or Full Astern. However, there are none included in this area in the answers from pilots. This may be because that they think that ship motion cannot be controlled when the astern engine is activated under such speed. Therefore, this area is defined as the amber color speed range because of the necessity of caution.

2. Control of Lateral Shift Motions in Approach Maneuvers

2.1 Risk of Damaging the Berth Structure Due to Excess Speed

In berthing maneuvers of larger ships, they are once stopped parallel to the face of the berth remaining about two ship beams from it, and then pushed to the berth by tugs. If lateral shift speed at this moment is excessive, not only the fenders or berth structures but also one's own hull may be damaged by the shock of contact. Therefore, persons engaged in ship handling must pay special attention to keep the proper speed of lateral shift corresponding to the distance between the berth adjusting the tug's thrust.

The shock to the berth may vary corresponding to the kinematic energy which is affected by the added mass of a ship, the distance between the contact point of the hull to the berth and the center of gravity of the ship in the final contact to berth and final contacting speed to berth etc. The fenders fitted to the berth face are designed and arranged so they can absorb such shock. The characteristic and capacity of fenders vary by their type and if excessive shock is given beyond their elasticity limit, they may be buckled or broken. Caution should be paid to maintain safe contacting speed taking the strength of fenders into consideration.

2.2 Estimation of the Level of Risk of Damaging the Berth Structure

(1) Definition of Safety Margin

The level of risk in the final lateral shift speed to the berth can be evaluated according to the same concept as the headway speed discussed in the previous section.

As shown in Fig. 3.1.22, when the lateral shift speed of a ship is "V" at a position where distance from the berth is "D" and it can be stopped at the position where the remaining distance to the berth is "A" by the tugs' brake force, the safety margin in this case can be deduced by the equation (3.1.8).

$$\text{Safety margin} \quad R = \frac{A}{D} \quad \cdots\cdots\cdots\cdots (3.1.8)$$

where

A : remaining distance between the ship where it is stopped and the berth

D : distance between an arbitrary position of the ship and the berth

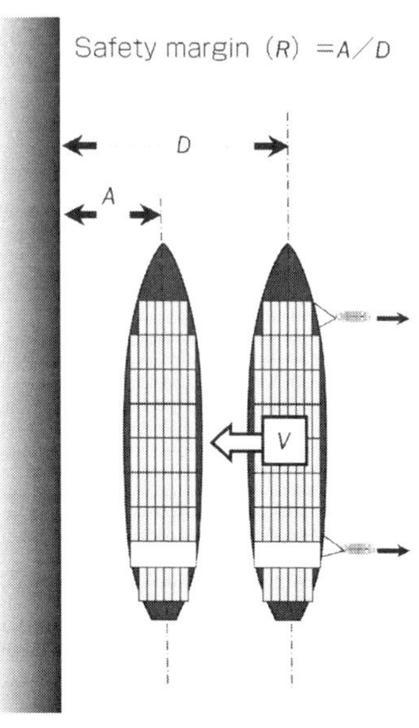

Fig. 3.1.22 Definition of safety margin (berthing maneuvers)

The safety margin can be represented by numerals from 0 to 1. The safety margin of "1" means that the ship can be stopped promptly after the activation of the tug's brake force. While the safety margin of "0" means that the ship can be stopped exactly at a predetermined stopping point.

(2) Deduction of Safety Margin

In deducing the safety margin, the tug's brake force (astern thrust) is the parameter assuming that there are two tugs available and they exert their thrusts for brake force by Full, Half and Slow.

The graphs of Fig. 3.1.23 illustrate the result of calculations in the case of safety margin zero for VLCCs, LNG carriers, PCCs and container ships, taking parameters of the two tug's brake force in Full, Half and Slow into account, μ and $\pm\sigma$ in the graphs respectively show the average figure of the frequency of answers and the standard deviation deduced from the questionnaires for pilots of VLCCs, LNG carriers , PCCs and container ships for the reduction of the speed of lateral shift.

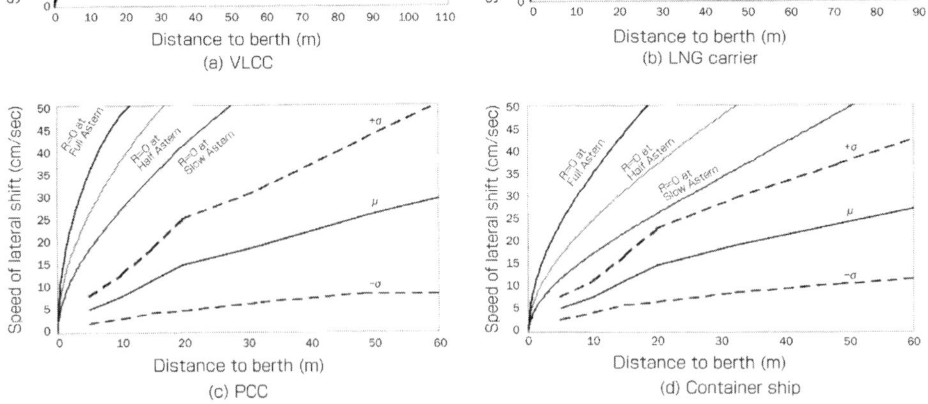

Fig. 3.1.23 Safety margin zero at every brake force by tugs (including pilots' views of safety margin)

The area between $\pm\sigma$ corresponds to the area of a safety margin of 0.3 to 0.6 when the brake force by two tugs is activated by Slow.

2.3 Guidelines to Control Lateral Shift Speed

If lateral shift speed is excessive, a ship will face the risk of collision with the berth. The view of most pilots engaged in maneuvering VLCCs, LNG carriers, PCCs and container ships to a safety margin of lateral shift speed by two tugs, lies in the range of 0.3 to 0.6 by slow astern thrust in which ships can be stopped with enough residual distance from the berth front.

The practical guidelines as shown in Fig. 3.1.24 are made taking into consideration the view of pilots to the safety margin for lateral shift speed and additional safety factors as described below.

(1) Available Speed Range

All areas below the curve that indicate a safety margin of "0" at Slow Astern of two tugs, are defined primarily as available speed range, which is where most responses about reducing lateral shift speed from pilots in the survey lie.

(2) Recommendable Speed Range

From the results of the questionnaire, in light of their experience, two-thirds of the pilots answered thinking that a safety margin between 0.3 and 0.6 is reasonable. The fact is that since most of the answers belong to this category it suggests that this range reflects the most standard concept among pilots.

Therefore, this range is defined as a recommendable speed range. However, it should be noted that the answers of safety range 0 to 0.3 at Slow Astern include 'audacious' type while safety range below 0.6 include 'prudent' type. Therefore, both sides of the recommended area are not defined as a recommendable, but just as available ranges.

(3) Red Color Speed Range

The curve to the utmost left in each graph shows the zero safety margin by Full Astern. If residual speed falls in the area to the left side of this curve, it means that even if Full Astern thrust of two tugs is activated at such residual speed, the ship could not be stopped within a predetermined stopping point and would collide with the berth.

Therefore, such excessive speed must not be maintained within a predetermined stopping point. This area where utmost caution is necessary is defined as the red color speed range.

(4) Amber Color Speed Range

In the area between the red color and available range, theoretically ships can be stopped before a predetermined stopping point by Half or Full Astern. However, there are none included in this area in the answers from pilots. This may be because that they consider possibilities of unexpected impedance such as snapped tug lines or sudden effect of wind or current.

Therefore, this area where utmost caution is neccessary is defined as the amber color speed range.

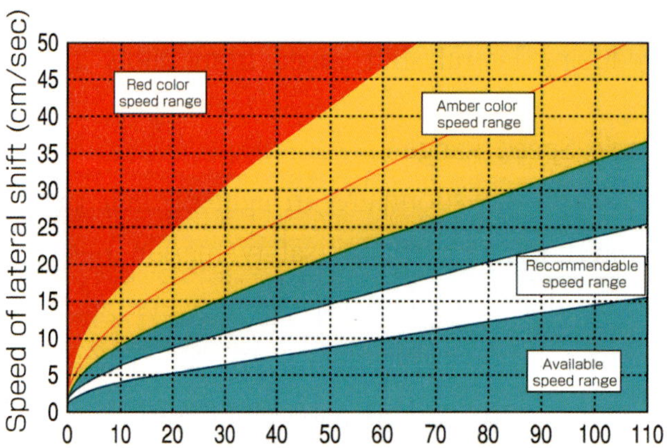

(a) VLCC

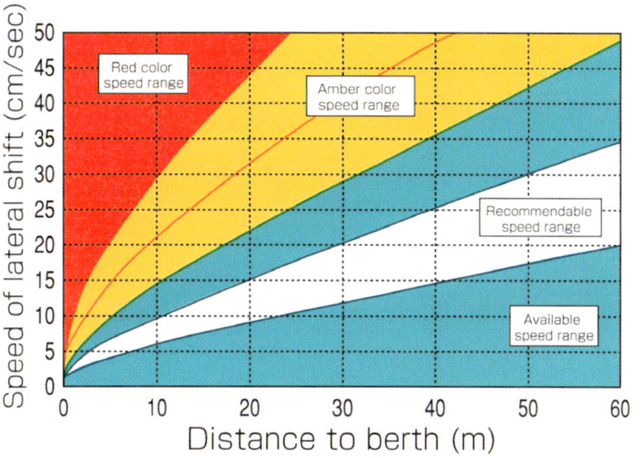

(b) LNG carrier

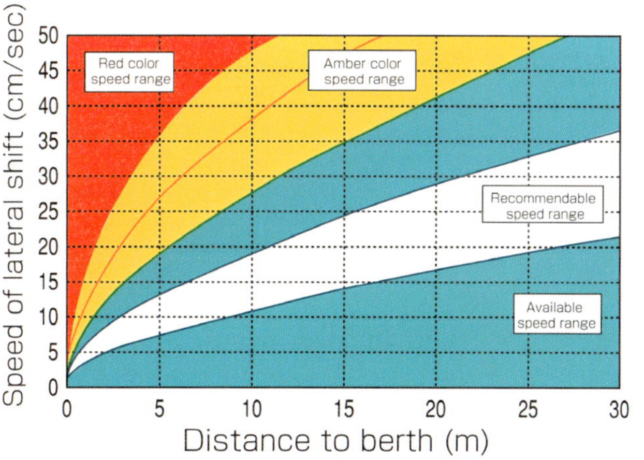

(c) PCC

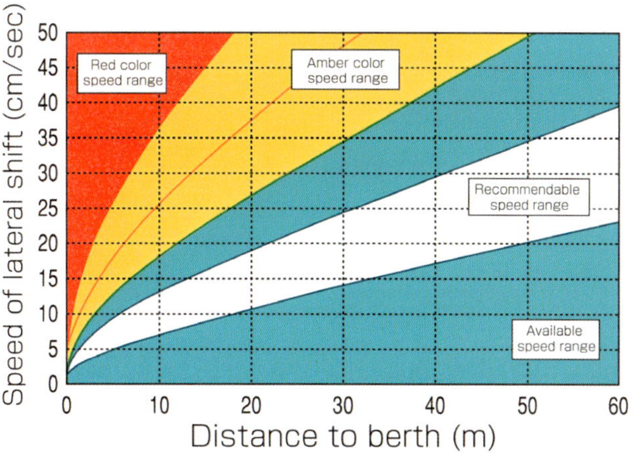

(d) Container ship

Fig. 3.1.24 Guidelines for reducing lateral shift speed

Chapter 6 Maneuvering in an Emergency

1. Maneuvering to Rescue Persons Overboard

1.1 Immediate Action after an Accident

If person overboard should happen while the ship is under way, the following actions must be taken immediately.

① Any person seeing a person overboard must throw in a life buoy, self-activating smoke signal, or self-igniting light close at hand and then report it immediately to the officer on watch.

② Keep an eye on the person overboard so as not to lose sight of them.

③ The officer on watch must make maneuvers to rescue the person.

1.2 Maneuvering to Rescue Persons Overboard

(1) **Single Turn** (Maneuver immediately after the accident)

As shown in Fig. 3.1.25, take hard over helm to the side where the person fell overboard to turn toward the person in distress to rescue them.

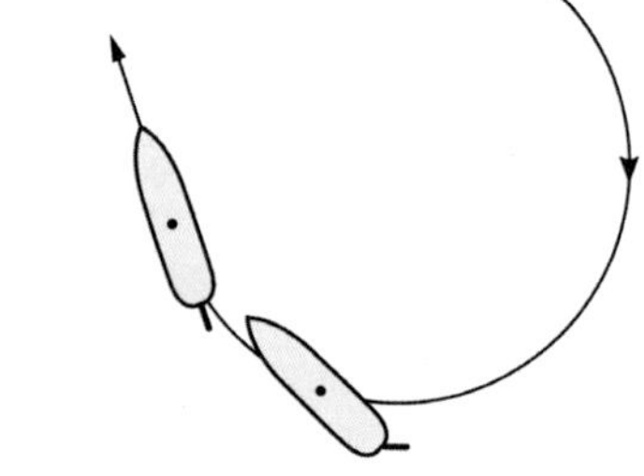

Fig. 3.1.25 Single turn

(2) **Williamson's Turn** (Maneuvering back to a point previously passed when the accident happens at night time or in restricted visibility)

As shown in Fig. 3.1.26, at first take hard over helm to the side where the person fell overboard to turn until the ship's heading turns 60° to 90° from its original course. Then, turn the helm to the opposite board until the ship's heading turns 180° from its original course to search for the person overboard.

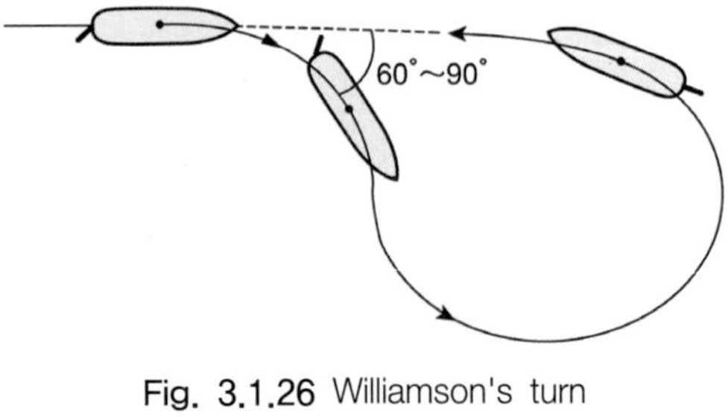

Fig. 3.1.26 Williamson's turn

Timing of changing helm to return the ship's heading back to the original course may vary depending on the maneuvering characteristics of each ship. Persons engaged in ship handling are required to be familiar with the maneuvering characteristics of their own ship and how to turn the helm in an emergency as well as how to return the ship back to its original course.

(3) Approaching the Persons Overboard

In approaching the person overboard, it would be better for larger ships to approach the distressed from its windward side as to ensure calm seas for the safe launching of a rescue boat and rescue operation. While for smaller ships, approaching the distressed from the leeward side is recommended to avoid the risk of the distressed being caught under the bottom of the ship while the ship is tossing up and down the waves.

2. Response and Procedures to an Accident

Collision, grounding, damage to the ship's hull or structure, trouble with engine or machinery, flooding, fire, capsizing and persons overboard are the main accidents at sea related to ship operation. To deal with these accidents in a composed manner, it is important for persons engaged in ship handling to have themselves acquainted with the processes and procedures of how to rescue someone in distress or prevent further development of damage to human life, the ship, and its cargo depending on the nature of accident. For this purpose, persons engaged in ship handling need to be trained through drills of an emergency muster list, and preparedness and responsiveness in an emergency.

The followings are the fundamental processes and procedures in an emergency.

① Suspension of navigation: Kill engine to reduce headway as quickly as possible.

② Confirmation of situation: Confirm the current situation, conceivable development of damage and necessary countermeasures

③ Immediate Action: To rescue human life first, the ship's hull and its cargo, countermeasures to prevent further development of damage

④ Possibility for continuation of navigation: Request for assistance or necessary deviation depending on capability of the ship

⑤ Report to all parties concerned: In compliance with emergency information and communication system guidelines

3. Report of Accident, Rescue System and Organization

If an accident cannot be dealt with within the ship's own capacity, request for assistance or a report to another ship nearby or the appropriate and nearest coastal administration or organization available is necessary by means of ordinary satellite communication system, VHF, air whistle, smoke signal and signal flags whichever available according to the relevant manual of the company.

When reporting the accident to the administration or organization of the nearest coastal state, the following items are generally required.

① the ship's name

② the number of crew members

③ location

④ type of accident

⑤ life-saving apparatus and equipment available

⑥ type and size of ship

⑦ other additional information

The area and responsible administration or organization for search and rescue and broadcasting information related to navigational safety including the ship's communication system in an emergency are defined by international conventions such as SOLAS (International Convention for the Safety of Life at Sea) and SAR (International Convention on Maritime Search and Rescue) conventions.

4. Types of Accidents and Countermeasures

4.1 Contact with Floating Object or Fishing Net

Although an officer on watch should pay attention to any obstacles around the ship that might interfere with safe navigation and avoid them, when any damage to the ship's hull or machinery is in doubt by contact with floating objects such as driftwood, fishing nets or ropes which could be caught by the ship's propeller or garbage which could possibly choke the cooling water inlets, the following actions must be immediately taken.

① Stop engines and check for damage
② Continue the voyage if the damage does not affect sea or cargo worthiness.
③ If damage is detrimental to sea or cargo worthiness, appropriate assistance needs to be requested.
④ If fishing nets are found to be damaged, whether it is minor or serious notification to the owner of them must be made.

4.2 Collision

In the case of collision, the following actions must be taken as quickly as possible to prevent further development of damage of life and/or property.

① Kill engines to reduce headway as quickly as possible. Control the ship in very slow speed because astern engines must not be used until it is confirmed that there is no possibility of ingress of water and consequential sinking.
② Investigate possible injuries to passengers and crew, and damage to the ship and its cargo.
③ Both ships involved in a collision must cooperate and endeavor to rescue life, the ships and cargo of one another and make a request for external assistance if necessary.
④ If there is water ingress, take countermeasures to prevent it including discharge of flooded water.
⑤ If there is a risk of capsizing or sinking, deliberate grounding or evacuation from the ship must be considered.
⑥ Prepare for carrying out any important documents and valuables.
⑦ Record date, time, position, courses of both ships, and weather and sea conditions.
⑧ Exchange information of the ship's name, port of registration, port of departure and destination to each other.

4.3 Grounding

Caution is necessary with the fact that in grounding, the load may be concentrated on the grounded part and shock of waves may aggravate damage to the hull.

① Stop engine immediately and check the following situation in grounding, extent of damage, water ingress, the ship's motion, list, depth, bottom quality, the contour of the sea bottom, topography of the surrounding shore, sea level, tidal times, tidal range, tidal current, swell, wind force and direction, the weather forecast, and capacity of one's own engine.

② Before using engine, be sure not to enhance the damage of the grounded part. It must be confirmed that if the propeller or rudder are clear from obstacles on the sea bed and there is no risk for the cooling water to intake mud and sand.

③ When it is practical to refloat the ship by its own ability taking sea level, weather and sea conditions into consideration, drop its anchor as appropriate preparing for clearing off from the shoal or reef.

④ If refloating by the ship's own ability is not possible, request external assistance and make every effort to secure the ship using anchor, anchor chain and mooring ropes appropriately.

4.4 Damage to the Rudder

Damage to the rudder causes the inability of a ship to keep its heading and may lead to grounding or collision. If a ship is provided with twin propellers, it is theoretically possible to steer her by adjusting the thrust of the two propellers. However, for ships with a single propeller, there is no practical means available except for preparing a temporary substitute or trying to use some available resistant object.

4.5 Damage to the Propeller System

Engine trouble or damage to propeller involves the risk of drifting and consequential grounding. Immediate action needs to be taken to restore them or immediate request for external assistance is necessary. Up until engines or propellers are restored or until rescue forces arrive, the ship needs to make every effort to minimize its drift or steady its posture appropriately using available means such as anchor or sea anchor.

4.6 Fire

Constant effort for prevention of fire is necessary as a safe haven for the crew is quite limited in the case of fire on board and excessive spraying of water involves a risk of capsizing. If a fire breaks out, the following actions must be taken immediately.

① The person who detects a fire break out must report it immediately to the officer on watch in order to muster fire stations and engage fire fighting at an early stage by any available means close at hand.

② Turn the ship to bring the source of the fire on the leeward side and stop the engine.

③ Shut off ventilation to the quarter where the fire has broken out and start the fire extinguisher system depending on the nature of a fire.

④ If the fire can not be controlled, evacuation from the ship and deliberate grounding should be considered so that spraying water would not sink the ship.

4.7 Flooding

In case of flooding from a breach in the hull, the following actions must be taken immediately.

① Make efforts to discharge the flood water and confirm the source of the flood and estimate the flood rate.

② Take appropriate means to prevent further flooding.

③ Close doors of the flooding quarter to prevent further development of flooding.

④ If flooding can not be controlled, deliberate grounding should be considered.

5. Maneuvering under Tow

5.1 Preparation for Towing

When a ship is unable to continue sailing under its own capacity due to a trouble with the engine, propeller, or rudder, assistance for towing is necessary. The following are fundamental aspects of the knowledge for towing. Persons engaged in ship handling should confirm these points before towing.

① Option of towing head first or stern first: If the condition of the towed ship permits, head first towing would be better in view of hydraulic resistance.

② Strength, properties, and length of towing lines: The strength of towing lines should be determined by the total resistance of the towed ship. Generally a line comprised of chain and fiber rope or fiber rope and wire rope as appropriate is used so that it may absorb shocks while towing. The length of tow lines should be 1.5 to 2 times the sum of the lengths of the towing ship and the towed ship.

③ Towing speed: It may vary depending on the strength of the towing line(s). It is better to keep it safe considering sea and weather conditions.

④ Securing method of towing line: Securing equipment must be strong enough to support towing the load. Between the tow line and ship's structures, appropriate buffers must be applied to prevent friction.

5.2 Precautions while under Tow

① Initiation of towing: Initiation of towing must be made in a manner so as not to give excessive shock to the towline. Towing thrust must be increased gradually according to the ahead inertia of the towed ship.

② Under tow: Alteration of course must be made gradually and incrementally. Tow lines must be prepared for quick release in the event of an emergency. Watch persons for tow lines must be stationed constantly.

6. Maneuvering in Rough Seas

It is important to alter course to avoid rough sea areas or take shelter in a timely manner, well in advance according to weather and sea state forecasts. However, if a ship should be caught in storm, maneuvering to lessen the shock of wind and waves to prevent structural damage must be taken.

6.1 Decision of Timing to Take Countermeasures

When a ship's motion becomes violent and the ship's speed decreases accompanied by frequent scooping seas over the deck and propeller racing, the person engaded in ship handling is required to take necessary countermeasures such as the alteration of course and/or the reduction of speed to mitigate shock by waves.

6.2 Selection of Course to Mitigate a Ship's Motion

It is said that the course at an angle of 20° to 30° to waves is most effective to reduce the shock of waves in rough seas because beam or quarter waves may acts synchronously to the ship's natural rolling period and bow waves may amplify pitching.

6.3 Steering and Alteration of Course in Rough Seas

In rough seas, turning by large helm angle or large angle of course alteration should be avoided to prevent lurch. When such a large angle of alteration of course is required, it must be made gradually starting with small angles in low speed depending on the roughness of the waves and the ship's motion.

6.4 Reduction of Speed to Mitigate the Shock of Waves

By reducing speed, shock of waves, a ship's synchronous motion and the frequency of scooping seas over the deck and propeller racing can be mitigated. And large speed reduction at an earlier stage is recommended.

6.5 Heave to

This is a method for a ship caught in stormy seas letting the storm pass by keeping the ship's least steerageway and keeping heading at an angle of 20° to 30° against the waves. The reason a ship must maintain a certain speed in this method, is to keep its posture and not to be laid in beam seas.

6.6 Scudding

In typhoons in the northern hemisphere, wind in the right semicircle is accelerated by general air flow while wind in the left semicircle is weakened by general air flow offsetting it. For this reason, the right semicircle is called dangerous semicircle while the left semicircle is called navigable semicircle.

As shown in Fig. 3.1.27, when a ship is in the right semicircle within the storm zone, maneuvering a ship that is running before the wind receiving wind and waves from her starboard quarter to pass in front of the typhoon to a navigable semicircle is called scudding. This is one of the effective means for a ship caught in the right semicircle of typhoon to get out of it quickly.

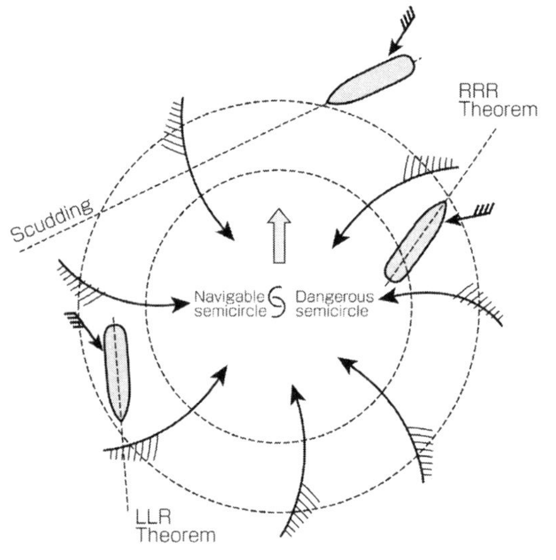

Fig. 3.1.27 Maneuver to avoid typhoon

6.7 RRR Theorem

In the northern hemisphere, when the direction of wind where a ship is running changes clockwise (R), it means that the ship is in the right semicircle (R). In such cases, if the ship continues to run receiving wind on her starboard bow (R), she can get away from the center of the typhoon.

6.8 LLR Theorem

By contrast, also in the northern hemisphere when the direction of wind changes counterclockwise (L), the ship is in the left semicircle (L) (navigable semicircle). In such cases, if the ship runs before the wind receiving wind and waves from her starboard quarter (R), she can get away from the center of the typhoon.

Ⅱ
Mooring and Anchoring

Chapter 1 Mooring

1. Function of Mooring Lines

For mooring ropes for ships, generally synthetic fiber ropes or wire ropes are used. Although wire rope is difficult to handle due to its weight and lack of flexibility, it is suitable to restrain a ship's subtle shift while ship is moored at berth because of its innate strength and small elasticity. For this reason, wire ropes are generally used for mooring lines mainly as spring lines for larger ships.

By contrast, synthetic fiber rope is widely used as ship's mooring lines except for spring lines as it is light, buoyant and superior in elasticity to absorb shock loads. The shortcomings of fiber rope are that they are liable to be damaged by friction and their larger elasticity against a heavy load in securing a ship's mooring position.

Fig. 3.2.1 shows the general arrangement of a ship's mooring lines. Spring lines are used to restrain a ship's fore and aft movement while breast lines are to secure a ship's side to the berth firmly. The functions of head and stern lines are to secure a ship's fore and aft position and to restrain yawing while ships are at berth.

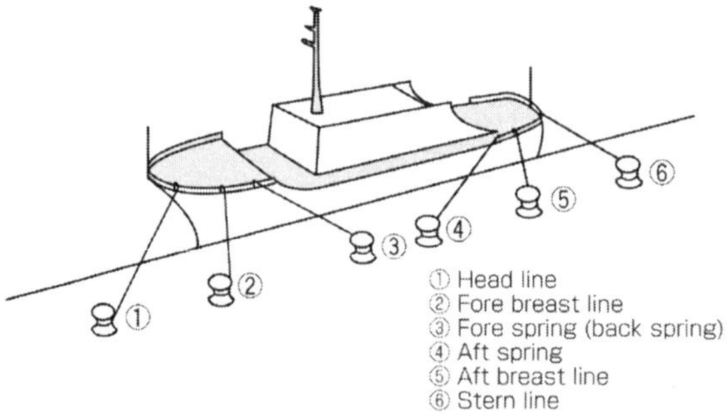

Fig. 3.2.1 Arrangement of mooring lines

To secure a ship firmly to the berth, it is ideal to take all mooring lines fore and aft symmetrically, not only head and stern lines but also spring lines including breast lines and they must be evenly tightened to minimize the ship's movement while at berth.

Caution is necessary as ship's movement while at berth may often be affected by the elastic properties of rubber fenders fitted to the face of berth, e.g. hard rubber could cause excessive reaction force against a ship's hull, which could enhance the ship's lateral movement.

2. Securing Force by Mooring Lines

When wind and wave forces have affect on a moored ship, tension created on each line acts to restrain the ship's movement. Such tension created on each line is called ***mooring force***. As illustrated in Fig. 3.2.2, horizontal and vertical components of mooring force of these lines can be derived from the equations (3.2.1). Where its vertical and horizontal angles and the tension of the rope are respectively "α", "β" and "T".

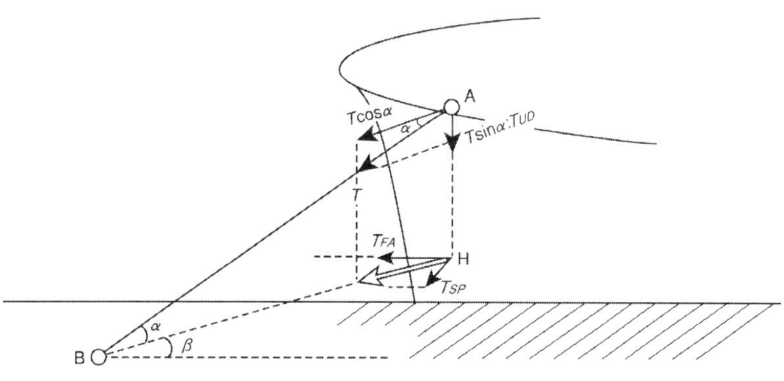

Fig. 3.2.2 Mooring force of rope

Let's look at the imaginary vertical surface made by drawing lines from the points A, B (both ends of the rope) and H in the Fig. 3.2.2 .Where horizontal and vertical components of the tension of the rope "T" is $T\cos\alpha$ and $T\sin\alpha$ respectively, the following equations stand.

$$
\left.\begin{array}{ll}
\text{Fore and aft Mooring Force} & : T_{FA} = T\cos\alpha \cdot \cos\beta \\
\text{Lateral Mooring Force} & : T_{SP} = T\cos\alpha \cdot \sin\beta \\
\text{Vertical Mooring Force} & : T_{UD} = T\sin\alpha
\end{array}\right\} \cdots\cdots (3.2.1)
$$

As you can see from equation (3.2.1), to gain optimum longitudinal mooring force, it is necessary to minimize the angles of "α" and "β" as much as possible. That is to say, it would be better to take the line close to parallel of the ship's longitudinal line and as horizontal as possible to gain optimum fore and aft mooring force. In contrast, to gain optimum lateral mooring force, it is necessary to minimize "α" and maximize "β". In other words, it is necessary to be taken as close to the beam and as horizontal as possible. Further, when obtaining maximum vertical mooring force, it is necessary to take the rope as close to vertical as possible.

However, arrangement of mooring lines in practice is solely dependent on that of shore bitts and bollards. In such circumstances, persons engaged in ship handling are required to make endeavors to flexibly achieve the optimum mooring effect by arranging each rope to share vertical, horizontal and lateral mooring force equally.

There are two kinds of bitts, straight bitts and curved bitts. Curved bitts are usually arranged along the edge of the wharf for general mooring purposes, while straight bitts are arranged further inshore to effectively gain extra horizontal mooring force preparing for rough weather and are sometimes called storm bitts.

Chapter 2 Anchoring

1. Anchor and Anchor Chain

1.1 Basic Functions and Performance Required to Anchor

Ships are secured to wharfs, quays, piers or jetties by mooring ropes or wires. To secure a ship to such berths within their operational limit under the effects of wind and waves, a sufficient number of mooring lines with the proper strength to offset such external disturbances and to absorb consequential kinematic energy is necessary corresponding to the ship size and weight.

On the other hand, in anchoring, such kinematic energy can be diffused by free movement of the ship around its dropped anchor. Therefore, the role of anchor dropped and anchor chain paid out is only to support the remaining load within their holding powers. In this regard, it may well be said that anchoring is a flexible mooring method with the mooring force provided by one (or two) anchor(s) and one (or two) chain(s).

Anchoring is used temporarily while waiting for a berth outside the ports and sheltering the ship from rough weather such as a typhoon. For instance, in ports opened for international trade on the Japanese coast, when a typhoon is approaching, most of ships lying inside port will be required to evacuate from their in-port berths to out-port anchorage or other appropriate sheltering places according to recommendations or orders of evacuation. In such case, to get reliable holding power of anchor(s) and anchor chain(s) is essential.

After the anchor is dropped, its condition in the water is invisible; thus keep it in mind beforehand that the following properties of anchor(s) are essential when a ship is obliged to lie at anchor.

① **Holding Performance** : The anchor fluke can grip the sea bed firmly when it is dropped.

② **Holding Power** : Maintenance of adequate holding power during anchoring.

③ **Stability of Anchor** : Even if the anchor is dragged, it can not be overturned.

1.2 Types and Weight of Anchor

Today AC14, JIS and Danforth type anchors as shown in Fig. 3.2.3 are widely used. Among them, AC14 and JIS types are used for general merchant ships and the Danforth type is used for smaller crafts. Size and minimum weight of anchor to be fitted to ships is defined by each ship's ***equipment number***. When an AC14 type anchor is fitted, 25% of its regulatory weight can be deducted because of its approved high holding characteristics.

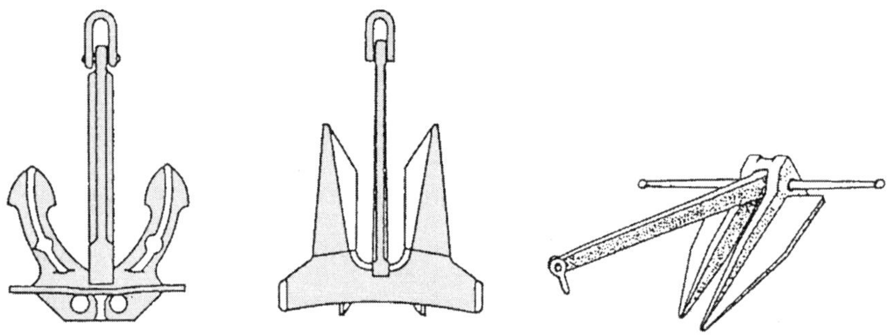

JIS type AC14 type Danforth type

Fig. 3.2.3 Typical types of anchor

For your reference, for small crafts, though there is no particular regulation to define the weight of the anchor to be fitted, anchors with proper holding power, strong enough to moor the ship are necessary.

Table 3.2.1 Examples of provisions of anchor and anchor chain

	Ship size & type	Weight of anchor	Diameter of anchor chain or rope	
			Rope	Chain
General merchant ship	230,000 DWT VLCC	18.7ton		102mm
	6,000 unit PCC	10.5ton		87mm
	25,000 GT TEU container ship	8.3ton		81mm
	5,000 GT general cargo	2.7ton		56mm
Small craft	Ship length 10ft 3.1m	1.8kg	6.4mm	4.8mm
	20ft 6.1m	3.6kg	6.4mm	4.8mm
	30ft 9.2m	4.5kg	9.5mm	6.4mm
	40ft 12.2m	8.2kg	11.1mm	7.9mm
	50ft 15.3m	18.1kg	12.7mm	9.5mm

The following table shows examples of anchor and anchor chain fitted to regular merchant ships and small crafts sorted by ship type and size.

1.3 Length of Chain to be Paid Out

Although chain connected to the anchor is very heavy and difficult to handle, it can mitigate load and shock to the anchor when they are laid out on the sea bed to a certain length by their own weight and holding power.

Therefore, it is important that anchor chain should be paid out appropriately corresponding to the depth and sea conditions, so that it may absorb shock due to the ship's swinging motion and anchor may be pulled horizontally over the sea bed, keeping its fluke firmly in the soil, which is not critical in calm seas but it is under strong wind and waves.

Caution is necessary if the anchor chain paid out is too short, the anchor shank may be raised at the anchor ring when it is pulled and adequate holding power cannot be expected.

For safe anchoring, always adequate length of chain is recommended to be paid out according to the depth of the water. The practical standard suggests that under normal weather, the chain length of $(3d+90)$m (d: water depth) should be necessary. However in rough seas, the longest chain allowed by the capacity is desirable. For your reference the followings standards are presented.

In calm sea : 5 times the water depth
Ordinary : 7 to 8 times the water depth
In rough weather : More than 10 times the water depth

2. Anchoring

2.1 Methods of Anchoring

Fig. 3.2.4 illustrates typical anchoring methods, riding with a single anchor and riding with two anchors. The method shown in the figure on the left, ***riding to a single anchor*** is most frequently and widely used while waiting for a berth or temporary stay under fair weather and sea conditions.

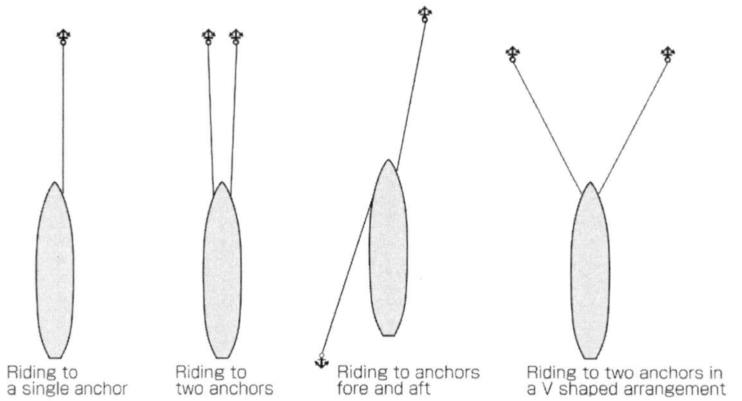

Fig. 3.2.4 Typical anchoring methods

The second figure from the left shows a ship ***riding to two anchors*** dropped simultaneously in the same direction with the same length of anchor chains paid out in parallel. In this method two times of holding power can be expected.

In the second figure from the right, each anchor is set fore and aft of the ship and the ship will be ***riding to either anchor alternately*** depending on tidal currents or wind direction even though two anchors are dropped.

And in the figure on the right side, the ship is riding midway between both anchors forming a ***letter V shape*** (***open moor***) which may be, in a broad sense, also included in riding two anchors.

2.2 Methods of Anchoring in Rough Weather

It is certain that if possible riding to two anchors in rough weather is safer than riding to a single anchor. In the letter V shaped arrangement riding to both anchors with an angle of 45° to 60° between the two chains, the swing of ship can be restrained against head wind. However, caution is necessary as their holding power will be decreased when the wind veers. This type of mooring is called ***open moor*** where a certain angle between two anchor chains is maintained against head wind and is an effective method under head wind.

In open mooring, optimum holding power can be attained at an angle of 45° to 60°between the two chains. However, in riding to both anchors in which the angle between the chains is not optimum, almost the same holding power as that of riding to a single anchor can be expected if enough length of the chain is paid out.

In a case where the angle between the two chains is 0°, it corresponds to riding to two anchors. In this method, about two times the holding power compared to that of riding to a single anchor can be expected if enough length of chain is paid out although caution to fouling anchors is necessary.

For larger ships today, they are often obliged to continue riding to a single anchor even when weather worsens, because when dropping a second anchor, certain angles between two chains cannot be afforded due to limited flexibility of the ship's movement. In such cases, it is recommended to use the second anchor as a ***snubber*** with chain length of about 1.5 times of depth, to restrain ship's swinging motion by almost half.

3. Methods of Dropping and Weighing Anchor

3.1 Anchoring Going Astern

If there is no particular restriction, two anchors should be used alternately. Before an anchoring point, the anchor is usually lowered a little above the water keeping it in a stand-by position to be dropped (***cock bill***). It is ideal to give astern thrust when the ship is at one ship's length before the anchoring point to stop the ship at that point. Drop anchor when the ship begins to go astern slowly and pay out chain using astern inertia.

The engine should be adjusted so as not to give excessive sternway. Anchor chain must be checked periodically so as to have the anchor grip the sea bottom. When the planned amount of chain length has been paid out, tighten the brake and wait for a while to see if the anchor grips the sea bed firmly (***brought up***). Bringing up of the anchor can be confirmed from the phenomenon that once tightened, the chain will slack again and ship's stern will turn.

When anchoring going astern, although ships are required to approach their anchoring points at exceptionally slow speed and are liable to be affected by the effects of wind and currents, it is a universal method for merchant ships.

3.2 Anchoring Going Ahead

Anchoring going ahead is often used by navy ships which are acting in a fleet and are required to do precise anchoring as designated. In such case, each ship lets its anchor go keeping headway and paying out the chain until its planned length. In this method, since there is a risk of damaging a ship's hull by friction between the chain running out or the chain itself, which could be snapped due to excess tension, cautious adjustment of headway is important.

3.3 Deep Sea Anchoring

In anchoring where depth is over 20m or in anchoring of larger ships fitted with heavy anchor, if an anchor is dropped from its ordinary stand-by state, there is a risk of damaging the anchor and chain caused by the impact of hitting the seabed hard or the chain bouncing violenty. Consequently the anchor and chain may not be laid over the sea bed properly. To avoid such cases, it is recommended to ***walk back*** the anchor by reversing the windlass following a deep sea anchoring criteria.

Following the advent of larger size ships, water depth of the water area for anchoring has been increased from 30m to 40m. Therefore, we had better think that deep sea anchoring is an ordinary anchoring procedure for these types of ships. When depth is about 50m, the anchor should be lowered to about 10m above the sea bed by walking back of the windlass, then drop the anchor as aforementioned in anchoring going astern. When the depth is more than 50m, the anchor should be walked back almost to the sea bed, and then dropped while going astern. In such cases, it is essential to keep the ship's operational speed to a minimum to restrain hydraulic resistance to the anchor and chain at a minimum.

Although the limit of deep sea anchoring depends on the capacity of each windlass, it is generally deemed up to around 100m in relation to the rated capacity of windlass. A person engaged in ship handling is required to have a thorough knowledge about a weight capacity of the windlass of their own ships and the limit of the depth it can be used in relation to the total weight of anchor and chain to be paid out.

3.4 Weighing Anchor

For preparation of weighing anchor, set the clutch and release the brake of the windlass to heave in the anchor chain. Then flush the hawse pipe washing water. It should take 3 minutes to heave in 1 shackle of chain based on the rated winding speed of a windlass at 9m/min. If in this process it seems that the windlass is over loaded or the chain is over tensioned due to the chain crossing under the ship's bottom or other conditions, the engine and rudder should be used appropriately to ease excessive tension on the chain.

In weighing anchor, the following particular terms are used. The chain hauled short to about 1.5 times of depth is called ***short stay***, the state that the anchor hanged right down from the hawse pipe is called ***up-and-down*** and the moment when the anchor is off the sea bed is called ***anchors aweigh***. In the moment of anchors aweigh, the chain becomes taut hanging right down from the hawse pipe as the whole weight of anchor and chain is loaded to the windlass. The moment of anchors aweigh is the start of the voyage. If the anchor is hoisted up over the sea surface, it is called ***clear up anchor*** and if it is entangled with other objects, its state is called ***foul anchor***. If the anchor is hoisted up in clear state, the weighing anchor will be complete and it can be restored to its original position.

4. Holding Characteristics of Anchors

4.1 Holding Power Coefficient

The element to define the performance of an anchor and its holding power can be derived from the following equation (3.2.2), assuming the holding power increases with the weight of an anchor.

$$H = kW \qquad (3.2.2)$$

where

H : holding power of anchor
W : weight of anchor
k : holding power coefficient

Although ***holding power coefficient*** "k" is a proportional constant and defined by the type of anchor and the quality of the sea bed, the holding power of anchor may vary depending on weight of anchor, hardness of soil, gripping condition of anchor fluke and dragging speed.

4.2 Values of Holding Power Coefficient

(1) Definition of Static Holding Power Coefficient

Static holding power means the holding power of the anchor when it rests in the same position with its fluke gripped firmly into the soil. The value of holding power can be obtained from a substantial amount of sample data of the relevant model tests according to the following conditions.

① As shown in Fig. 3.2.5, a model anchor is pulled at a constant speed repeatedly under the same bottom quality conditions.

② Fig. 3.2.6 shows the graph of holding power obtained by the test above. From the graph, the maximum holding power of the anchor when it is

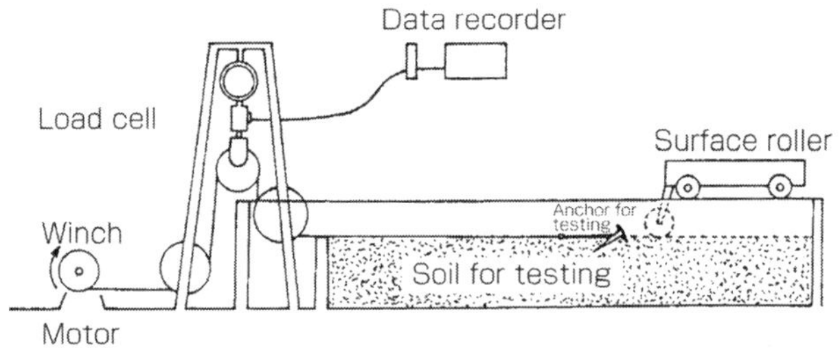

Fig. 3.2.5 Anchor tension testing arrangement

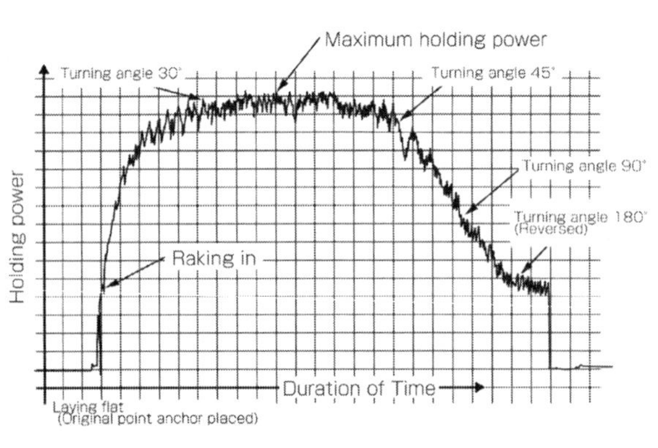

Fig. 3.2.6 Characteristic curve of holding power

dragged with its fluke firmly gripping the ground can be measured. The test was repeated varying pulling speed to obtain as many sample combinations of the speed and the maximum holding power as possible.

③ Fig. 3.2.7 and Fig. 3.2.8 respectively show the plots of the data obtained from the tests of 10 kg JIS type model anchor and 10 kg AC14 type model anchor on sandy bottom conditions. On each graph, the horizontal axis represents pulling speed "V" and the vertical axis represents maximum holding power coefficient H_D/W (H_D: maximum holding power, W: weight of anchor).

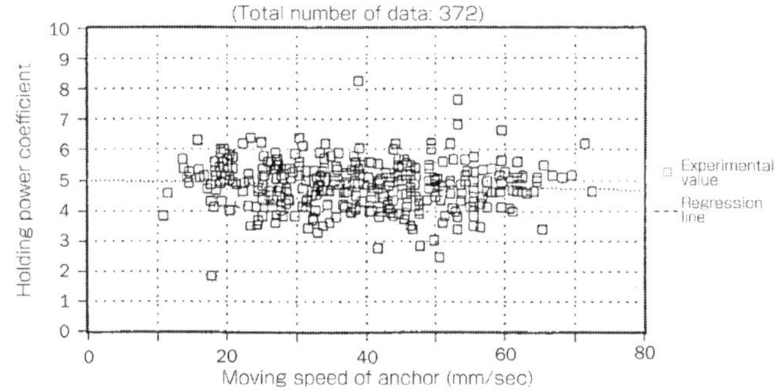

Fig. 3.2.7 Relation between maximum holding power coefficient and pulling speed (JIS type anchor, 10 kg, bottom quality : sand)

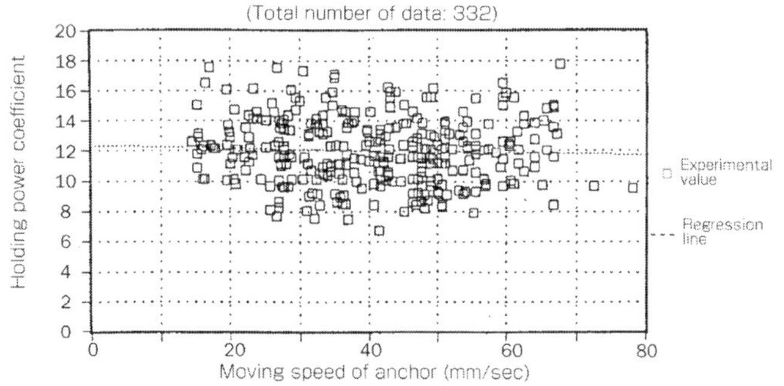

Fig. 3.2.8 Relation between maximum holding power coefficient and pulling speed (AC14 type anchor, 10 kg, bottom quality : sand)

④ Assuming the linear correlation between the maximum holding power coefficient H_D / W and pulling speed "V", the relation as shown in equation (3.2.3) holds.

$$\frac{H_D}{W} = aV + b \quad \cdots\cdots (3.2.3)$$

The point on the vertical axis where this regression line crosses represents the maximum holding power coefficient "b" at $V = 0$. This can be deemed as the static holding power coefficient as it indicates holding power in the resting state of the anchor.

⑤ As far as the results shown in Fig. 3.2.7 and 3.2.8, there is no significant correlation between values of the maximum holding power coefficient and pulling speed and the static coefficient and kinematic coefficients seems to be almost the same.

(2) Irregularity of Maximum Holding Power Coefficient

There is no significant correlation between values of the maximum holding power coefficient and pulling speed. And the static and kinematic coefficients, as well as their irregularities seem to be almost same.

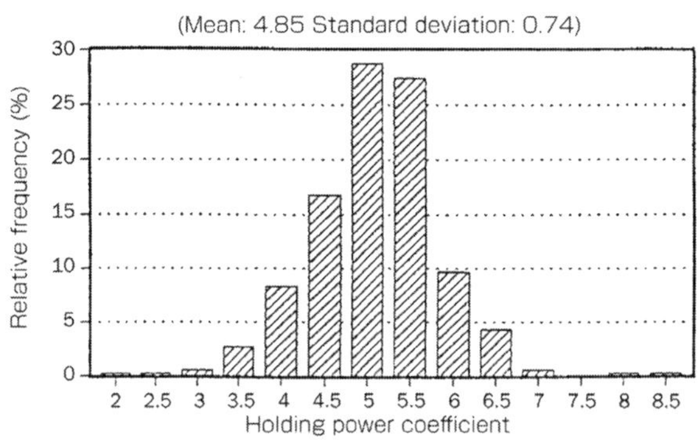

Fig. 3.2.9 Irregularities of maximum holding power coefficient
(JIS type anchor, 10 kg, bottom quality : sand)

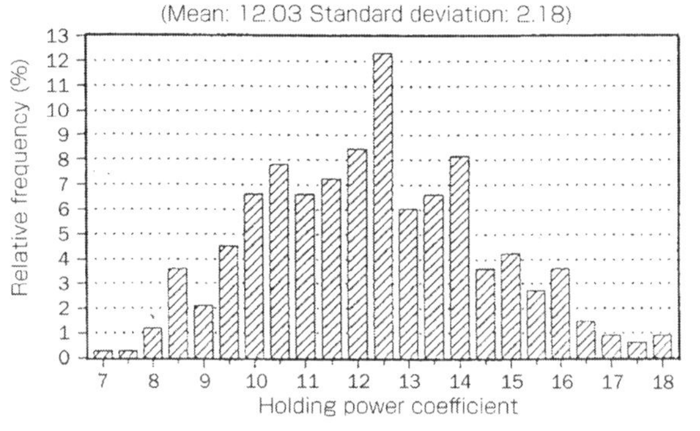

Fig. 3.2.10 Irregularities of maximum holding power coefficient
(AC14 type anchor, 10kg, bottom quality : sand)

Table 3.2.2 Statistical values of maximum holding power coefficient
(Bottom quality : sand)

Bottom quality sand	Mean	Standard deviation	Variable coefficient	Total sample data
JIS type	4.85	0.74	15%	372
AC14 type	12.03	2.18	18%	332

Table 3.2.3 Statistical values of maximum holding power coefficient
(Bottom quality : mud)

Bottom quality mud	Mean	Standard deviation	Variable coefficient	Total sample of data
JIS type	4.82	0.99	21%	77
AC14 type	6.58	1.61	24%	29

Figs. 3.2.9 and 3.2.10 illustrate the irregularity of the data plotted in Figs. 3.2.7 and 3.2.8. However, among the values of the maximum holding power coefficient obtained from multiple experiments, there exists the irregularities which can be deemed as a normal distribution. This means that the holding power of the same anchor may vary during each anchoring operation depending on various indeterminate elements. Table 3.2.2 illustrates the irregularity of statistical data of tests done by 10kg JIS type and AC14 type anchors under the condition that the bottom quality is sand, and Table 3.2.3 illustrates the irregularity of statistical data of tests done by 20kg JIS type and AC14 type anchors under the condition that the bottom quality is mud.

(3) Determination of Holding Power Coefficient

We can imagine that the holding power of the actual anchor may vary further, affected greatly by many indeterminate elements on the sea bed than that of the model tests. From this point of view, the holding power should be considered as a random variable. However, it would be practical to consider this as a determined value based on the understanding that this value may vary. Methods on how to define standard holding power coefficient are presented below based on the aforementioned assumptions.

① The lighter the weight of the model anchor, the larger the value of holding power tends to appear, and the heavier the weight of the model anchor, the smaller the value of holding power. However, when the weight of the model anchor is beyond 1 ton, the value of holding power almost becomes constant. Therefore, the holding power coefficient of an actual anchor is defined by the mean of holding power of the model anchor of which the weight exceeds 1 ton. Table 3.2.4 shows the mean values of maximum holding power coefficient by type and by bottom quality.

Table 3.2.4 Mean value of maximum holding power coefficient

AC14 type	Sand	Mud	JIS type	Sand	Mud
Mean (λ)	8.8	12.6	Mean (λ)	3.9	3.7

② The determined values of holding power coefficient "Λ" were defined by adopting the statistical characteristic values derived from abundant model test data in the Tables 3.2.2 and 3.2.3 for the irregularity of maximum holding power coefficient, taking its random deviation into account. And then the practical standard values on the side of prudence were gained according to the following equation (3.2.4).

$$\left.\begin{aligned} \Lambda &= \lambda - \gamma \\ \gamma &= 0.6745\sigma \end{aligned}\right\} \qquad (3.2.4)$$

where

λ : Mean Value
γ : Random Deviation
σ : Standard Deviation

③ The reason that $\Lambda = \lambda - \gamma$ is deemed as the standard value of holding power coefficient is as follows. The probability that the value of the holding power of an actual anchor falls below the standard value of "Λ" is not more than 0.25.
That is to say, although for three out of four times anchoring, the standard value may be warranted, for one time, the holding power may fall under the standard value.

④ The values of the standard holding power coefficient derived from the above process are shown in table 3.2.5 together with the statistical characteristic values to determine the standard values. As shown in the table, the standard holding power coefficients of JIS type anchors are 3.5 in sand and 3.2 in mud. While those of AC14 type are 7.7 in sand and 10.6 in mud.

Table 3.2.5 Determined standard holding power coefficients

JIS type	Sand	Mud	AC14 type	Sand	Mud
Mean (λ)	3.9	3.7	Mean (λ)	8.8	12.6
Variable parameter (C)	15%	21%	Variable parameter (C)	18%	24%
Standard deviation (σ)	.59	.78	Standard deviation (σ)	1.58	3.02
Random deviation (γ)	.39	.52	Random deviation (γ)	1.07	2.04
Standard holding power coefficient (Λ) $= \lambda - \gamma$	3.5	3.2	Standard holding power coefficient (Λ) $= \lambda - \gamma$	7.7	10.6

4.3 Anchor and Chain under Dragging

(1) Holding Power Coefficient of Anchor under Dragging

The holding power of an anchor overturned on the sea bed with its fluke upside down when it is pulled over the sea bed solely depends on the friction between the sea bed; for its standard value, 1.5 for JIS type and 2.0 for AC14 type in both sand and mud can be usually applied.

(2) Holding Power Coefficient of Anchor Chain under Dragging

The holding power of anchor chain sliding over the sea bed can be represented by the ratio of friction force obtained by measuring actual friction of a unit length of anchor chain sliding over seabed to weight of chain per unit length. For its standard value, 0.75 in sand and 0.6 in mud can be usually applied.

(3) Behavior of Anchor under Dragging

The posture of JIS type anchors under dragging conditions are unstable. When either side of the fluke digs deeper in the soil while it is being pulled, they tend to be overturned suddenly. And once overturned, they continue sliding over the sea bed. Another firm grip cannot be expected because sand or mud often accumulates between the tripping palm and shank.

On the other hand, AC14 type anchors sink into the soil almost at the same time as when they grip the soil. And when they are not overturned by pulling, they will be dragged keeping their original posture. Then they rise up toward the surface of the soil when they start to turn. And they react to the sea bed when they are turned about 90° and dragged obliquely towards the pulling direction. When they make a left-handed turn as seen from behind, they move to the right. And when making a right-handed turn, they move to the left. This is because the lift component of the pressure from the seabed acts on the fluke.

An AC14 type anchor will restore its posture and move straight and grip soil firmly again if dragging is continued. The process is, sinking into soil → listing → oblique dragging → overturn (restore original posture) → straight dragging, → gripping in soil again. As you see from the above, an AC14 type anchor has the characteristic to grip soil again even if it is once dragged and overturned. In this respect, an AC14 type anchor is fundamentally different from the JIS type anchor.

5. Anchoring in Rough Weather

5.1 Swinging Motion

A ship lying at anchor under strong wind moves, swinging as shown in Fig. 3.2.11 drawing a figure 8 (***figure 8 motion***). When a ship is laid to the wind at an angle, lift affects the hull. This lift causes the ship to move laterally. Following the ship's lateral shift, restraining force by the anchor chain at the bow increases and the ship's lateral movement reaches its limit. When the ship is swung to this limit, the hull is pulled forward towards the windward by the reaction force of the tightened chain until its tension is released.

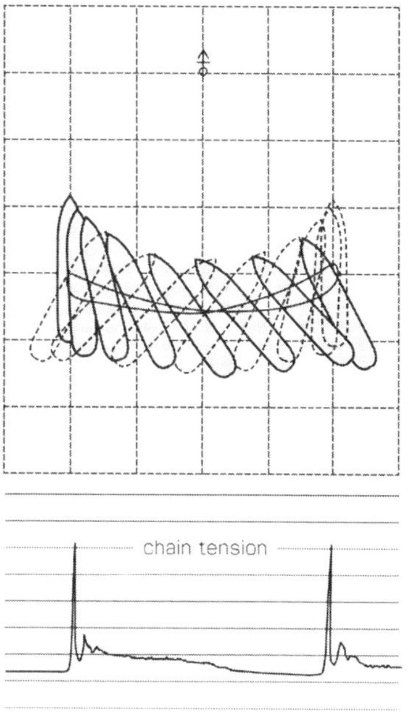

Fig. 3.2.11 Swinging motion of a ship at anchor

However, swing of the stern and forward movement of the hull continues for some time after the tension at the bow is released and the ship, after it receives wind towards the bow, changes to its windward board. At this phase the movement of the ship becomes most sluggish. Then, suddenly the bow is blown down toward the leeward and the direction of the chain and the ship's longitudinal center line are placed on the same line. At around this instance, maximum load is created on the anchor and anchor chain.

In the process of this swinging motion, constant and impact loads have alternate effects on the anchor. A constant load created while the ship moves laterally does not exceed the holding power of the anchor as long as it is laid out correctly. However, when an impact load which is created at the extreme position of its swinging motion on the anchor is beyond its holding power, it gradually moves the anchor towards the ship.

Even if the anchor begins to start sliding over the sea bed, the ship will be able to continue riding to the anchor as long as proper holding power can be maintained thereafter. However, if the anchor is overturned and unable to restore its holding power again, dragging the anchor is unavoidable.

5.2 Relationship between Swinging Motion and the Length of the Catenary Part of Chain

The total length of chain paid out composed of the length of the cable laid on the sea bed and the length of the catenary part can be calculated according to the following equation (3.2.5) as long as the proper length of chain is paid out according to the depth of water, assuming that the anchor, anchor chain and ship are placed on the same line as shown in Fig. 3.2.12.

Let the total chain length paid out "L" be the sum of the length of the cable laid on the sea bed "l" and the length of the catenary part "s".

$$\left.\begin{aligned} L &= l + s \\ s &= \sqrt{y^2 + 2\left(\frac{T_x}{w_c}\right)y} \end{aligned}\right\} \quad \cdots\cdots (3.2.5)$$

where

L : total chain length paid out
l : length of chain laid on the sea bed

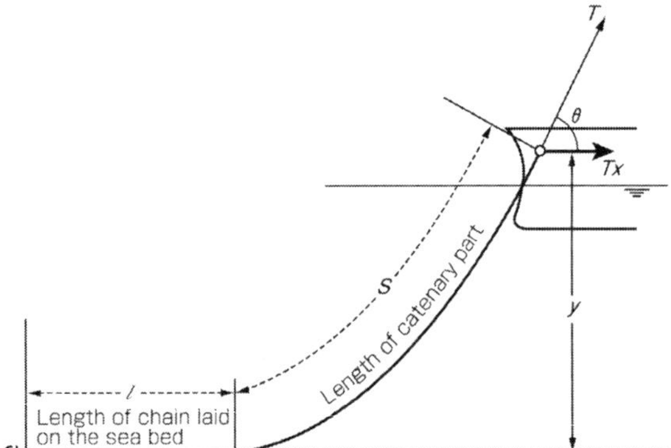

Fig. 3.2.12 Length of catenary part and total chain length paid out

s : length of catenary part
y : height from bell-mouth to sea bed
T_x : horizontal component of chain tension "T"
w_c : weight of chain in water

The anchor chain paid out is comprised of the part lying on the sea bed and the catenary part. The chain paid out as a whole absorbs the kinematic energy of the ship's swinging motion, changing the shape of the catenary curve and transmitting its tension to the anchor. The smaller the load delivered to the anchor, the longer the length of chain lying on the sea bed becomes and the catenary part will be slackened and the bigger the load delivered to the anchor, the shorter the length of chain lying on the seabed becomes.

The chain lying on the sea bed also contributes to holding power corresponding to the length paid out. Fig. 3.2.13 illustrates how the length of the part of

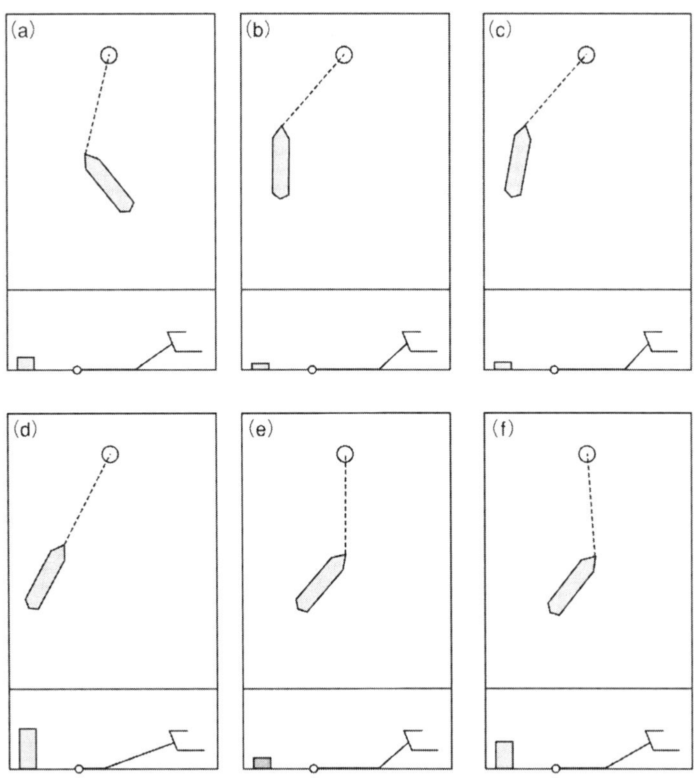

Fig. 3.2.13 Relationship between swinging motions and catenary part of the length of chain

the chain laid on the sea bed changes the following swinging motions. While the ship moves (a) right to left, (e) and (f) left to right following ship's lateral movement, the tension on the chain is gentle and the length of the part of the chain laid on the sea bed is almost unchanged under the constant load affecting the chain. At (b), due to the extreme of the ship's swing the restraining force by the anchor increases and the catenary begins to slack.

At (c) the ship begins to move forward against the wind for a moment then changes its windward board. At this moment the tension on the chain is slackened most and the length of the part lying on the sea bed is maximized. At (d) the once slackened catenary part abruptly becomes taut and at around the moment that the anchor chain and the ship's longitudinal line are placed on the same line, a large impact on the anchor will be created. At this moment, the chain is pulled taut and length of the part of chain laid on the sea bed becomes shortest.

5.3 Safety Countermeasures in Maneuvering

The important point in achieving safe anchoring is to prevent dragging the anchor. Dragging the anchor often occurs when the load to the anchor is increased corresponding to the strength of the swinging motion. Therefore, it is important to take appropriate countermeasures to mitigate such motion as much as possible.

Generally, the swinging motion appears more intense on a ship in which the ratio of projected area above water to that under water is larger. For instance, the motion of fat ships such as VLCCs is not so serious and box type ships such as barges and dredgers are almost free from such motion. While ships like PCCs show quite violent swinging motions. In view of the design of the superstructure, it can be said that the further backward the center of the wind pressure on its superstructure is placed, the slower the swinging motion becomes.

Table 3.2.6 provides an overview of practical maneuvering for how general merchant ships should adjust their ship's draft, trim, the length of chain to be paid out, use of a second anchor, bow thruster and main engine to mitigate swinging motion effectively.

Table 3.2.6 Countermeasures to mitigate swinging motions

Countermeasures	Efficiency	Remark
To increase draft	Swinging motion can be lessened corresponding to the increased weight of hull.	
To make trim even or by the head if possible	Swinging motion can be restrained as the center of wind pressure moves toward the stern.	About 1.5m trim by the head will be effective to mitigate swinging motion.
To increase the length of chain to be paid out	It can increase the total holding power due to the increased friction between the chain laid on the sea bed and the shock absorbing effect of elongated catenary.	Effective regardless ship type and size.
To use the second anchor as a snubber	It is effective to cut swinging motion short by about 50%. In addition, the load to the first anchor can be lessened 30 to 40% lower than that without a snubber.	Effective under a certain wind force.
To ride to two anchors, by dropping anchors simultaneously with chains paid out equally	Holding power can be increased.	Caution is necessary because of the possibility of fouling anchors depending on the change of wind direction.
To ride to both anchors which chains are paid out equally at an appropriate angle	If two anchors are dropped at an angle of 45° to 60°, swinging motion can be lessened and the load to each anchor can be decreased by 60%.	Caution is necessary according to the change of wind direction, as the load of the anchor on either side may be increased.
To use bow thruster	The effect of wind which causes swinging motion or tension on the cable can be largely mitigated by keeping the bow headed towards the wind. If thrust force of 80% of front wind pressure is available, about 40% of the range of swinging motion and shock load can be decreased.	
To put engine S/B ready to use	It is effective if the ship's heading is kept right against the wind using slow ahead thrust and rudder.	Caution is necessary as excessive headway may cause an increased impact to the chain once slackened when the ship is drifted down wind again, which could cause dragging of the anchor. Sternway is also effective to restrain swinging motion to some extent; however, excessive sternway has an affect opposite to holding power in addition to the wind pressure. It should be noted that considerable effort is needed to maintain appropriate thrust.

6. Risk of Dragging Anchor

6.1 Deducing the Risk of Dragging Anchor

(1) Probability of Dragging Anchor

When a ship is lying at anchor under strong wind and the load to the dropped anchor exceeds its holding power, the anchor will be dragged over the sea bed down-wind. At that moment, an unstable anchor is liable to be turned to some extent around the shank. And the turning angle of the anchor will increase in the process while it is being dragged. And when its angle of list becomes around 45°, the anchor will lose almost all its holding power and the dragging of the anchor will be accelerated.

For how long the anchor will be moved until dragging of the anchor starts when it has been turned to 45° and most of its holding power has been lost can be given by normal random variables derived from the following empirical equation (3.2.6). In the equation "μ_Y" and "σ_Y" respectively represent nondimensionalized mean values and standard deviation divided by the height of the anchor "h".

$$f(Y)=\frac{1}{\sqrt{2\pi(\sigma_Y)}}\exp\left[-\frac{1}{2}\left(\frac{Y-\mu_Y}{\sigma_Y}\right)^2\right] \quad \cdots\cdots (3.2.6)$$

where

$$\mu_Y=\frac{\mu_y}{h}=4.94$$

$$\sigma_Y=\frac{\sigma_y}{h}=1.40$$

When shocks act intermittently on the anchor at both extremes of the ship's swinging motion, the distance a/h in which the anchor moves at every shock can be deduced by the following equation (3.2.7) corresponding to the scale of impact force "T_D".

However, when the acting shock is smaller than the holding power, the anchor will not be moved. In this equation, let "W_A", "a" and "h" respectively be the weight of the anchor in the air, moving distance per shock, and the height of the anchor. From this equation (3.2.7), an accumulated moving distance "X" of the anchor for a certain period of time under strong wind conditions can be obtained corresponding to the number and the scale of the shocks for that period.

$$\frac{a}{h} = 0.158 \times 10^{-2} \exp\left[0.0896 \frac{T_D}{W_A}\right] \quad \cdots\cdots (3.2.7)$$

If we let the limit of the moving distance before dragging of the anchor starts be "Y" and if the anchor moves "X" for a period of "t", then the probability of dragging anchor "P_{yf}" after the time duration of "t" will be defined as the following equation (3.2.8).

$$P_{yf} = Prob[\,Y - X \leqq 0] \quad \cdots\cdots (3.2.8)$$

In the above, if both "Y" and "X" are given as normal random variables, the probability density function of the difference between "Y" and "X" randomly picked out from each population $Z = Y - X$ will be a normal distribution. Therefore,

$$f_z(Z) = \frac{1}{\sqrt{2\pi(\sigma_Z)}} \exp\left[-\frac{1}{2}\left(\frac{Z - \mu_Z}{\sigma_Z}\right)^2\right] \quad \cdots\cdots (3.2.9)$$

where

$$\mu_Z = \mu_Y - \mu_X$$
$$\sigma_Z^2 = \sigma_Y^2 + \sigma_X^2$$

Accordingly, the probability of dragging anchor is given as follows.

$$P_{Zf} = Prob[\,Z \leqq 0]$$
$$= \int_{-\infty}^{0} f(Z)\,dZ \quad \cdots\cdots (3.2.10)$$

When the distance moved for a certain amount of time is given as a definite value, the probability of dragging the anchor can be obtained from the distribution of "Y" and the definite value "X" according to the following equation (3.2.11).

$$P_{yf} = \int_{-\infty}^{X} f(Y)dY \quad \cdots\cdots (3.2.11)$$

Fig. 3.2.14 illustrates the flow chart of the calculation process to derive the probability of dragging the anchor.

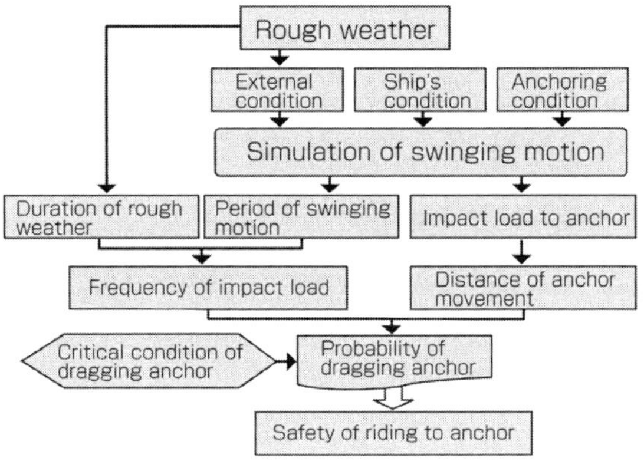

Fig. 3.2.14 Calculation process of dragging anchor probability

At first, carry out the simulation of the ship's swinging motion by inputting the ship type, size and external conditions such as water depth, bottom quality, sheltering condition of anchorage as well as weather and sea conditions. Next, calculate the accumulated moving distance of the anchor under rough weather based on the load acting on the anchor and the period of ship's swinging motion. Then, deduce the cumulative movement of the anchor and the probability of the dragging anchor according to the above process.

Fig 3.2.15 illustrates the probabilities of dragging the anchor of a 6,000 unit

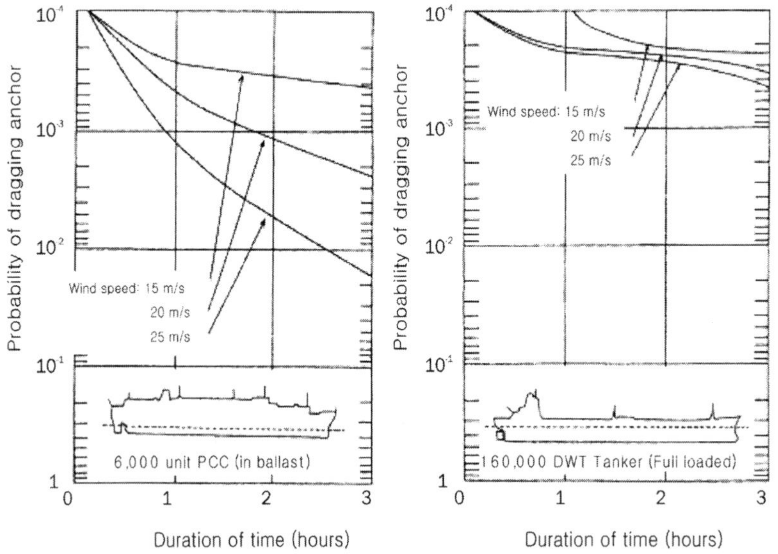

Fig. 3.2.15 Result of calculation on dragging anchor probability

PCC in ballast conditions and a 160,000 DWT tanker in full load on the conditions that each ship is riding to a single anchor together with 10 shackles of chain under a wind speed of 15 m/sec, 20 m/sec and 25 m/sec. You can understand from the graphs that a risk of dragging the anchor for a tanker does not increase corresponding to the duration in time but for the PCC, a risk of dragging the anchor increases rapidly corresponding to the duration in time.

(2) Risk Index of Dragging Anchor

Another means of deducing the risk of dragging the anchor is using the risk index of dragging the anchor. In this method, forces acting on the anchor are brought into focus if they are capable of moving the anchor. When a force acts beyond the holding power of the anchor and the chain paid out, it is considered that there is a risk of dragging the anchor and its capability will be made indexation according to the concept of ***stress-strength-model.***

Let the force acting on the anchor and holding power of the anchor and chain be random variables of "S" and "R" respectively. The anchor can be moved when "S" is larger than "R" ($S > R$). And its occurrence probability "P_r" can be expressed as shown in the following equation (3.2.12).

$$P_r = Prob\,[S - R > 0] \quad \cdots\cdots (3.2.12)$$

When probability distribution function "F_S" of acting force "S" is considered as the occurrence frequency of "S" and has been calculated as shown in Fig. 3.2.16, the risk index of dragging the anchor can be deduced as the ratio of the occurrence of acting force "S" beyond the defined holding power of the anchor and chain "R_0". That is to say, the risk index of dragging the anchor can be expressed as a probability that "S" exceeds the defined value of "R_0" in random distribution functions as shown in the following equation (3.2.13).

$$P_r = 1 - F_S(R_0) \quad \cdots\cdots (3.2.13)$$

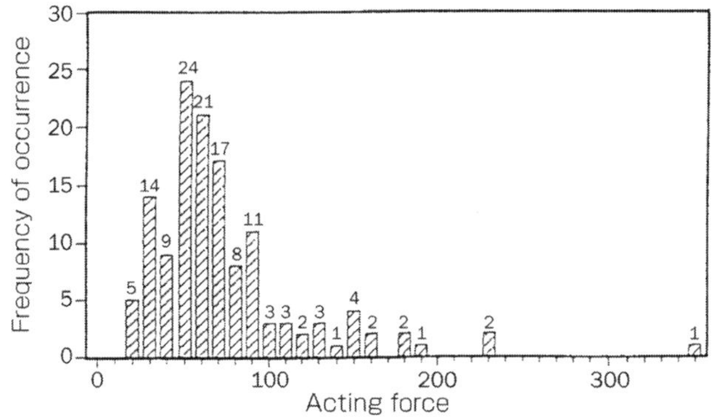

Fig. 3.2.16 Occurrence distribution of acting force to anchor

In respect to the risk index of dragging the anchor deduced from the above process, the following can be said. When a ship's swinging motion and consequential acting forces on the anchor are small and do not exceed the holding power of the anchor and chain, the value of the risk index of dragging the anchor is zero and there is no risk of dragging the anchor. On the other hand, when more than a half of the occurrence of acting force on the anchor and chain exceeds the holding power of them, the risk index of dragging the anchor will be 50% and risk of dragging the anchor is impending.

It can be said that the concept of the risk index of dragging the anchor is to present a similar probabilistic indication of rainfall of the weather forecast in deciding if one should carry an umbrella when they are going out to work.

(3) Correlation between Risk Index of Dragging Anchor and Probability of Dragging Anchor

Fig. 3.2.17 shows the correlation plotted between the probability of dragging anchor and the risk index of dragging anchor after 2 hours at every varying wind speed, length of chain paid out and water depth derived from the simulation of a 6,000 unit PCC in ballast condition fitted with a JIS type anchor under the conditions that the bottom quality is sandy. From the graph the followings can be said.

In the area where risk index is less than 30%, the probability of dragging anchor is also low. And where the risk index of dragging anchor exceeds 70%,

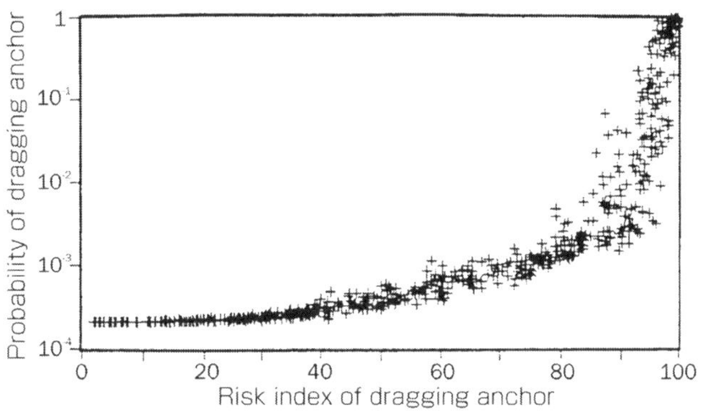

Fig. 3.2.17 Correlation between risk index of dragging anchor and probability of dragging anchor

reliability of the anchor will be lost rapidly and risk of dragging anchor will be impending. Therefore a risk index of under 30% can be said as the safety range for anchoring conditions while an index of over 70% should be considered as dangerous for anchoring.

6.2 Length of Anchor Chain Required to be Paid out when Riding to a Single Anchor in View of the Risk Index of Dragging Anchor

Fig. 3.2.18 shows the length of the anchor chain for a 6,000 unit PCC in ballast condition riding to a single anchor where the bottom quality is sandy and water depth is 20m under varying wind speeds maintaining its risk index of dragging

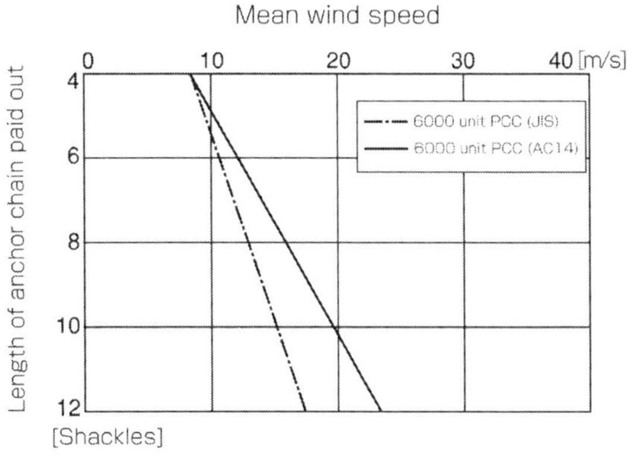

Fig. 3.2.18 Relation between length of chain to be paid out and wind speed (Risk index of dragging anchor : 30%)

anchor at 30%. The graph includes the figures of both JIS and AC14 types based on the standard holding power coefficients in Table 3.2.5.

According to the graph in Fig. 3.2.18, the subject PCC needs 10 shackles of anchor chain for a JIS type anchor and 8 shackles for an AC14 type anchor to retain the risk index of dragging anchor under 30% when she is riding to a single anchor under 15m/sec of wind.

7. Countermeasures to Prevent Dragging Anchor

7.1 Behavior of Ship under Dragging Anchor

(1) Swinging Motion when Anchor is Dragged

When an anchor is dragged under a certain wind pressure, the ship will drift leeward repeating swinging motion (distorted figure 8).

Within the range of 'A' in Fig. 3.2.19, the ship repeats a regular swinging motion being pressed downwind. In the range of 'B', the ship drifts downwind with the meandering anchor and continuing a distorted swinging motion corresponding to increasing wind pressure.

Then, when wind pressure increases further, the ship will drift downwind at a constant speed drawing a gentle curve without a swinging motion any longer as shown in the range of 'C'. This phenomenon tells us that we would be too hasty if we judge the ship is not under dragging anchor conditions because she regularly repeats swinging motion. We should keep it in mind that the anchor could be dragged even when a ship has such a motion.

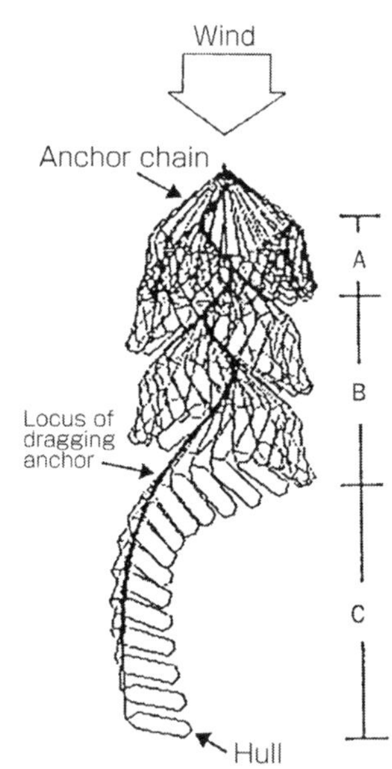

Fig. 3.2.19 Ship's swinging motion under dragging anchor

(2) Drifting Speed and a Ship's Posture while Dragging Anchor

Even if an anchor is placed on the sea bed without gripping firmly into the sea bed, the ship will not drift downwind as long as the total frictional force of the anchor and chain exceeds external pressure acting on the ship. Table 3.2.7 shows the critical wind speed in relation to the total frictional force of an anchor. When wind speed slightly exceeds the critical wind speed, dragging anchor under distorted swinging motion will start. The ship will drift downwind dragging its anchor at a constant speed as wind speed increases further.

Fig. 3.2.20 shows the ship's speed of drift downwind after the anchor has been dragged. From the graph, the PCC yields to dragging anchor accompanying distorted swinging motion under wind speed 16 to 18m/sec. If the wind becomes stronger, the ship will drift downwind at a constant rate. Under wind speeds of 20~25m/sec, the ship's speed of drift downwind will not be less than 3~4 knots.

Table 3.2.7 Critical wind speed of dragging anchor in view of frictional holding power of an anchor

	Critical wind speed of dragging anchor in view of frictional holding power of an anchor
5,000 GT general cargo ship	30 m/sec
160,000 DWT tanker	26 m/sec
700 TEU container ship	24.2 m/sec
6,000 unit PCC	16.5 m/sec

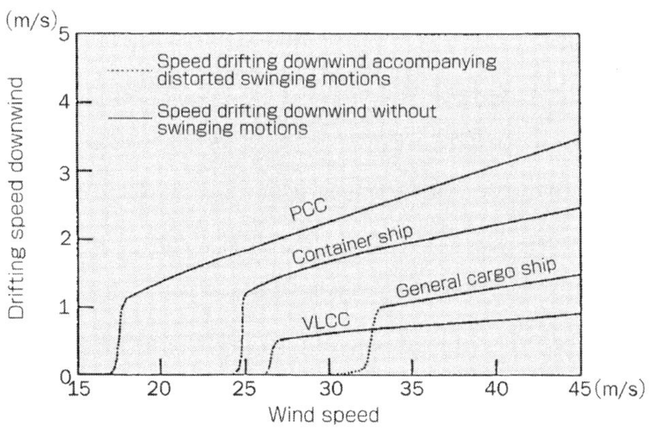

Fig. 3.2.20 Relation between drifting speed downwind and wind speed

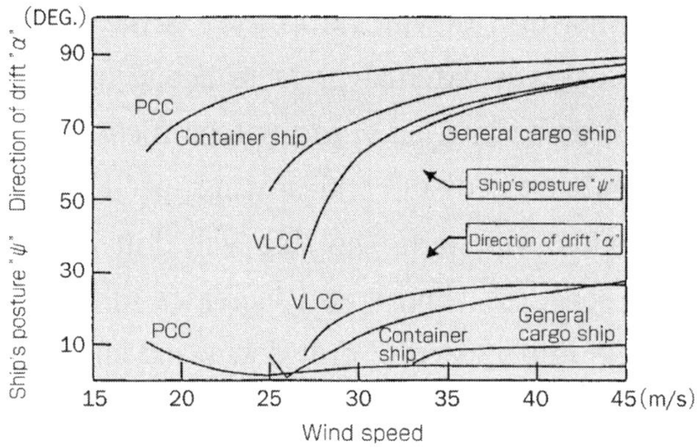

Fig. 3.2.21 Ship's posture and direction of drift while dragging anchor

Fig. 3.2.21 shows the posture and direction of a drifting ship dragging anchor. The ship's posture is represented by an angle between ship's longitudinal center line and wind direction, while the direction of drift is represented by an angle between the direction of wind and the drifting direction of a ship's center of gravity. For instance, when the anchor of a PCC is dragged under 30m/sec of wind speed, it will be carried straight downwind lying abeam to the wind. While a VLCC will drift in a direction 20° to the wind and lying 60° to the wind.

7.2 Countermeasures and Effect

(1) Extra Chain Length and Use of the Second Anchor

It is difficult to stop dragging the anchor once it has started. This is because the inertia of a drifting ship at a certain speed is too large to restrain its motion by the holding power of only the anchor and chain. Although at the initial stage of dragging anchor before its drifting inertia fully develops, the effect of paying out extra length of chain and/or dropping the second anchor to restrain dragging the anchor may be of some use, it should be noted that for a ship drifting at a constant speed after its anchor has been dragged, the effect of paying out extra length of chain and/or use of the second anchor may be almost useless.

(2) Use of Bow Thruster

Bow thrusters if fitted, are one of the useful means to maintain the ship's heading towards the wind even when it is drifting downwind at a constant speed. However, for this purpose, it is necessary for the bow thrusters to be provided with at least equal power to frontal wind pressure. For instance, when a PCC is drifting under a wind of 20m/sec, 22 tons or more of thrust will be needed to offset the moment by wind pressure.

(3) Use of Main Engine and Rudder

When the main engine and rudder are used to correct a ship's heading towards the wind and maintain its posture once its anchor has been dragged, it is required to use at least slow ahead of the engine against a wind of 20m/sec, half ahead engine against a wind of 25m/sec and S/B full ahead engine against a wind of 30m/sec in addition to hard over rudder. Today, if engine and rudder are used, it is possible for a ship to maintain its heading under fairly strong wind as the output of engines has been increased.

However, sometimes it is difficult to maintain the continuous high performance of the engine due to propeller racing by the ship's motions in rough seas. Therefore, it should be noted that the output of a ship's engine as a counter-measure to stop dragging the anchor may be restricted under rough weather conditions.

However once the ship is back under control and the use of the rudder and main engine is effective in keeping the ship's posture, immediate action is necessary to avoid accidents while attempting to start anchoring over from the beginning or to switch to drifting until rough seas calm down.

7.3 Detection of Dragging Anchor

It is imperative for a person engaged in ship handling on anchor watch to detect dragging anchor of their own ship as early as possible. The followings are points to detect and discern dragging the anchor at an early stage.

① Check the ship's position regularly and see if its position is found to be outside of the ship's movable range beyond the length of paid out chain.

② When the ship's heading does not face directly against the wind.

③ When the ship remains lying abeam to the wind, not changing its board from wind received.

④ When the anchor chain does not slack at the extreme point at both sides of the figure 8 motion (swinging motion).

⑤ When abnormal vibrations are felt on the chain.

8. Preparation for Safe Anchoring

8.1 Concepts to Ensure Safety

Most accidents that occur to a ship lying at anchor follow a pattern of dragging anchor → drift→ collision or grounding. Table 3.2.8 presents the fundamental concept to prevent such accident.

It is essential for persons engaged in ship handling to select the optimum anchorage taking topography, contour of the sea bed, bottom quality, water depth, available water area sheltered condition, ship congestion of water area and other related information into account. It is also important to maintain safe distance from ships lying in anchor, shoals and other offshore structures.

Also indispensable elements in deciding anchorage are, sea and weather conditions, such as wind direction, wind speed, wave height, direction and period

Table 3.2.8 Concept to ensure safe anchoring

Considerations required before dropping anchor	Technical measures for anchor watch	Measures for management and security
• Select the area precluding dragging anchor. • Secure enough ship-to-ship distance or ship-to-shore distance that precludes being involved in a marine casualty even in the possible event of dragging anchor.	• Technical measures capable of preventing the occurrence of dragging anchor. • Prediction or early detection of dragging anchor. • Post-incident measures following dragging of the anchor.	• Assistance as a means to support an anchor watch management system. For instance, providing sheltering information, management of anchoring water, individual passage guidance, risk information about possible dragging, and consolidating rescue systems and organization, etc.

of waves, and the direction and speed of tidal streams and the height of tide, along with one's own ship's conditions, such as ship type, size, draft, trim and the characteristics of the anchor including the type and weight of the anchor, unit weight and size of the anchor and chain and the length of chain to be paid out.

Further, it is important to establish means to detect dragging anchor as early as possible and countermeasures for reacting promptly if it should occur. Prepare yourself by collecting information to study post-incident technical measures organizing such knowledge and scrutinize their effectiveness.

8.2 Safe Distance between Objects and Surrounding Obstructions

It is important to maintain a safe distance from other ships, structures, shoals and obstructions in the vicinity, taking all environmental conditions and the ship's capacity into consideration to prevent a marine casualty in the case of dragging anchor as discussed in the previous subsection.

Although there are no quantitative standards or established theories for a safe distance to be maintained between these objects or surrounding obstructions, considering the necessary space for a ship after its anchor has started dragging, which should be detected immediately, until the time a ship is maneuverable again with its engine and rudder, according to one hypothetical proposition, a ship will need at least a 0.5 mile of space.

Based upon this distance, for this hypothetical situation, in addition to taking the length of chain paid out and the length of one's own ship into account, at least about one mile will be necessary for a safe anchoring for larger ships.

However, it is actually difficult to maintain such an ideal safe distance because it is often intruded upon by smaller ships which need only about half the size of the space than that needed for big ships to keep a safe distance.

This reality teaches us that it is indispensable for the safety of the water area for ships lying in anchor cannot be achieved without systematic management and control, eg: designate water area by ship type and size or provide ships looking for an anchoring spot with port information as to how congested it is, etc..

Volume 4

Ship Handling Characteristics Data Base

Ⅰ. Maneuverability

Ⅱ. Ship Handling Characteristics

I
Maneuverability

Chapter 1 List of Trial Ships

Ships that persons engaged in ship handling operate are quite versatile in size and type. Today, ships have become increasingly specialized and have increased in size corresponding to the needs of international trade. For persons engaged in ship handling, in addition to the basic and general knowledge of ship handling, knowledge of the maneuvering characteristics for the individual ship which they will handle are indispensable. In other words, each person engaged in ship handling is required to familiarize and master the maneuvering characteristics of their ship before they are commissioned for duty and for safe operation of the ship.

This volume is compiled with the purpose of assisting persons engaged in ship handling to study preliminary knowledge of ship maneuvering by presenting the database of maneuvering characteristics of typical ships by size and type.

Table 4.1.1 Specifications of ships selected for maneuvering simulation

Ship type	Condition	Lpp	B	df	da	C_b
260,000 DWT tanker						
VLCC (full)	Fully loaded	311	58.0	20.0	20.0	0.83
VLCC (half)	Half loaded			14.6	14.6	0.75
VLCC (ballast)	Ballast			9.2	10.7	0.75
10,000 DWT tanker						
Tanker (full)	Fully loaded	108	18.4	7.9	7.9	0.75
127,000 m^3 LNG carrier						
LNG (full)	Fully loaded	273	43.9	10.8	10.8	0.70
6,000 TEU container ship						
CNTN (full)	Fully loaded	301	42.8	14.0	14.0	0.63
6,000 unit PCC						
PCC (full)	Fully loaded	190	32.3	9.5	9.5	0.52

Table 4.1.2 Engine revolution (RPM) and speed (telegraph speed)

Ship type	Main engine power output	Main engine (RPM)					Speed (kts)				
		N/F	S/BF	Half	Slow	D.S	N/F	S/BF	Half	Slow	D.S
260,000 DWT tanker VLCC (full)	28,000PS	Ah'd 69	58	48	39	29	Ah'd 14.3	12.0	10.0	8.0	6.0
		Ast'n ___	58	48	39	29					
10,000 DWT tanker Tanker (full)	4,560PS	Ah'd 200	172	138	106	78	Ah'd 13.1	11.9	10.0	7.8	5.6
		Ast'n ___	172	138	106	78					
127,000 m^3 LNG carrier LNG (full)	43,000PS	Ah'd 86	54	46	36	25	Ah'd 19.6	12.3	10.4	8.2	5.6
		Ast'n ___	54	46	36	25					
6,000 TEU container ship CNTN (full)	42,500PS	Ah'd 125	92	70	49	34	Ah'd 22.8	17.3	13.2	9.1	6.4
		Ast'n ___	92	70	50	33					
6,000 unit PCC PCC (full)	19,200PS	Ah'd 106	58	45	37	30	Ah'd 20.0	11.4	8.9	7.4	5.9
		Ast'n ___	49	46	40	30					

Bridge mockup

Wing mode maneuvering

Operation room

Simulator Dome

Fig. 4.1.1 Ship handling simulator

The ships in this database have been selected out of all modern and typical ships by size and type based on "C_b" (***block coefficient***) for readers' easy reference. Table 4.1.1 shows specifications and Table 4.1.2 shows engine revolution and telegraph speed of the selected ships

The data for the maneuvering characteristics of the selected ships as shown in Fig. 4.1.1 are collected from the results of simulations made by the ship handling simulator system owned by the ship handling laboratory of the faculty of maritime sciences, Kobe University (manufactured by Japan Marine Science Inc.).

Chapter 2 Steering Characteristics

1. Zigzag Maneuver Test

The indices of responsiveness to the helm "T" and turning ability "K" are deduced from the results of simulations carried out on 7 conditions with 5 typical ship types listed in Table 4.1.1 according to the test methods and procedures of zigzag maneuver tests as discussed in Section 1, Zigzag Maneuver Test (Volume 2, [I], Chapter 4, Section 1). In simulation, 15° and 35° zigzag tests were carried out under the conditions of water depth to ship draft ratio $h/d=\infty$.

Fig. 4.1.2 shows the chronological relation among helm angle "δ", ship's heading "ψ" and rate-of-turn "r". And Table 4.1.3 shows the analytical results of maneuverability indices "T" and "K".

260,000 DWT tanker (Full)_15°Z

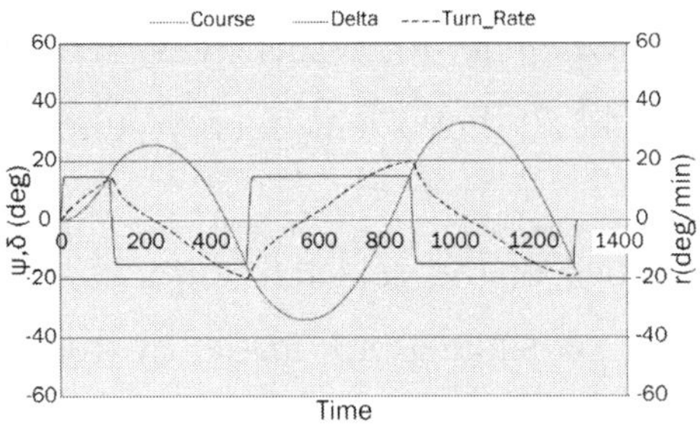

260,000 DWT tanker (Full)_35°Z

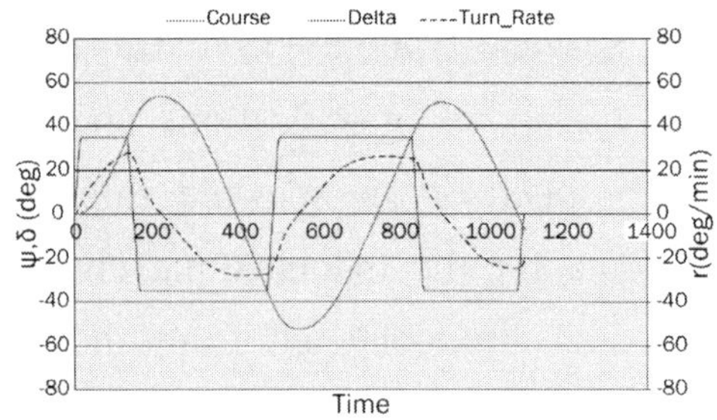

260,000 DWT tanker (Half)_15°Z

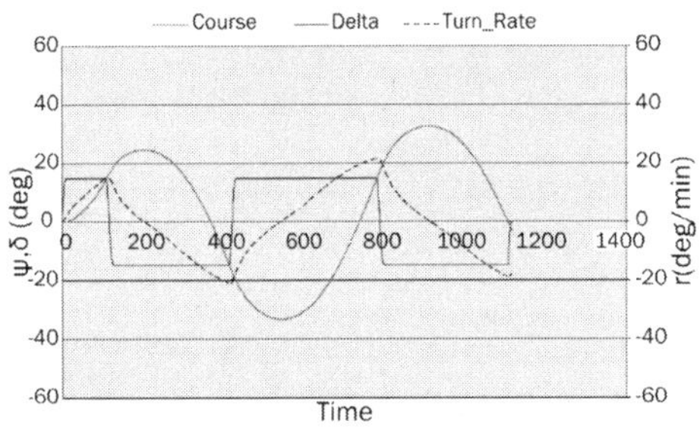

260,000 DWT tanker (Half)_35°Z

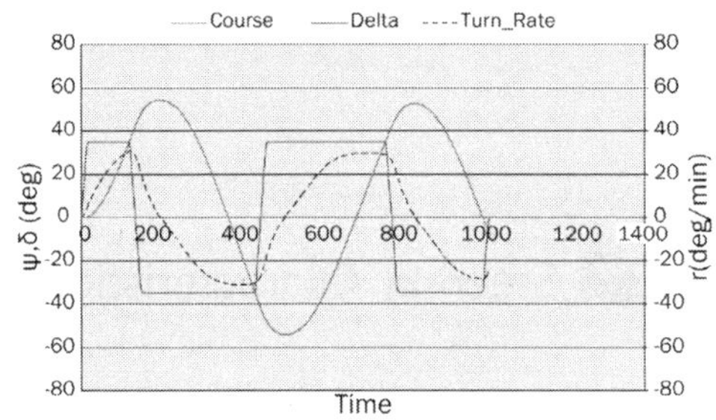

260,000 DWT tanker (Ballast)_15°Z

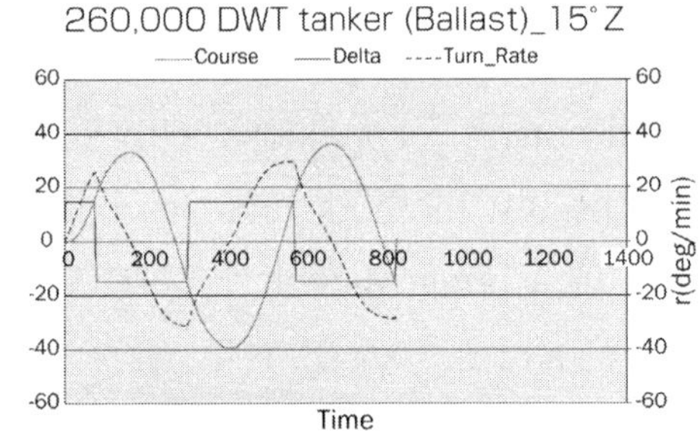

260,000 DWT tanker (Ballast)_35°Z

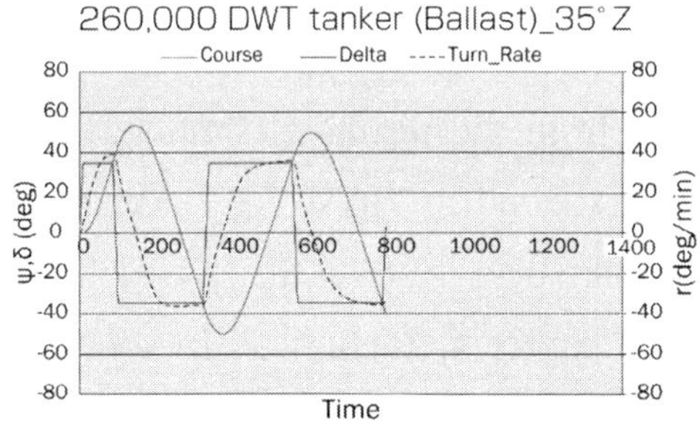

10,000 DWT tanker (Full)_15°Z

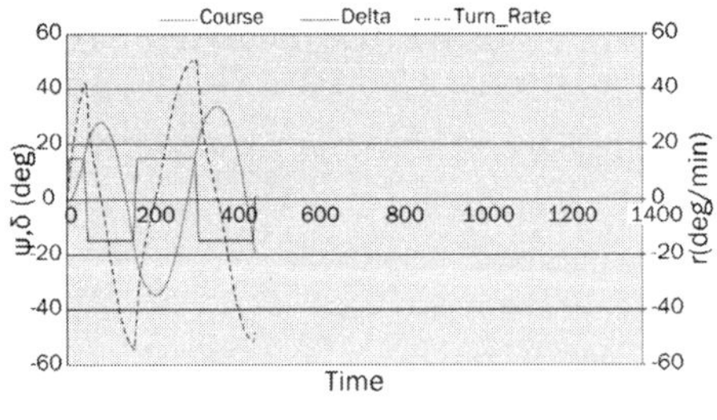

10,000 DWT tanker (Full)_35°Z

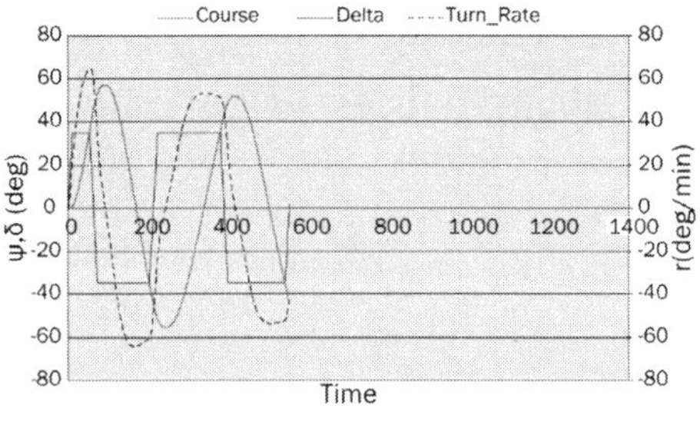

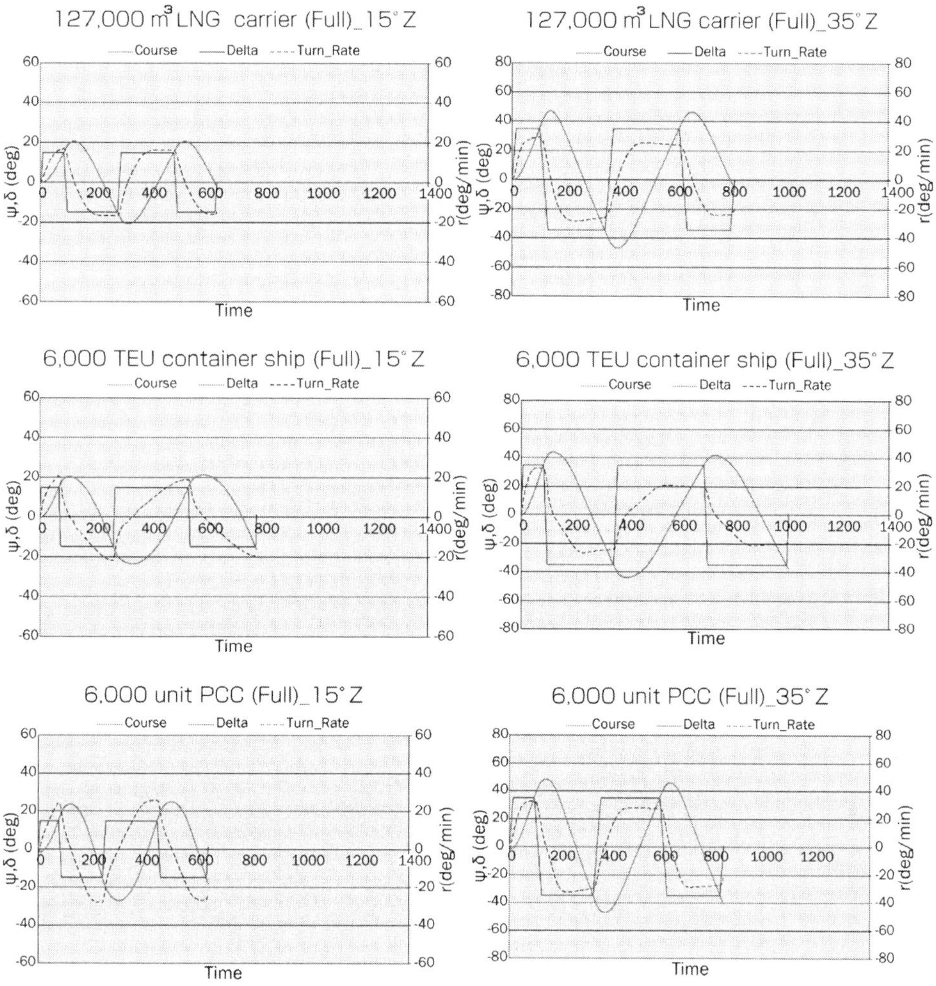

Fig. 4.1.2 Chronological data of zigzag test
(15° and 35° zigzag maneuver tests under h/d = ∞)

Table 4.1.3 Analytical results of maneuverability indices "T" and "K" (15° and 35° zigzag maneuver tests under h/d = ∞)

			Initial speed	Index of responsiveness to the helm	Index of turning ability	Non-dimensional maneuverability indices	
Ship type	Helm angle	Water depth condition	(kts)	T (sec)	K(sec^{-1})	T'	K'
260,000 DWT tanker VLCC (full)	15° Z test	h/d = ∞	11.99	237.17	0.03	4.70	1.59
	35° Z test			78.73	0.01	1.56	0.61
260,000 DWT tanker VLCC (half)	15° Z test		12.19	199.87	0.03	4.03	1.60
	35° Z test			75.22	0.01	1.52	0.68
260,000 DWT tanker VLCC (ballast)	15° Z test		13.11	207.87	0.06	4.51	2.79
	35° Z test			58.14	0.02	1.26	0.81
10,000 DWT tanker Tanker (full)	15° Z test		11.87	91.84	0.09	5.19	1.63
	35° Z test			37.78	0.03	2.14	0.46
127,000 m^3 LNG carrier LNG (full)	15° Z test		12.33	40.77	0.02	0.95	0.82
	35° Z test			21.77	0.01	0.51	0.55
6,000 TEU container ship CNTN (full)	15° Z test		16.6	73.54	0.02	2.09	0.75
	35° Z test			48.35	0.01	1.37	0.33
6,000 unit PCC PCC (full)	15° Z test		11.53	76.60	0.04	2.39	1.16
	35° Z test			36.02	0.01	1.12	0.43

2. Spiral Test

Spiral tests of a 260,000 DWT tanker and a 127,000 m^3 LNG carrier under full load for which specifications are listed in Table 4.1.1 were carried out under the conditions of the water depth to ship draft ratio as $h/d = \infty$ and initial speed of 12 knots, according to the test methods and procedures of spiral test as discussed in Section 2, Spiral Test (Volume 2, [I], Chapter 4, Section 2). Fig. 4.1.3 shows the $\delta - r$ curve that resulted from the spiral test.

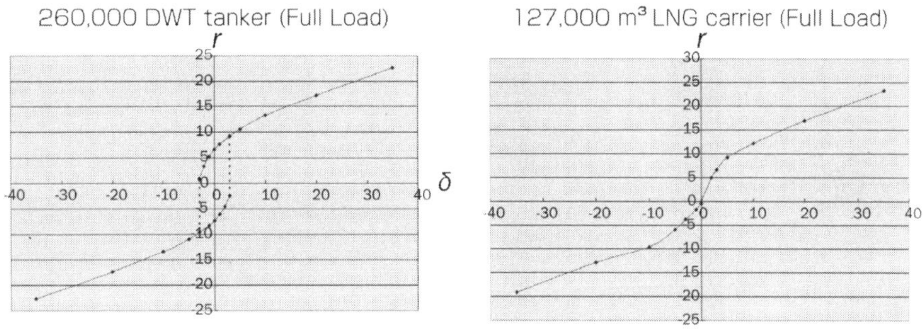

Fig. 4.1.3 Result of spiral test
(260,000 DWT tanker and 127,000 m^3 LNG, full load, initial speed 12 kts, h/d = ∞)

3. Turning Test

A 360° turning test of all the ships listed in Table 4.1.1 was carried out and analyzed at 15° starboard helm and 35° port helm under the conditions of $h/d=\infty$ and $h/d=1.3$ according to the test methods and procedures as discussed in turning test in Section 3, Turning Test (Volume 2, [I], Chapter 4, Section 3).

Fig. 4.1.4 illustrates the locus of the turning circle for each trial. And Table 4.1.4 shows the elements of the turning circle. Horizontal and vertical axis of each graph for Fig. 4.1.4 represent standardized value of ship length (Lpp) under the conditions of $h/d=\infty$ and $h/d=1.3$. The results of the tests under different h/d conditions are shown in comparison. And Table 4.1.5 shows speed reduction during turning tests.

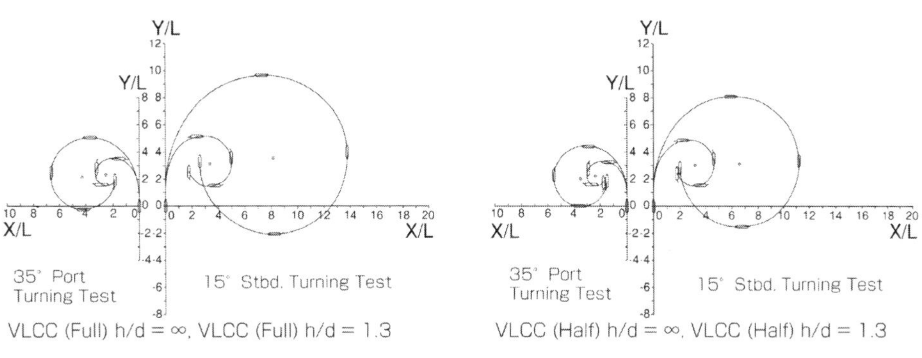

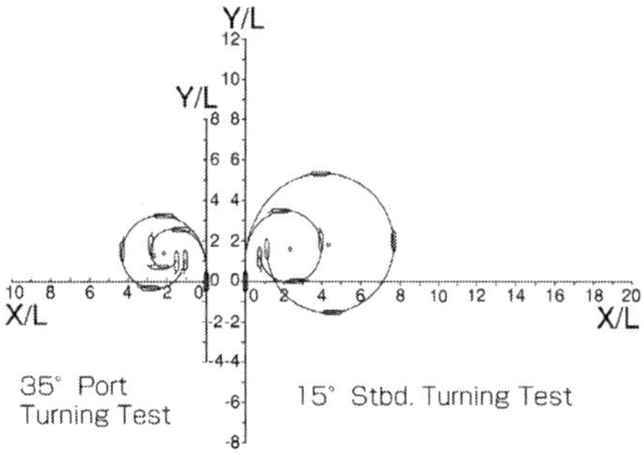

VLCC (Ballast) h/d = ∞, VLCC (Ballast) h/d = 1.3

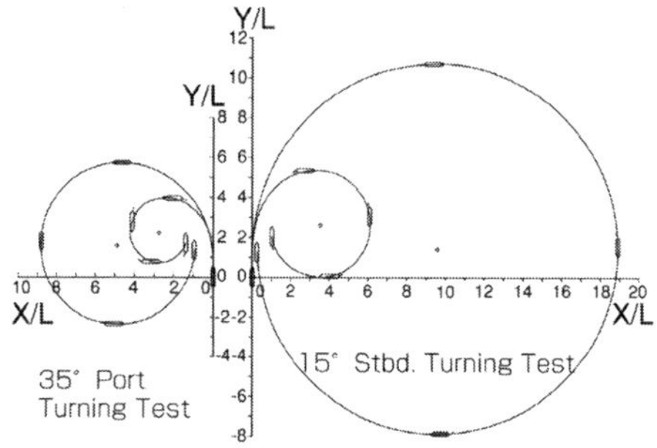

Tanker (Full) h/d = ∞, Tanker (Full) h/d = 1.3

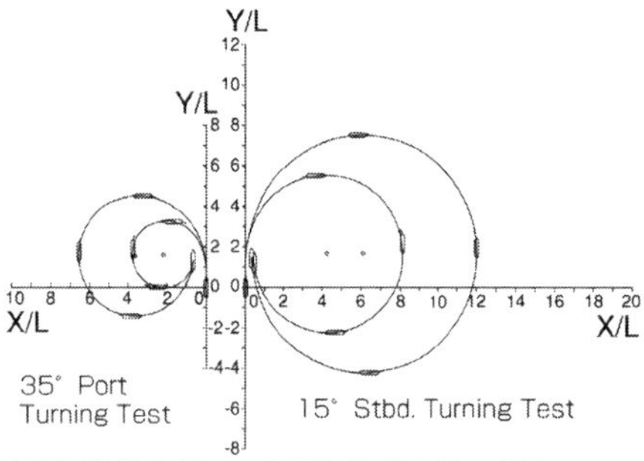

LNG (Full) h/d = ∞, LNG (Full) h/d = 1.3

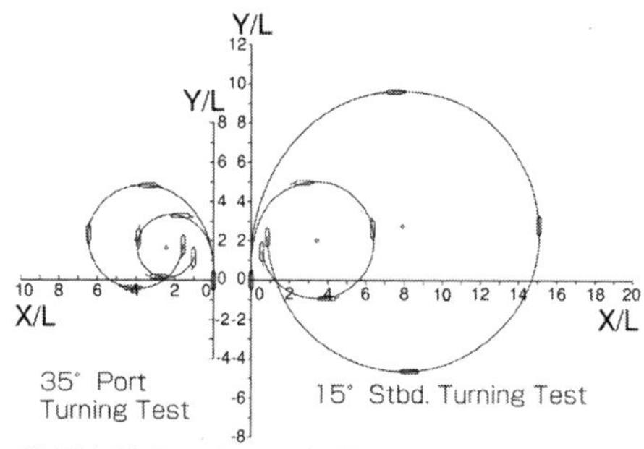

CNTN (Full) h/d = ∞, CNTN (Full) h/d = 1.3

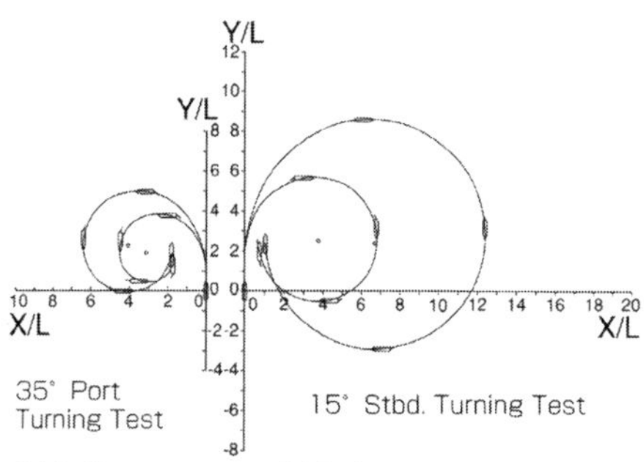

PCC (Full) h/d = ∞, PCC (Full) h/d = 1.3

Fig. 4.1.4 Locus of turning tests (standardized by ship's length)
(15° starboard helm and 35° port helm under the conditions of h/d = ∞ and h/d = 1.3)

Table 4.1.4 Elements of turning circle (standardized by ship length)
(15° starboard helm and 35° port helm under the conditions of h/d = ∞ and h/d = 1.3)

Ship type	Helm angle	Depth/Draft ratio	Initial speed (kts)	Advance	Transfer	Tactical diameter	Max. advance	Max. transfer	Final diameter	Reach	Drift angle at = 180°
260,000 DWT tanker VLCC (full)	35° port	h/d=∞	11.99	3.47	1.55	3.18	3.52	3.26	1.32	2.29	23.20
	15° stbd.		11.99	5.12	2.48	4.98	5.16	5.01	3.16	3.08	12.40
	35° port	h/d=1.3	11.57	5.09	3.57	6.61	5.09	6.62	4.75	2.20	1.80
	15° stbd.		11.53	9.69	7.36	13.76	9.69	13.76	11.13	3.55	1.40
260,000 DWT tanker VLCC (half)	35° port	h/d=∞	12.19	3.23	1.40	2.90	3.28	3.00	1.14	2.20	15.70
	15° stbd.		12.19	4.82	2.31	4.61	4.86	4.66	2.83	3.55	13.50
	35° port	h/d=1.3	11.68	4.45	2.99	5.55	4.45	5.55	4.04	2.01	1.70
	15° stbd.		11.68	8.12	5.95	11.18	8.13	11.18	9.17	3.10	1.60
260,000 DWT tanker VLCC (ballast)	35° port	h/d=∞	13.11	2.58	1.31	2.79	2.61	2.83	1.29	1.42	16.90
	15° stbd.		13.11	3.47	1.85	3.99	3.49	4.02	3.24	1.60	9.70
	35° port	h/d=1.3	12.23	3.23	2.24	4.33	3.24	4.33	3.25	1.27	3.60
	15° stbd.		12.10	5.33	3.96	7.65	5.33	7.66	6.59	1.79	1.80
10,000 DWT tanker Tanker (full)	35° port	h/d=∞	11.87	3.88	1.91	4.15	3.93	4.21	2.75	2.17	16.90
	15° stbd.		11.87	5.27	2.83	6.07	5.32	6.11	5.04	2.55	10.30
	35° port	h/d=1.3	10.28	5.71	4.53	8.80	5.71	8.80	7.88	1.57	1.30
	15° stbd.		10.28	10.69	9.47	18.86	10.69	18.86	18.57	1.38	0.60
127,000m³ LNG carrier LNG (full)	35° port	h/d=∞	12.33	3.22	1.75	3.73	3.24	3.75	3.10	1.56	9.10
	15° stbd.		12.33	5.48	3.75	8.10	5.50	8.12	7.62	1.60	6.40
	35° port	h/d=1.3	11.90	4.55	3.33	6.57	4.55	6.57	5.86	1.53	3.10
	15° stbd.		11.88	7.52	5.91	11.98	7.52	11.99	11.66	1.66	2.00
6,000 TEU container ship CNTN (full)	35° port	h/d=∞	16.60	3.26	1.70	3.85	3.32	3.91	2.83	1.63	15.90
	15° stbd.		16.58	4.94	2.79	6.36	5.00	6.41	5.79	2.01	11.00
	35° port	h/d=1.3	14.29	4.83	3.35	6.39	4.84	6.40	4.87	2.05	4.50
	15° stbd.		14.89	9.63	7.01	15.05	9.63	15.06	14.18	2.80	2.30
6,000 unit PCC PCC (full)	35° port	h/d=∞	11.53	3.73	2.02	4.48	3.79	4.55	2.74	1.87	16.40
	15° stbd.		11.53	5.60	2.99	6.73	5.65	6.78	5.98	2.47	10.40
	35° port	h/d=1.3	11.01	4.96	3.25	6.32	4.97	6.33	4.53	2.23	5.90
	15° stbd.		11.01	8.60	6.20	12.37	8.61	12.38	11.28	2.44	2.90

Table 4.1.5 Speed reduction while turning
(15° starboard helm and 35° port helm under the conditions of h/d = ∞ and h/d = 1.3)

Ship type	Helm angle	Depth/ Draft ratio	Initial speed (kts)	Duration time to turn 90° (sec)	Speed reduction at 90°	Duration time to turn 180° (sec)	Speed reduction at 180°	Duration time to turn 270° (sec)	Speed reduction at 270°	Duration time to turn 360° (sec)	Speed reduction at 360°
260,000 DWT tanker VLCC (full)	35° port	h/d=∞	11.99	239.00	0.64	446.00	0.35	688.00	0.24	940.00	0.18
	15° stbd.		11.99	346.00	0.74	607.00	0.53	896.00	0.43	1202.00	0.38
	35° port	h/d=1.3	11.57	428.00	0.73	779.00	0.62	1121.00	0.57	1458.00	0.55
	15° stbd.		11.53	837.00	0.76	1530.00	0.70	2203.00	0.67	2868.00	0.68
260,000 DWT tanker VLCC (half)	35° port	h/d=∞	12.19	219.00	0.63	403.00	0.35	616.00	0.24	897.00	0.22
	15° stbd.		12.19	319.00	0.74	553.00	0.53	811.00	0.42	1086.00	0.38
	35° port	h/d=1.3	11.68	366.00	0.73	659.00	0.63	943.00	0.58	1223.00	0.56
	15° stbd.		11.68	682.00	0.77	1240.00	0.71	1748.00	0.69	2823.00	0.68
260,000 DWT tanker VLCC (ballast)	35° port	h/d=∞	13.11	168.00	0.69	317.00	0.46	466.00	0.33	1223.00	0.26
	15° stbd.		13.11	217.00	0.80	405.00	0.65	609.00	0.56	2823.00	0.52
	35° port	h/d=1.3	12.23	258.00	0.73	485.00	0.61	716.00	0.54	951.00	0.50
	15° stbd.		12.10	435.00	0.77	809.00	0.77	1173.00	0.70	1543.00	0.69
10,000 DWT tanker Tanker (full)	35° port	h/d=∞	11.87	96.00	0.67	188.00	0.44	298.00	0.35	416.00	0.32
	15° stbd.		11.87	127.00	0.79	237.00	0.64	358.00	0.57	484.00	0.54
	35° port	h/d=1.3	10.28	196.00	0.76	379.00	0.70	562.00	0.69	745.00	0.68
	15° stbd.		10.28	370.00	0.84	727.00	0.83	1084.00	0.83	1441.00	0.83
127,000m^3 LNG carrier LNG (full)	35° port	h/d=∞	12.33	214.00	0.63	428.00	0.48	660.00	0.45	896.00	0.45
	15° stbd.		12.33	358.00	0.80	705.00	0.74	1059.00	0.73	1414.00	0.73
	35° port	h/d=1.3	11.90	357.00	0.65	705.00	0.60	1053.00	0.59	1402.00	0.59
	15° stbd.		11.88	546.00	0.82	1054.00	0.80	1564.00	0.80	2074.00	0.80
6,000 TEU container ship CNTN (full)	35° port	h/d=∞	16.60	191.00	0.52	426.00	0.34	710.00	0.27	1017.00	0.25
	15° stbd.		16.58	277.00	0.62	581.00	0.51	913.00	0.49	1251.00	0.48
	35° port	h/d=1.3	14.29	397.00	0.50	804.00	0.39	1236.00	0.36	1762.00	0.34
	15° stbd.		14.89	771.00	0.59	1533.00	0.57	2299.00	0.57	3065.00	0.57
6,000 unit PCC PCC (full)	35° port	h/d=∞	11.53	177.00	0.65	362.00	0.44	558.00	0.37	755.00	0.35
	15° stbd.		11.53	249.00	0.78	471.00	0.68	703.00	0.66	937.00	0.65
	35° port	h/d=1.3	11.01	264.00	0.69	504.00	0.54	747.00	0.49	989.00	0.47
	15° stbd.		11.01	441.00	0.84	815.00	0.81	1188.00	0.80	1560.00	0.80

4. Rudder Responsiveness

For a person engaged in ship handling, the development of rate-of-turn after the rudder has turned is one of the important elements to know about the characteristic of one's ship. The graphs of Fig. 4.1.5 illustrate the development of rate-of-turn of each ship listed in Table 4.1.1 at 15° starboard helm and 35° port helm under the conditions of $h/d=\infty$ and $h/d=1.3$.

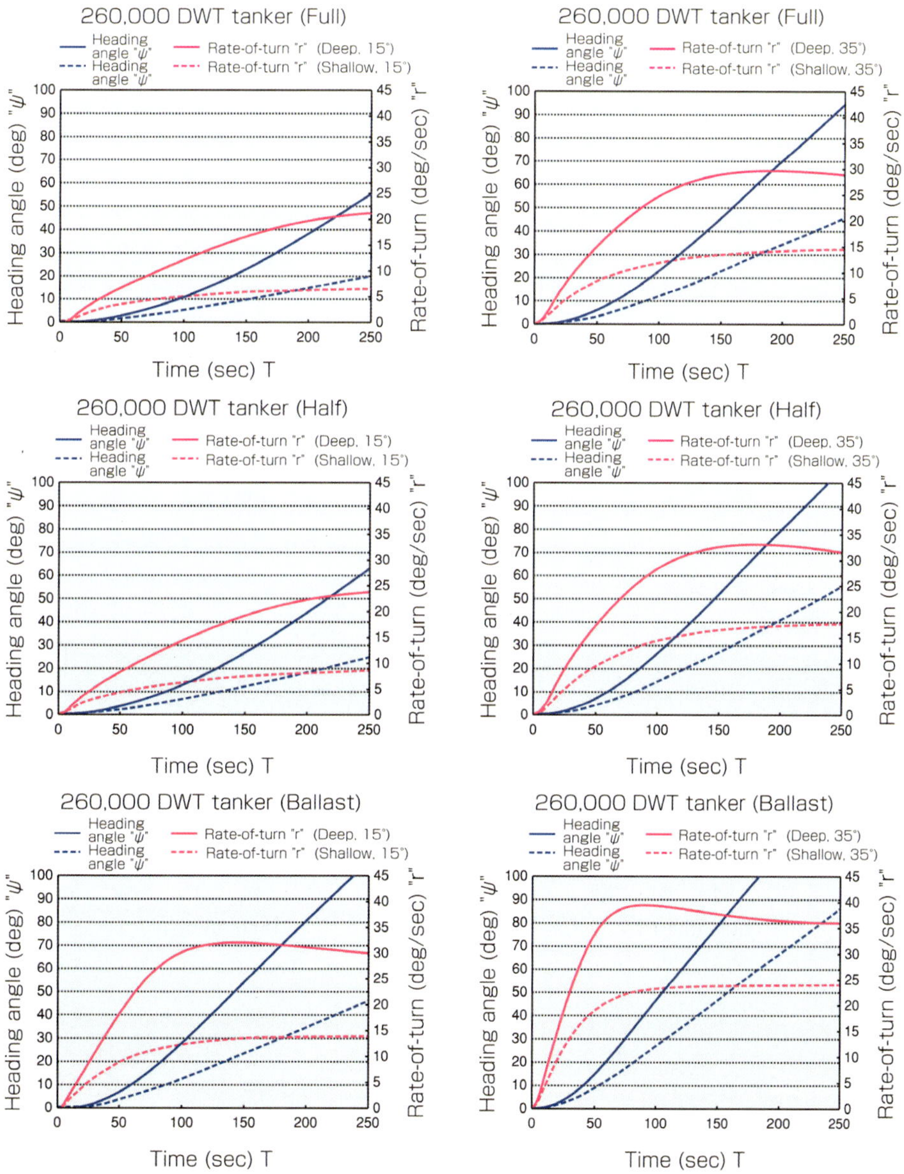

10,000 DWT tanker (Full)

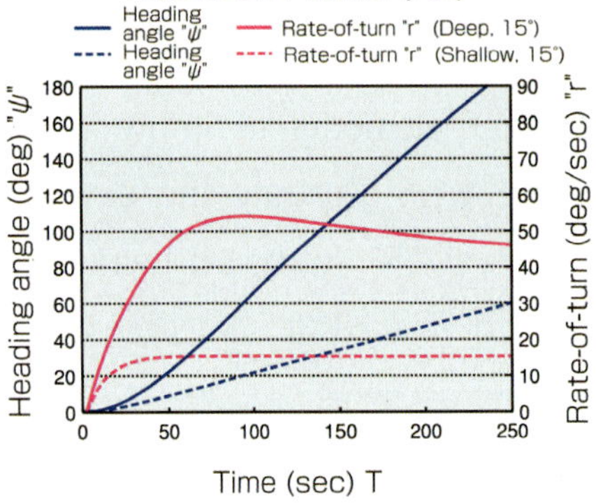

10,000 DWT tanker (Full)

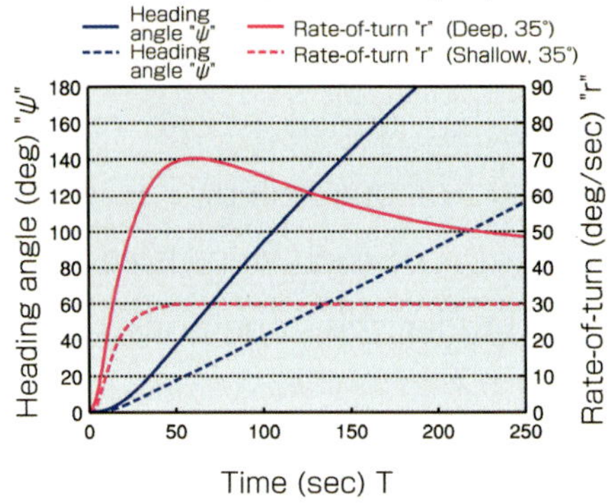

127,000 m³ LNG carrier (Full)

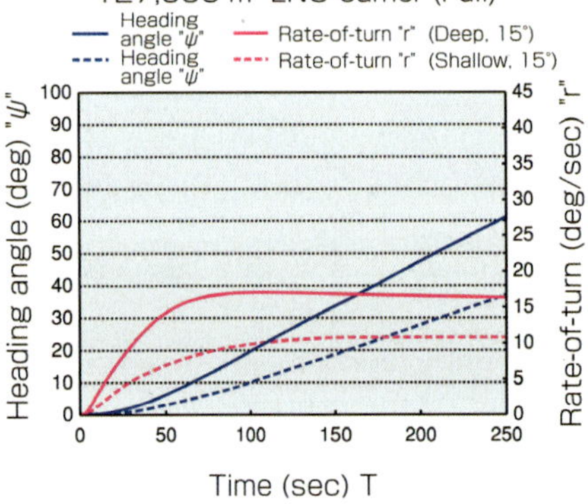

127,000 m³ LNG carrier (Full)

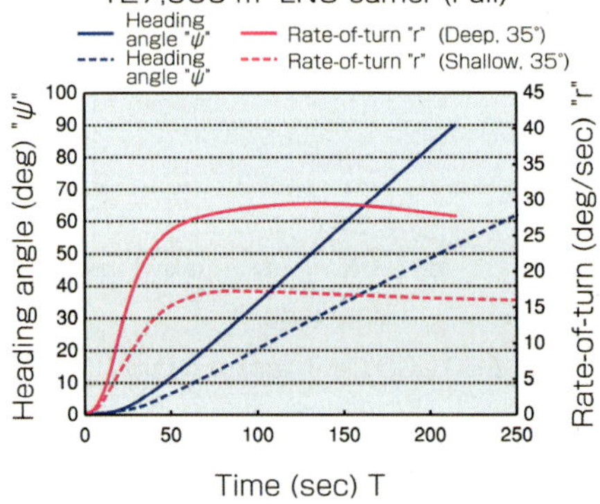

6,000 TEU container ship (Full)

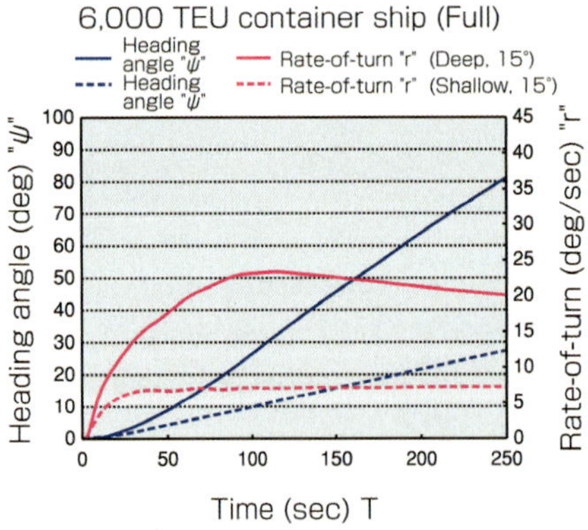

6,000 TEU container ship (Full)

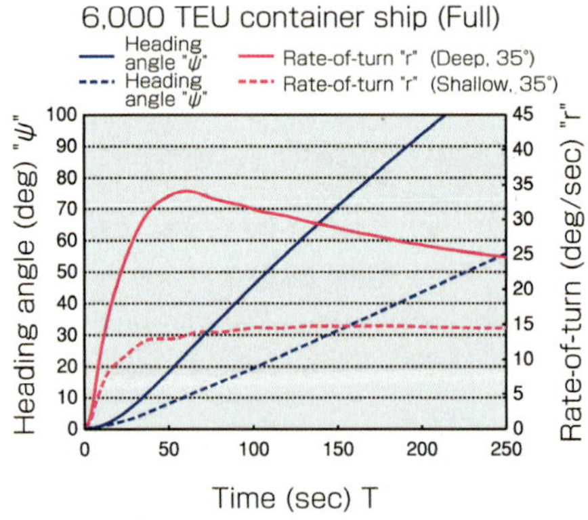

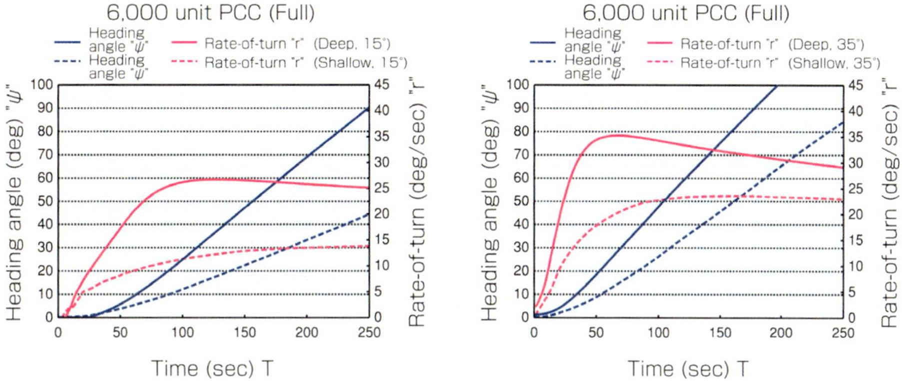

Fig. 4.1.5 Development of rate-of-turn after rudder is turned
(15° starboard helm and 35° port helm under the conditions of h/d = ∞ and h/d = 1.3)

Tables 4.1.6 and 4.1.7 show overshoot angles until the initial turning inertia can be completely offset after the helm is returned to midship and after 15° or 35° checking rudder is applied when the ship is turned at helm angle of 15° starboard or 35° port. These characteristics of rudder responsiveness to the helm are an important element to settle a ship's heading on a new intended course in alteration of course or collision avoidance maneuvers.

Note: blank columns in the first line in Table 4.1.6 where the ratio of water depth to ship draft is ∞ means that the time until the initial turning inertia can be completely offset was too long to observe.

Table 4.1.6 Overshoot Angle to midship helm (h/d = 1.3)

Ship type	Helm angle	Depth/Draft ratio	Rate-of-turn at helm midship (deg/sec)	Overshoot angle (deg)	Time required to stop turning (sec)
260,000 DWT tanker VLCC (full)	35° port	h/d = ∞			
	15° stbd.				
	35° port	h/d = 1.3	0.26	17.30	214.00
	15° stbd.		0.14	8.50	154.00
260,000 DWT tanker VLCC (half)	35° port	h/d = ∞			
	15° stbd.				
	35° port	h/d = 1.3	0.32	23.03	257.00
	15° stbd.		0.17	12.60	201.00

Ship type	Helm angle	Depth/Draft ratio	Rate-of-turn at helm midship (deg/sec)	Overshoot angle (deg)	Time required to stop turning (sec)
260,000 DWT tanker VLCC (ballast)	35° port	h/d=∞			
	15° stbd.				
	35° port	h/d=1.3	0.38	30.50	332.00
	15° stbd.		0.25	15.50	177.00
10,000 DWT tanker Tanker (full)	35° port	h/d=∞			
	15° stbd.				
	35° port	h/d=1.3	0.49	11.50	79.00
	15° stbd.		0.25	4.00	41.00
12.7000 m^3 LNG carrier LNG (full)	35° port	h/d=∞	0.38	41.70	708.00
	15° stbd.		0.25	20.50	308.00
	35° port	h/d=1.3	0.26	17.00	221.00
	15° stbd.		0.18	9.50	135.00
6,000 TEU container ship CNTN (full)	35° port	h/d=∞			
	15° stbd.				
	35° port	h/d=1.3	0.20	24.60	499.00
	15° stbd.		0.12	5.10	113.00
6,000 unit PCC PCC (full)	35° port	h/d=∞			
	15° stbd.				
	35° port	h/d=1.3	0.37	26.10	303.00
	15° stbd.		0.24	12.90	185.00

Table 4.1.7 Overshoot angle for checking rudder (h/d=∞)

			Overshoot angle		Overshoot angle		Overshoot angle	
Ship type	Checking helm angle	Depth/ Draft ratio	ψ(deg)	at (r) (deg/min)	ψ(deg)	at (r) (deg/min)	ψ(deg)	at (r) (deg/min)
260,000 DWT tanker VLCC (full)	15°	h/d=∞	10.49	14.07	−19.01	−19.68	17.97	19.71
	35°		18.39	27.49	−17.64	−27.53	15.68	25.04
260,000 DWT tanker VLCC (half)	15°		9.63	15.07	−18.20	−21.10	17.64	21.40
	35°		19.29	30.25	−19.41	−30.93	17.57	28.77
260,000 DWT tanker VLCC (ballast)	15°		18.20	25.18	−24.49	−31.36	21.07	29.46
	35°		18.47	39.13	−15.16	−36.00	14.70	35.63
10,000 DWT tanker Tanker (full)	15°		12.79	41.38	−19.75	−54.39	18.69	50.56
	35°		21.99	64.36	−20.50	−60.92	16.71	51.18

			Overshoot angle		Overshoot angle		Overshoot angle	
Ship type	Checking helm angle	Depth/ Draft ratio	ψ(deg)	at (r) (deg/min)	ψ(deg)	at (r) (deg/min)	ψ(deg)	at (r) (deg/min)
12,7000 m^3 LNG carrier LNG (full)	15°	h/d=∞	5.58	16.72	−5.78	−16.68	5.55	16.10
	35°		11.8	28.71	−12.07	−26.05	11.27	23.58
6,000 TEU container ship CNTN (full)	15°		5.31	20.66	−8.47	−22.02	6.25	18.77
	35°		8.85	31.58	−8.89	−24.34	6.85	19.42
6,000 unit PCC PCC (full)	15°		8.58	23.94	−12.11	−28.28	6.85	25.58
	35°		13.24	31.98	−2.30	−30.15	10.78	27.10

Three samples of data on overshoot angle in Table 4.1.7 are derived from the result of zigzag maneuver test as shown in Fig. 4.1.2.

Ⅱ Ship Handling Characteristics

Chapter 1 List of Trial Ships

In ports and harbors where water area for maneuvering is quite limited, ships are required to move at a limited speed so that they can be stopped by the smallest braking force possible and at the same time, give persons engaged in ship handling the time to take appropriate counter-measures against possible risks based on correct decisions. On the other hand, movement at slow speed means that the steerability decreases making it more difficult to keep a ship's posture and position, the slower the ship moves, the greater the effect of external disturbances grows and the harder ship handling becomes.

To handle a ship correctly based on precise data and objective judgment under such conditions, it is indispensable for each person engaged in ship handling to be conversant with the maneuvering characteristics of their own ship beforehand so that they could make the best decisions for maneuvering. The database of Ship Handling Characteristics of this volume has been created purposing to provide persons engaged in ship handling with database for practical and technical information for various handling of ships inside ports and harbors such as driving, stopping and crash astern inertia, the effect of external disturbances, shallow water effect, critical speed to maintain steerageway, stopping distance and deflection of heading under astern engine, turning ability by boosting, lateral shifting maneuver and stern kick.

The data was collected by means of simulation tests of the ship types for which specifications and telegraph speeds are listed in Table 4.2.1 and Table 4.2.2 respectively by fast time ship maneuvering simulation system of Japan Marine Science Inc. Selected trial ships were similar to those used for simulator tests in types and sizes.

Table 4.2.1 Particulars of test ships

Ship type	Condition	Lpp	B	df	da	C_b
300,000 DWT tanker VLCC (full)	Fully loaded	324	60.0	20.5	20.5	0.86
10,000 DWT bulker Bulker (full)	Fully loaded	107	19.4	8.0	8.0	0.75
135,000 m^3 LNG carrier LNG (full)	Fully loaded	276	46.0	10.8	10.8	0.72
6,000 TEU container ship CNTN (full)	Fully loaded	303	42.8	13.6	13.6	0.63
50,000 GT large cruise ship	Fully loaded	205	29.6	7.3	7.3	0.62
4,500 unit PCC PCC (full)	Fully loaded	170	32.3	8.8	8.8	0.58

Table 4.2.2 Telegraph speed

Ship type	Condition	Speed (kt)				
		N/Full	S/B Full	Half	Slow	D. Slow
300,000 DWT tanker VLCC (full)	Fully loaded	15.2	10.5	7.1	4.2	2.7
10,000 DWT bulker Bulker (full)	Fully loaded	13.6	9.6	7.4	5.1	3.6
135,000 m^3 LNG carrier LNG (full)	Fully loaded	18.9	12.0	9.0	6.3	4.3
6,000 TEU container ship CNTN (full)	Fully loaded	24.7	18.1	14.0	9.7	6.8
50,000 GT large cruise ship	Fully loaded	22.2	11.0	9.2	6.4	2.8
4,500 unit PCC PCC (full)	Fully loaded	19.9	12.2	10.1	8.0	5.9

Chapter 2 Inertial Characteristics

1. Driving Inertia and Stopping Inertia

In ports and harbors, persons engaged in ship handling are required to adjust their speed depending on each ship's inertial characteristics based on the knowledge of how it behaves when the ship speed increases or decreases. The following data is presented as a reference for those persons. Fig. 4.2.1 shows the results of driving and stopping inertia tests of the 6 typical types of ships listed in Table 4.2.1 under the conditions of the ratio of water depth to ship draft $h/d=\infty$ and $h/d=1.3$. In driving inertia tests, chronological change of speed when the engine is started S/B Full Ahead from the state that the ship is lying dead in the water with its engine stopped is measured. While in stopping inertia tests, chronological change of speed and distance run from the time the engine of a running ship at S/B Full Ahead speed is stopped to the time until it comes to a complete stop in water are measured. Thin and thick lines in the graphs represent change of speed and distance run respectively.

300,000 DWT tanker (Full)_deep
Driving inertia characteristics (Stop Eng.→S/B Full Ahead)

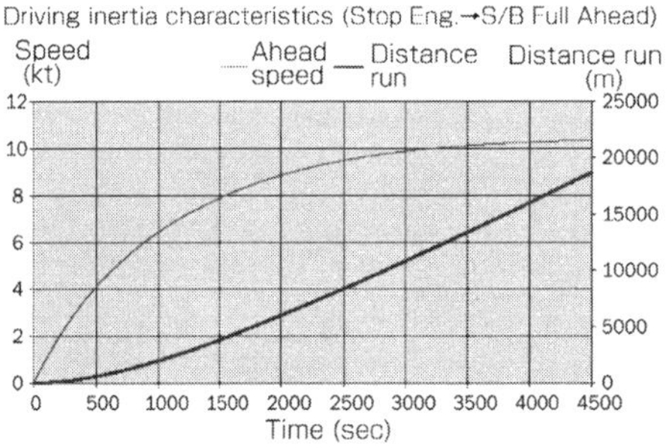

300,000 DWT tanker (Full)_shallow
Driving inertia characteristics (Stop Eng.→S/B Full Ahead)

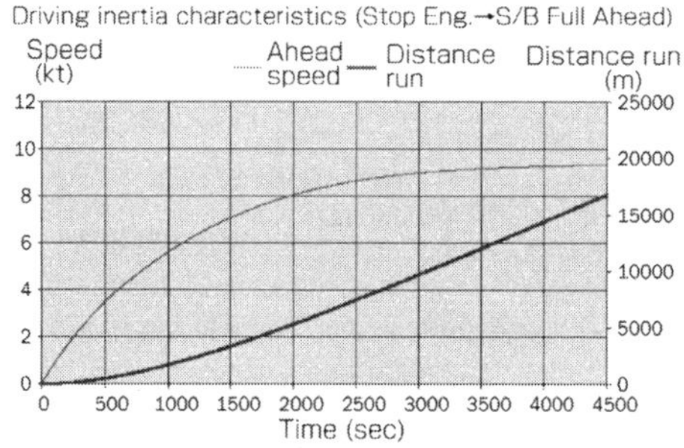

300,000 DWT tanker (Full)_deep
Stopping inertia characteristics (S/B Full Ahead→Stop Eng.)

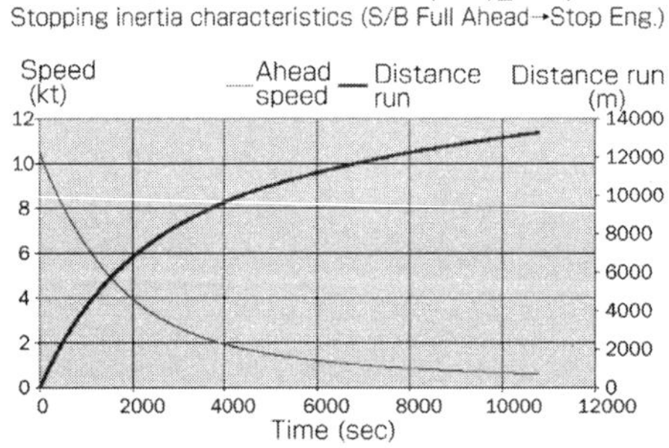

300,000 DWT tanker (Full)_shallow
Stopping inertia characteristics (S/B Full Ahead→Stop Eng.)

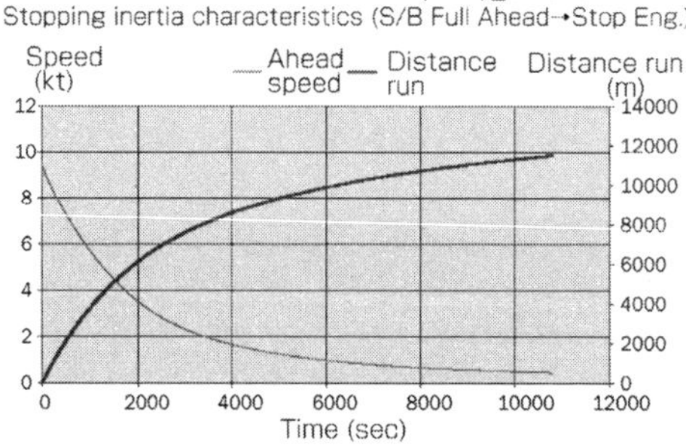

10,000 DWT bulker (Full)_deep
Driving inertia characteristics (Stop Eng.→S/B Full Ahead)

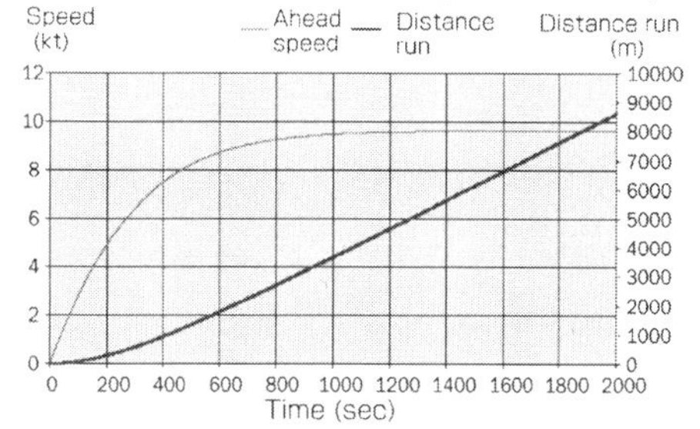

10,000 DWT bulker (Full)_shallow
Driving inertia characteristics (Stop Eng.→S/B Full Ahead)

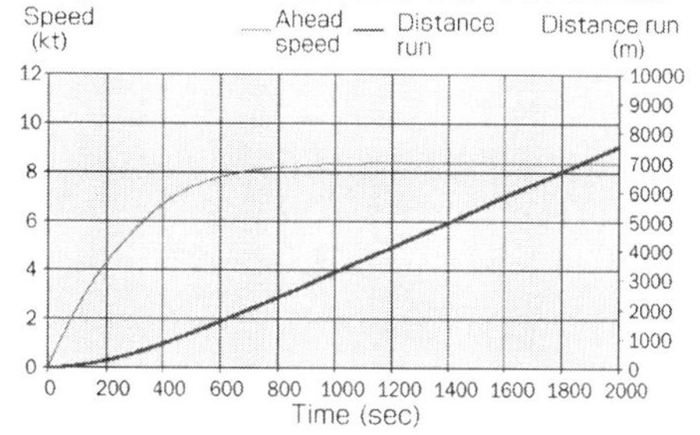

10,000 DWT bulker (Full)_deep
Stopping inertia characteristics (S/B Full Ahead→Stop Eng.)

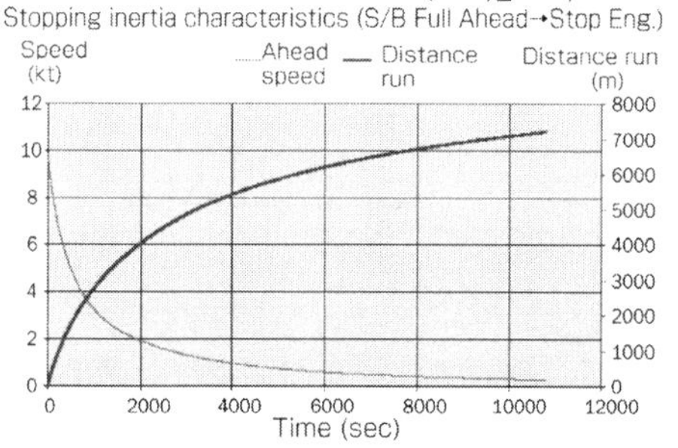

10,000 DWT bulker (Full)_shallow
Stopping inertia characteristics (S/B Full Ahead→Stop Eng.)

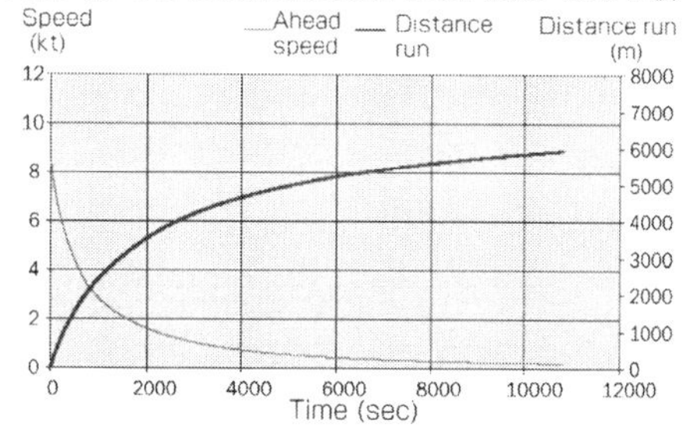

135,000 m^3 LNG carrier (Full)_deep
Driving inertia characteristics (Stop Eng.→S/B Full Ahead)

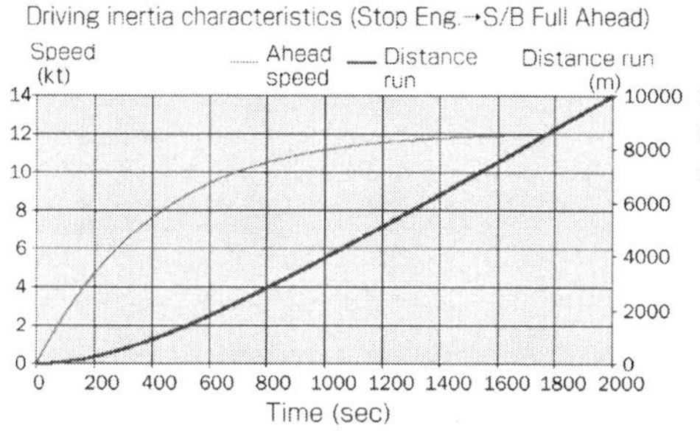

135,000 m^3 LNG carrier (Full)_shallow
Driving inertia characteristics (Stop Eng.→S/B Full Ahead)

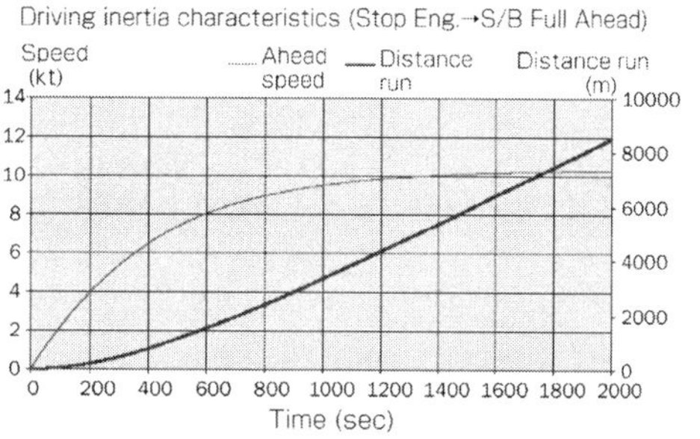

135,000 m^3 LNG carrier (Full)_deep
Stopping inertia characteristics (S/B Full Ahead→Stop Eng.)

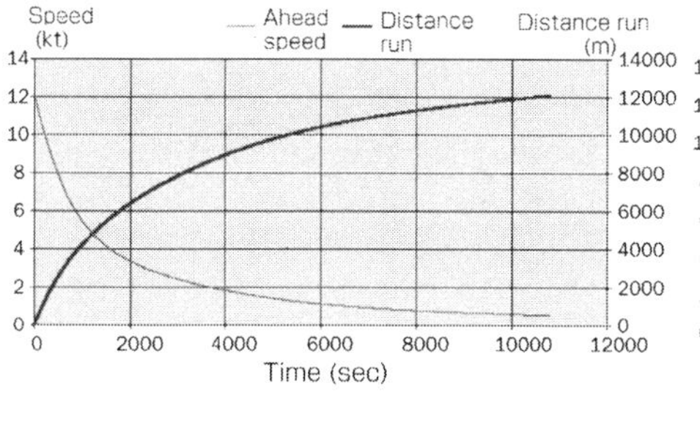

135,000 m^3 LNG carrier (Full)_shallow
Stopping inertia characteristics (S/B Full Ahead→Stop Eng.)

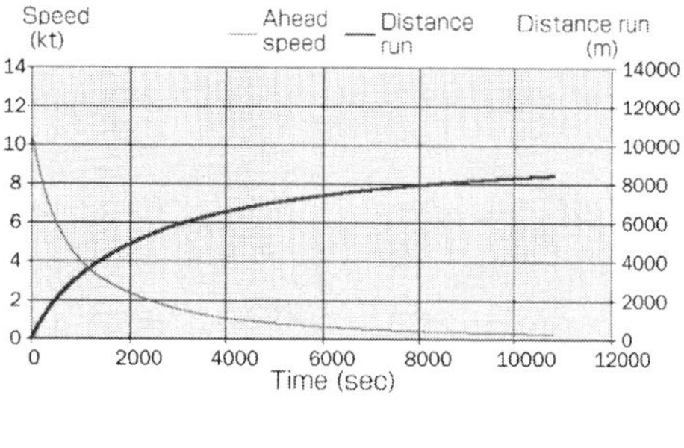

6,000 TEU container ship (Full)_deep

Driving inertia characteristics (Stop Eng.→S/B Full Ahead)

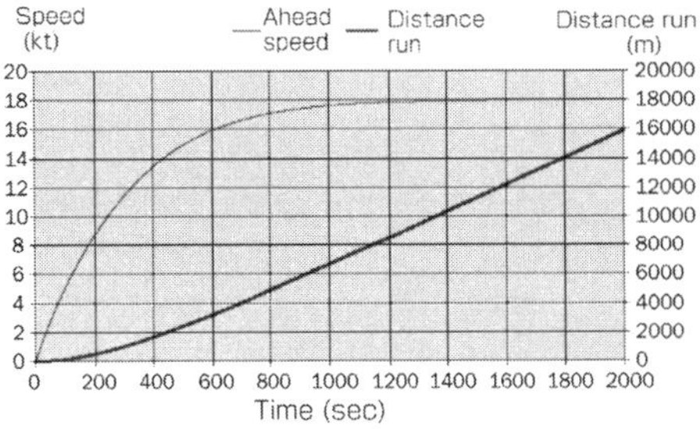

6,000 TEU container ship (Full)_shallow

Driving inertia characteristics (Stop Eng. → S/B Full Ahead)

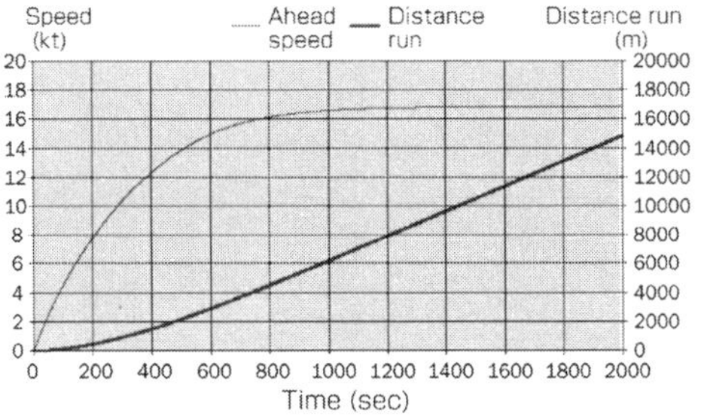

6,000 TEU container ship (Full)_deep

Stopping inertia characteristics (S/B Full Ahead→Stop Eng.)

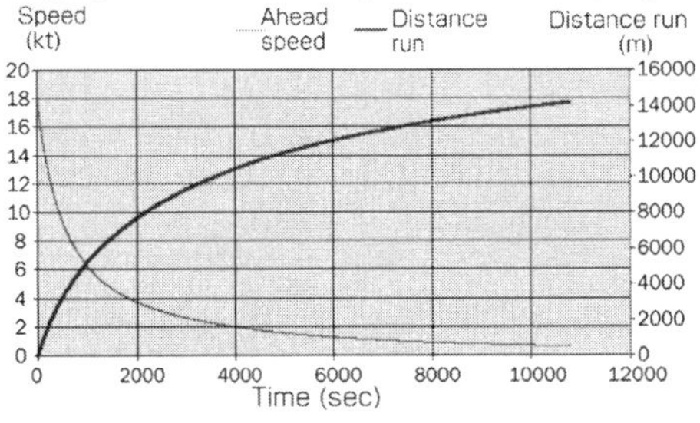

6,000 TEU container ship (Full)_shallow

Stopping inertia characteristics (S/B Full Ahead→Stop Eng.)

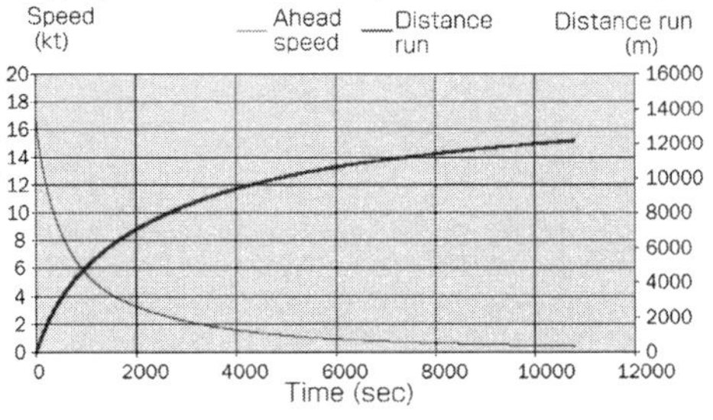

50,000 G/T large cruise ship (Full)_deep

Driving inertia characteristics (Stop Eng.→S/B Full Ahead)

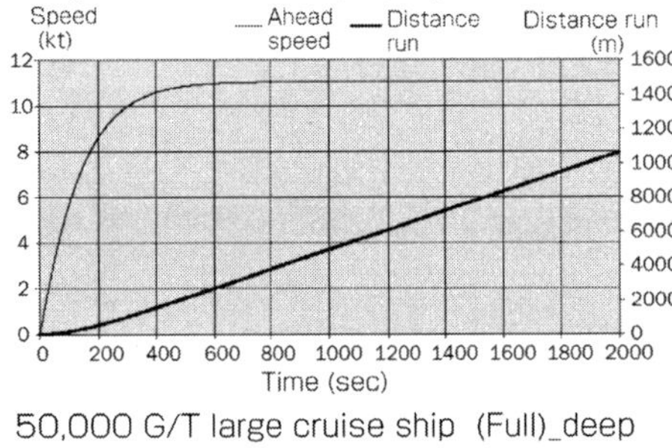

50,000 G/T large cruise ship (Full)_shallow

Driving inertia characteristics (Stop Eng.→S/B Full Ahead)

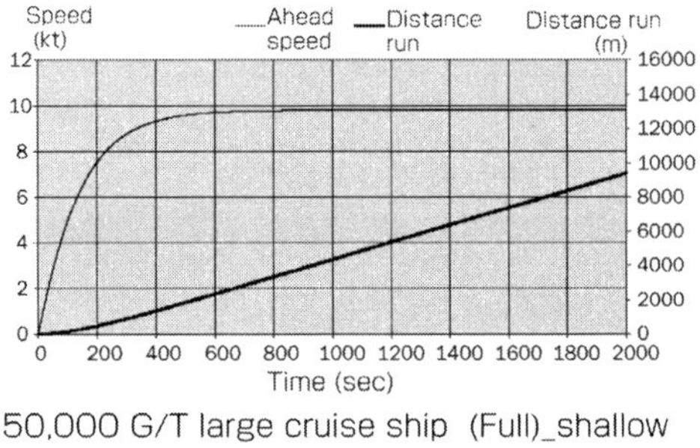

50,000 G/T large cruise ship (Full)_deep

Stopping inertia characteristics (S/B Full Ahead→Stop Eng.)

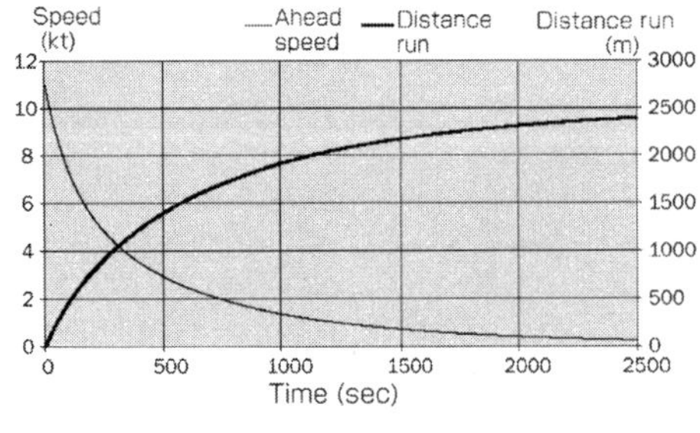

50,000 G/T large cruise ship (Full)_shallow

Stopping inertia characteristics (S/B Full Ahead→Stop Eng.)

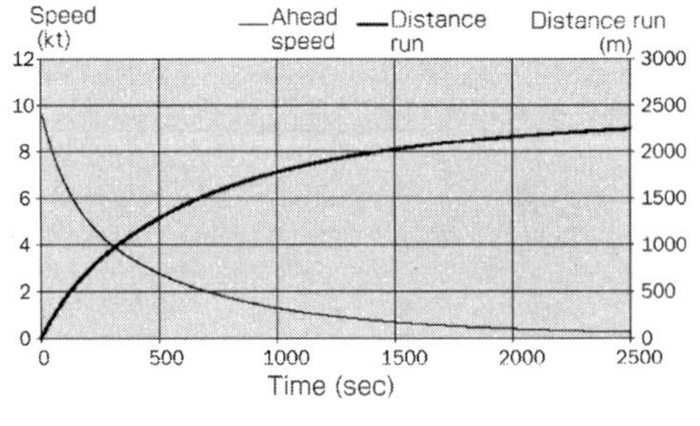

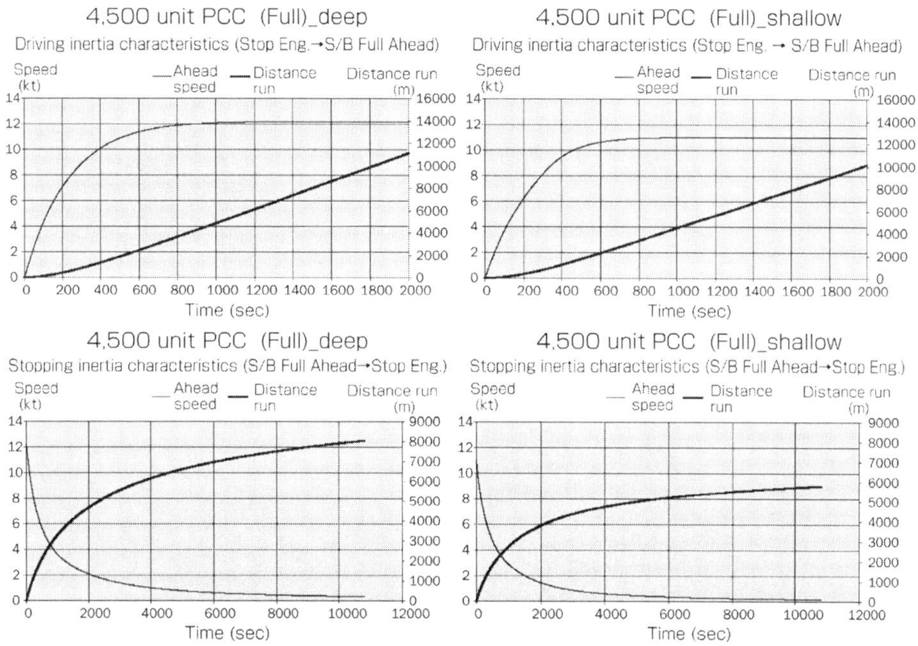

Fig. 4.2.1 Driving inertia and stopping inertia (h/d = ∞, h/d = 1.3)

2. Emergency Stopping Inertia

Persons engaged in ship handling are also required to know beforehand how long it will take to stop their ship and how far it will run when the ship is running at a certain speed and is stopped in an emergency. Fig. 4.2.2 shows the data for emergency stopping inertia of 6 typical types of ships listed in Table 4.2.1 when their engines are reversed to Full Astern from S/B Full Ahead speed under the conditions of the ratio of water depth to ship draft $h/d = \infty$ and $h/d = 1.3$. In addition to the inertia at this moment, deflection of the heading by accompanying astern engine is also an important element.

In the figure, changes of heading angle and rate-of-turn until the ship comes to a complete stop in the water are given together with the ship's track. In the upper graph, thin and thick lines respectively represent change of speed and distance run, while in the middle graph, they respectively represent the ship's heading and rate-of-turn. And thick lines in the graphs at the bottom represent the ship's track.

300,000 DWT tanker (Full)_deep
Emerg'y stop'g inertia characteristics (S/B Full Ahead→Full Ast'n)

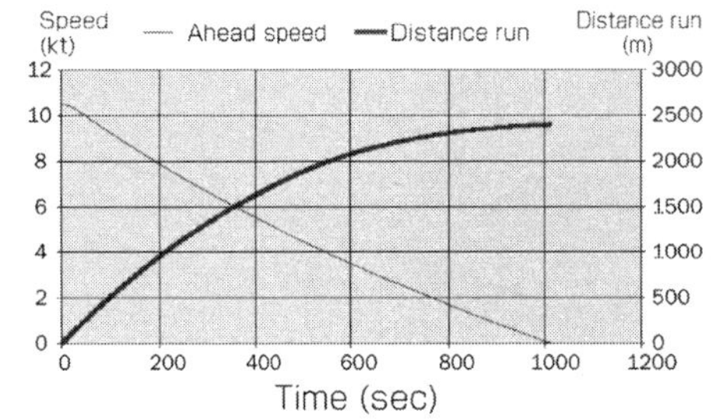

300,000 DWT tanker (Full)_shallow
Emerg'y stop'g inertia characteristics (S/B Full Ahead→Full Ast'n)

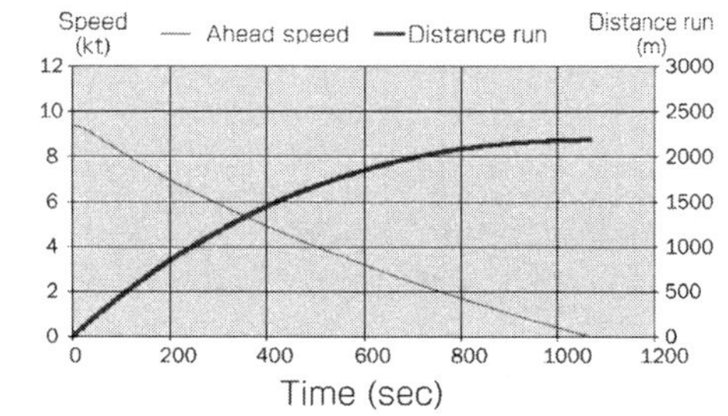

300,000 DWT tanker (Full)_deep
Emerg'y stop'g inertia characteristics (S/B Full Ahead→Full Ast'n)

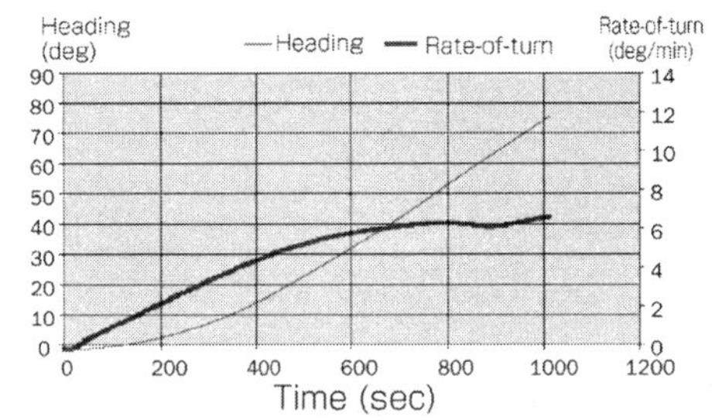

300,000 DWT tanker (Full)_shallow
Emerg'y stop'g inertia characteristics (S/B Full Ahead→Full Ast'n)

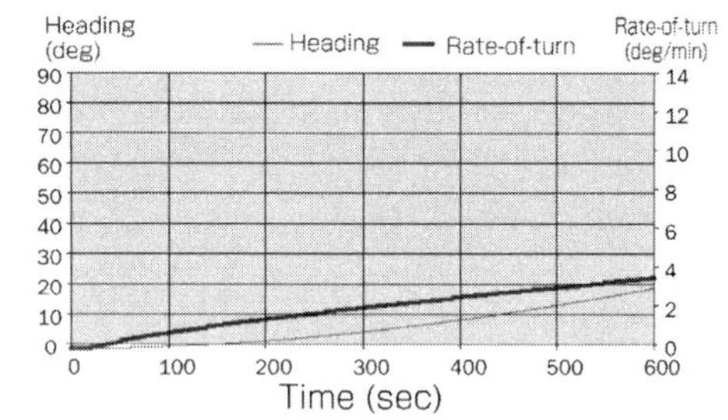

300,000 DWT tanker (Full)_deep
Emerg'y stop'g inertia characteristics (S/B Full Ahead→Full Ast'n)

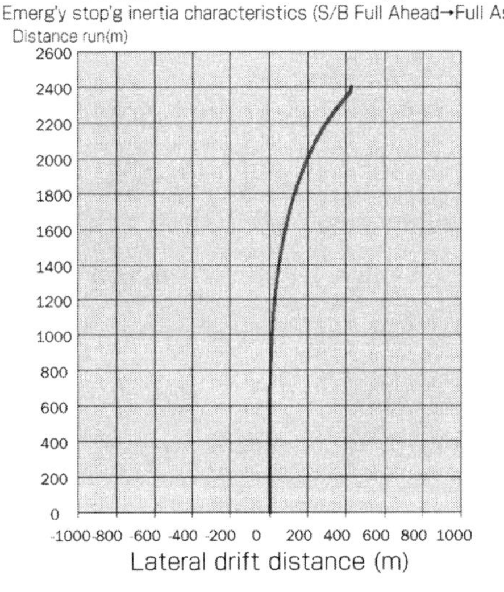

300,000 DWT tanker (Full)_shallow
Emerg'y stop'g inertia characteristics (S/B Full Ahead→Full Ast'n)

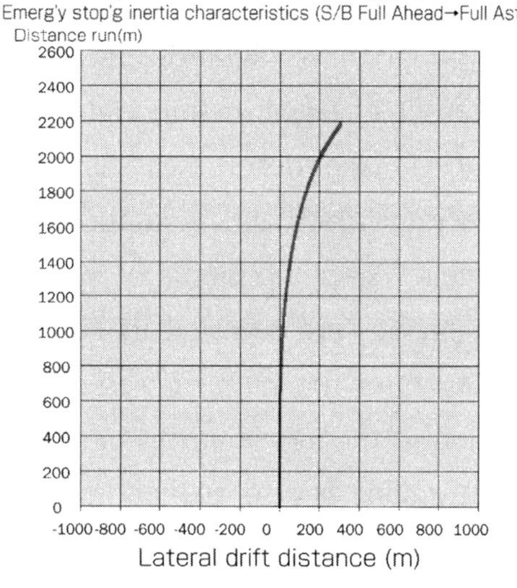

10,000 DWT tanker (Full)_deep
Emerg'y stop'g inertia characteristics (S/B Full Ahead→Full Ast'n)

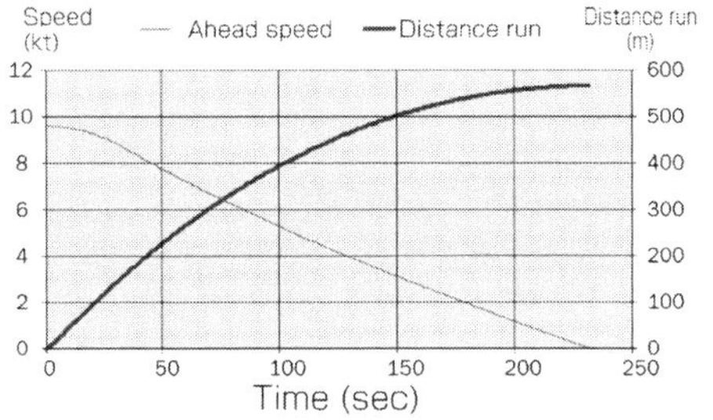

10,000 DWT tanker (Full)_shallow
Emerg'y stop'g inertia characteristics (S/B Full Ahead→Full Ast'n)

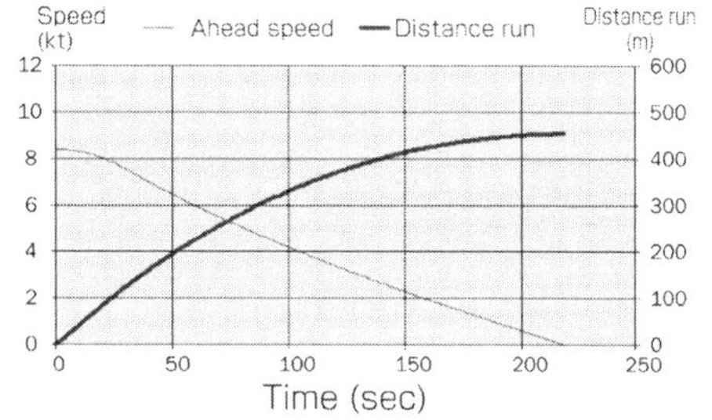

10,000 DWT tanker (Full)_deep
Emerg'y stop'g inertia characteristics (S/B Full Ahead→Full Ast'n)

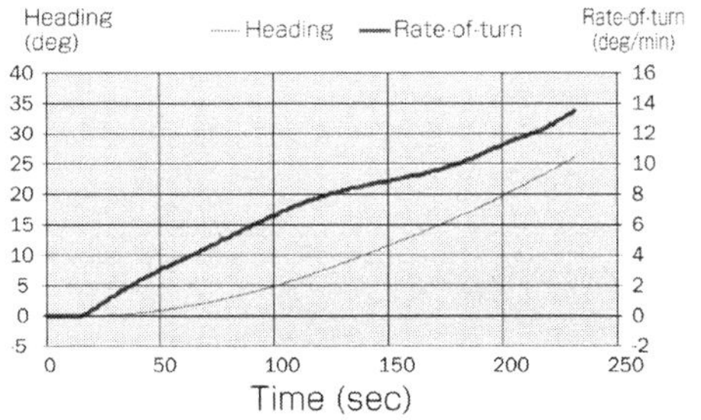

10,000 DWT tanker (Full)_shallow
Emerg'y stop'g inertia characteristics (S/B Full Ahead→Full Ast'n)

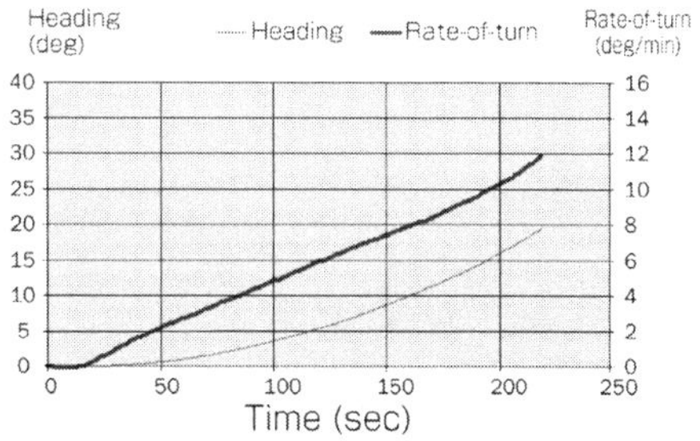

10,000 DWT tanker (Full)_deep
Emerg'y stop'g inertia characteristics (S/B Full Ahead→Full Ast'n)

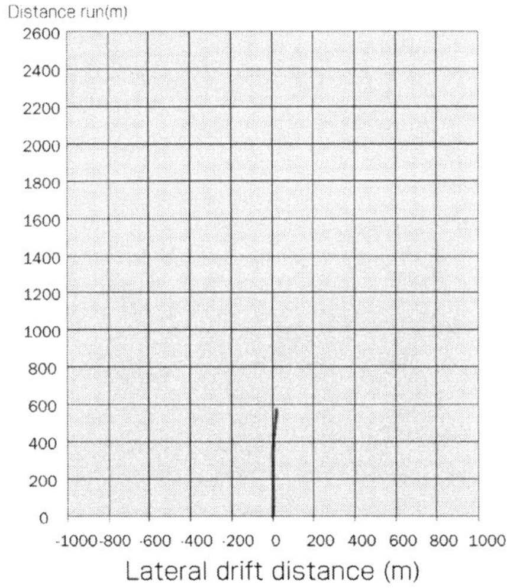

10,000 DWT tanker (Full)_shallow
Emerg'y stop'g inertia characteristics (S/B Full Ahead→Full Ast'n)

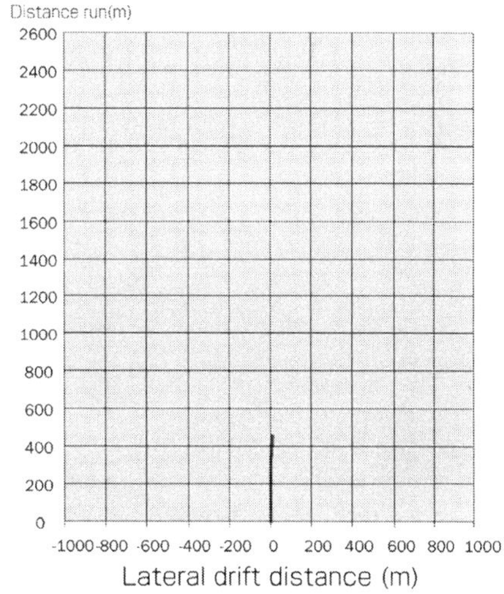

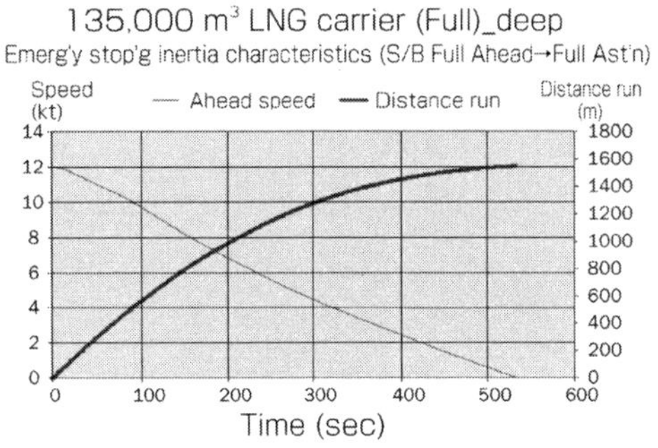
135,000 m³ LNG carrier (Full)_deep
Emerg'y stop'g inertia characteristics (S/B Full Ahead→Full Ast'n)
Speed (kt)
Ahead speed
Distance run
Distance run (m)
Time (sec)

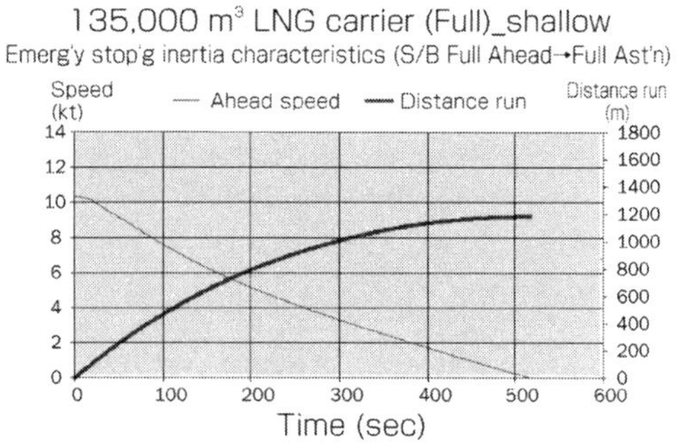
135,000 m³ LNG carrier (Full)_shallow
Emerg'y stop'g inertia characteristics (S/B Full Ahead→Full Ast'n)
Speed (kt)
Ahead speed
Distance run
Distance run (m)
Time (sec)

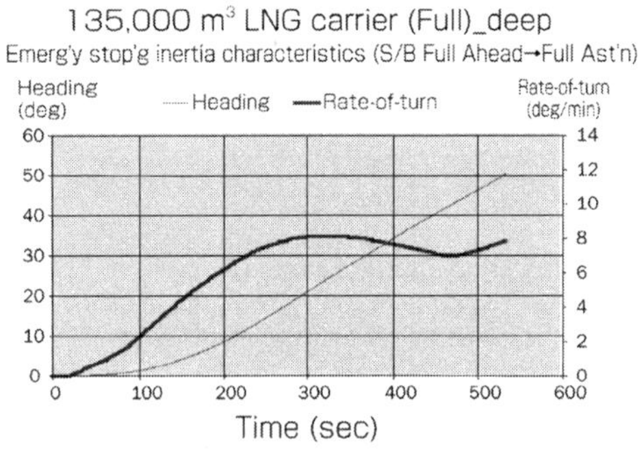
135,000 m³ LNG carrier (Full)_deep
Emerg'y stop'g inertia characteristics (S/B Full Ahead→Full Ast'n)
Heading (deg)
Heading
Rate-of-turn
Rate-of-turn (deg/min)
Time (sec)

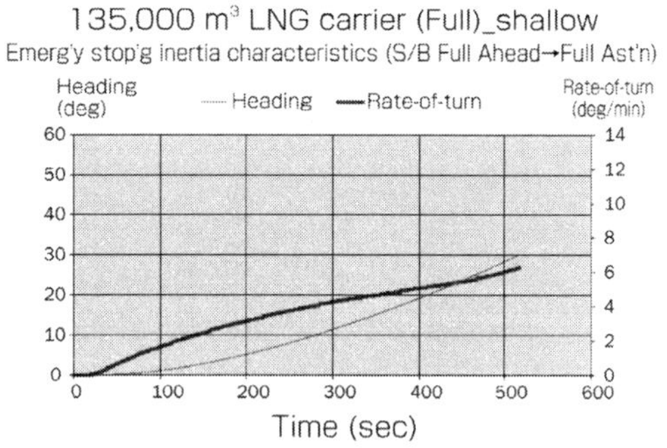
135,000 m³ LNG carrier (Full)_shallow
Emerg'y stop'g inertia characteristics (S/B Full Ahead→Full Ast'n)
Heading (deg)
Heading
Rate-of-turn
Rate-of-turn (deg/min)
Time (sec)

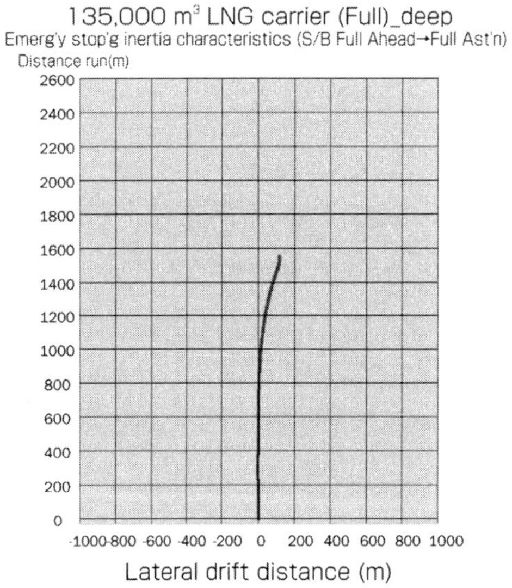
135,000 m³ LNG carrier (Full)_deep
Emerg'y stop'g inertia characteristics (S/B Full Ahead→Full Ast'n)
Distance run(m)
Lateral drift distance (m)

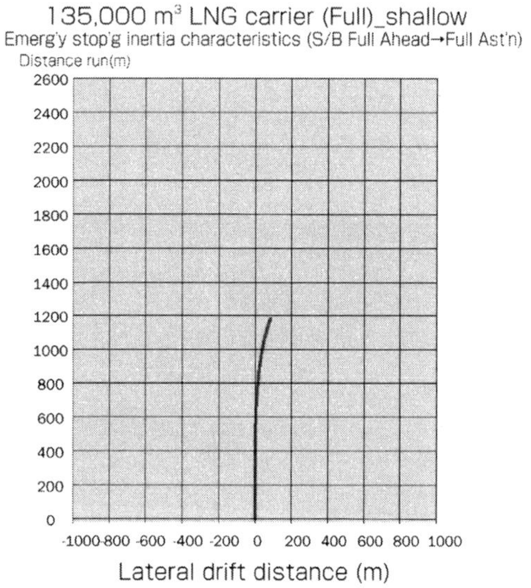
135,000 m³ LNG carrier (Full)_shallow
Emerg'y stop'g inertia characteristics (S/B Full Ahead→Full Ast'n)
Distance run(m)
Lateral drift distance (m)

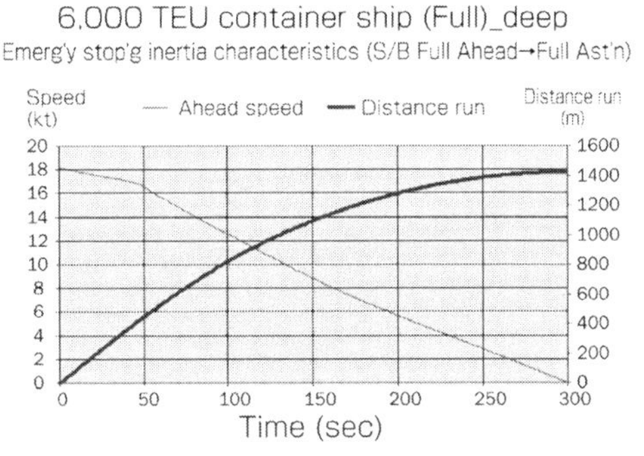
6,000 TEU container ship (Full)_deep
Emerg'y stop'g inertia characteristics (S/B Full Ahead→Full Ast'n)
Speed (kt)
Ahead speed
Distance run
Distance run (m)
Time (sec)

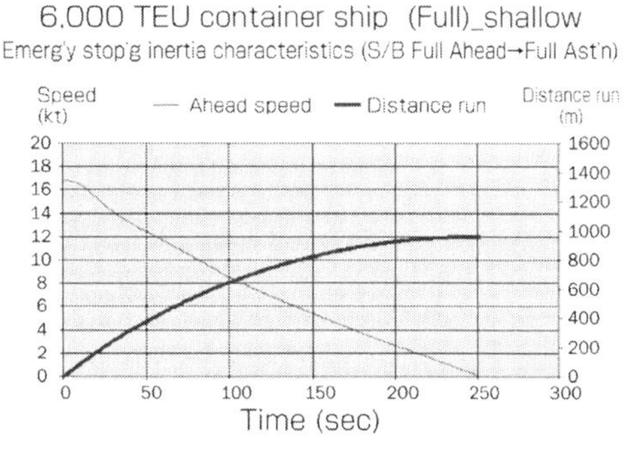
6,000 TEU container ship (Full)_shallow
Emerg'y stop'g inertia characteristics (S/B Full Ahead→Full Ast'n)
Speed (kt)
Ahead speed
Distance run
Distance run (m)
Time (sec)

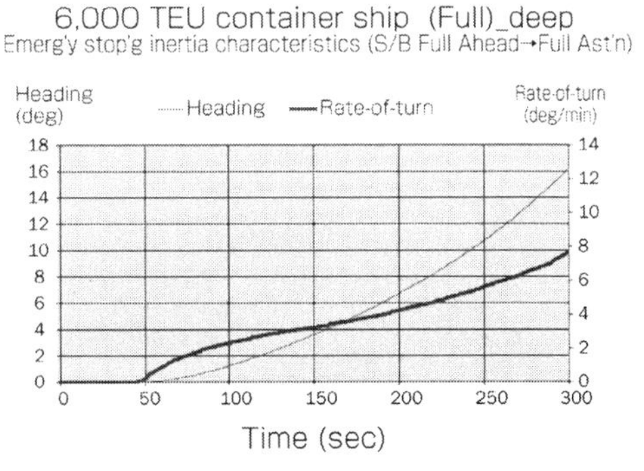
6,000 TEU container ship (Full)_deep
Emerg'y stop'g inertia characteristics (S/B Full Ahead→Full Ast'n)
Heading (deg)
Heading
Rate-of-turn
Rate-of-turn (deg/min)
Time (sec)

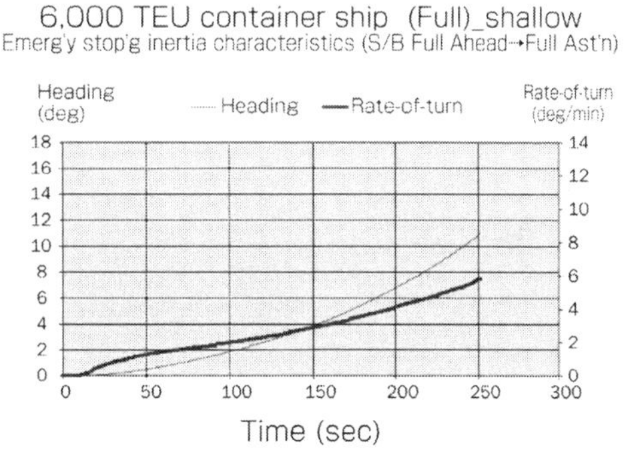
6,000 TEU container ship (Full)_shallow
Emerg'y stop'g inertia characteristics (S/B Full Ahead→Full Ast'n)
Heading (deg)
Heading
Rate-of-turn
Rate-of-turn (deg/min)
Time (sec)

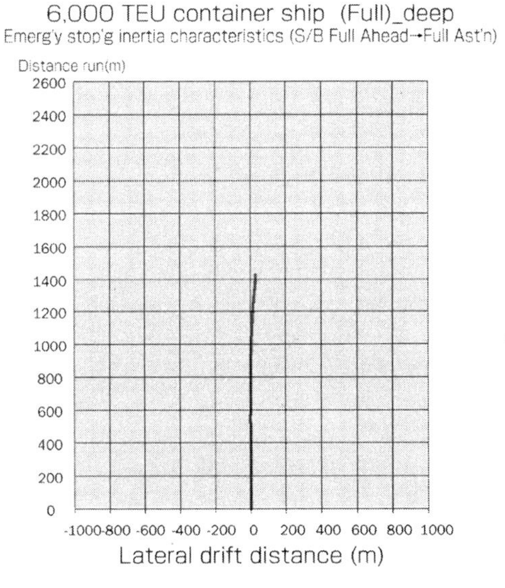
6,000 TEU container ship (Full)_deep
Emerg'y stop'g inertia characteristics (S/B Full Ahead→Full Ast'n)
Distance run(m)
Lateral drift distance (m)

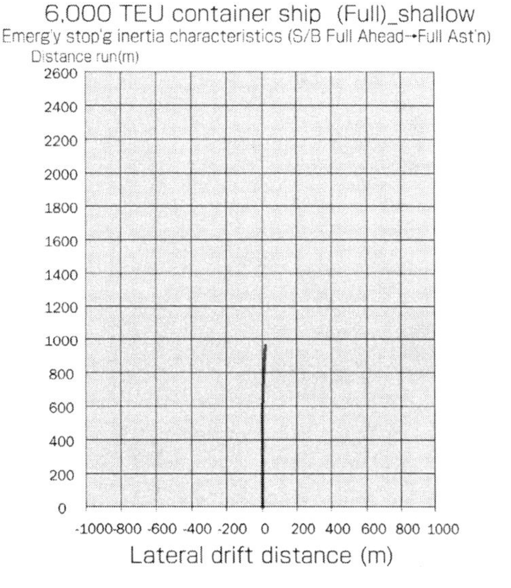
6,000 TEU container ship (Full)_shallow
Emerg'y stop'g inertia characteristics (S/B Full Ahead→Full Ast'n)
Distance run(m)
Lateral drift distance (m)

50,000 GT large cruise ship (Full)_deep

Emerg'y stop'g inertia characteristics (S/B Full Ahead→Full Ast'n)

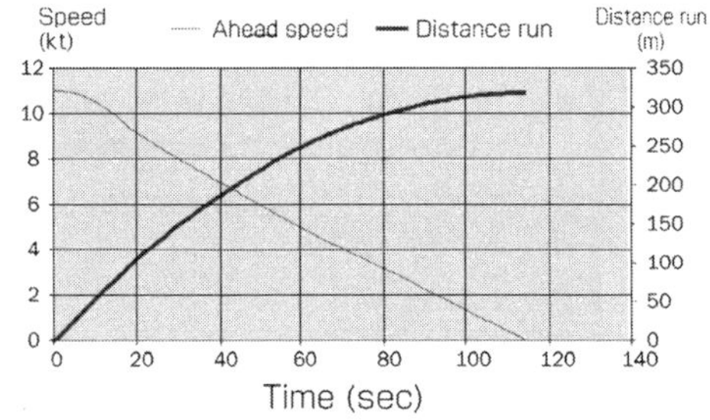

50,000 GT large cruise ship (Full)_shallow

Emerg'y stop'g inertia characteristics (S/B Full Ahead→Full Ast'n)

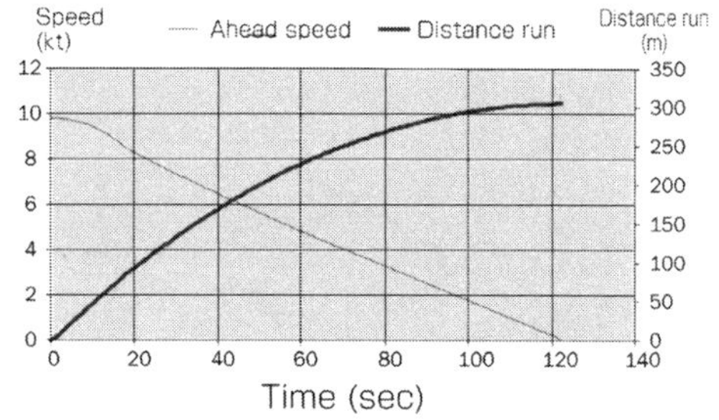

50,000 GT large cruise ship (Full)_deep

Emerg'y stop'g inertia characteristics (S/B Full Ahead→Full Ast'n)

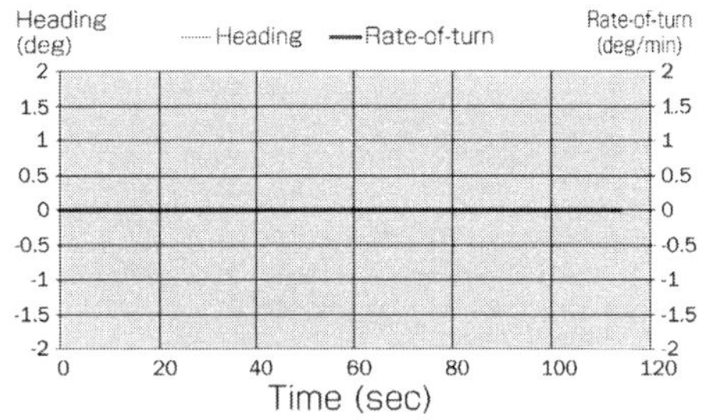

50,000 GT large cruise ship (Full)_shallow

Emerg'y stop'g inertia characteristics (S/B Full Ahead→Full Ast'n)

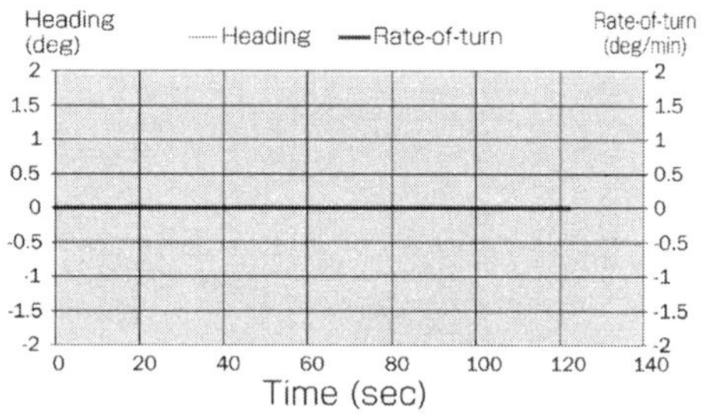

50,000 GT large cruise ship (Full)_deep

Emerg'y stop'g inertia characteristics (S/B Full Ahead→Full Ast'n)

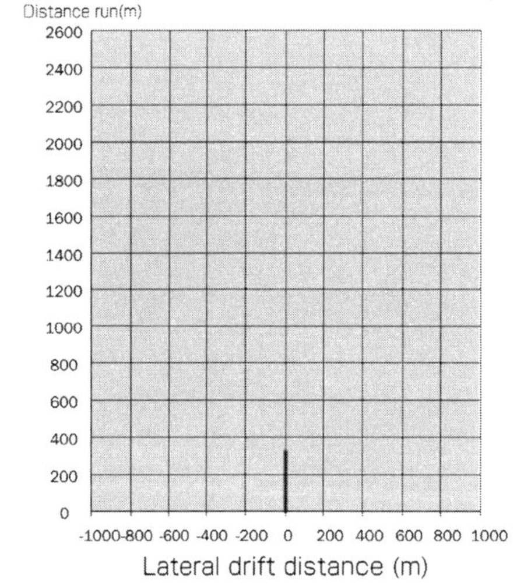

50,000 GT large cruise ship (Full)_shallow

Emerg'y stop'g inertia characteristics (S/B Full Ahead→Full Ast'n)

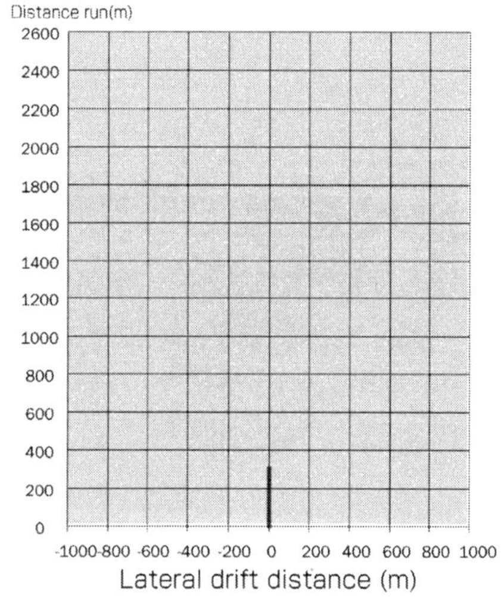

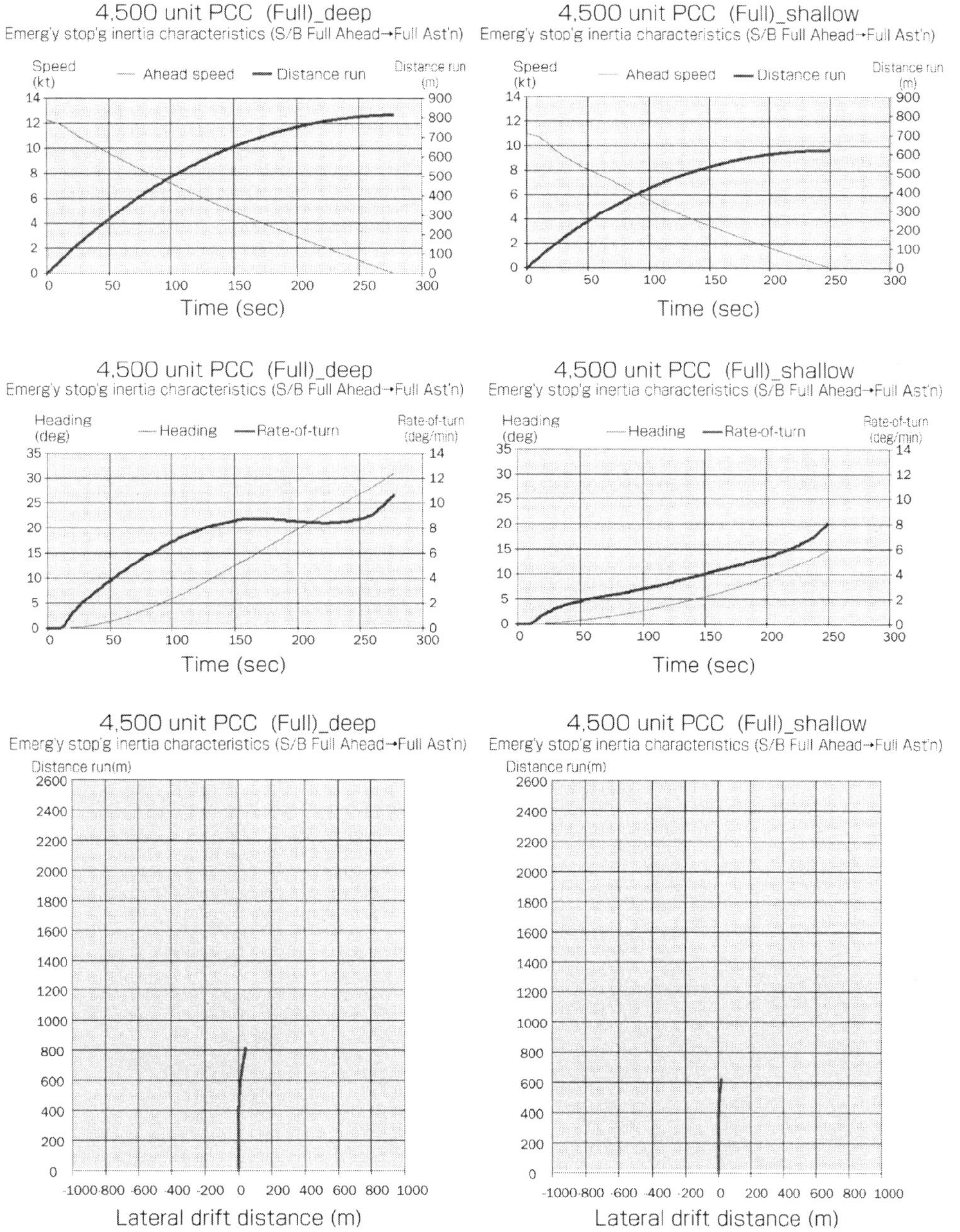

Fig. 4.2.2 Emergency stopping inertia (Initial speed: S/B Full ahead, h/d = ∞, h/d = 1.3)

Chapter 3 Crash Astern Maneuvering

Fig. 4.2.3 shows the data of crash astern stopping inertia of 6 typical ship types listed in Table 4.2.1; e.g., how long it will take to stop one's ship and how far it will run when their engines are reversed to Full Astern from Navigation Full Ahead speed under the conditions of the ratio of water depth to ship draft $h/d=\infty$. The data also includes changes of heading angle, rate-of-turn and ship's track until the ship comes to a complete stop in the water. In the upper graph, thin and thick lines represent change of speed and distance run respectively, in the middle graph, they represent the ship's heading and rate-of-turn respectively, and thick lines in the graphs at the bottom represents ship's track.

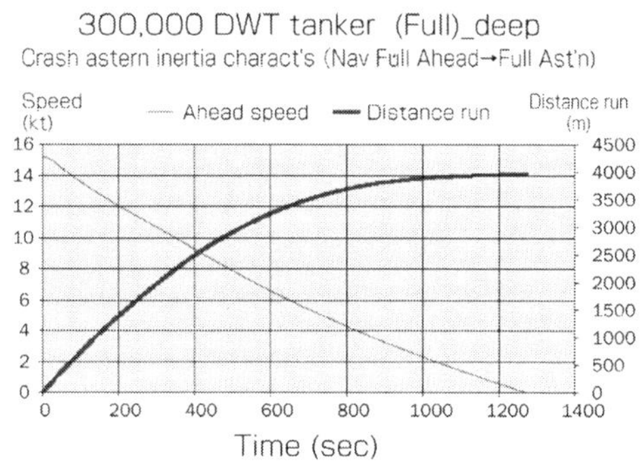
300,000 DWT tanker (Full)_deep
Crash astern inertia charact's (Nav Full Ahead→Full Ast'n)
Speed (kt)
Ahead speed
Distance run
Distance run (m)
Time (sec)

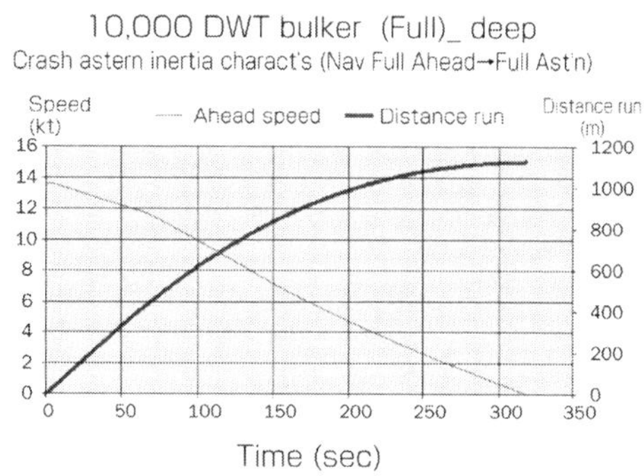
10,000 DWT bulker (Full)_ deep
Crash astern inertia charact's (Nav Full Ahead→Full Ast'n)
Speed (kt)
Ahead speed
Distance run
Distance run (m)
Time (sec)

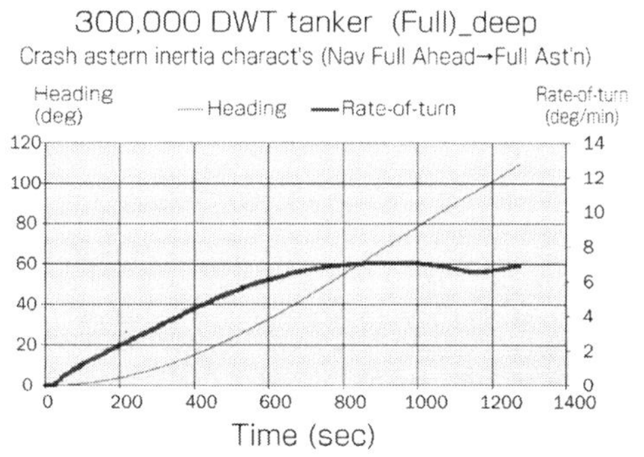
300,000 DWT tanker (Full)_deep
Crash astern inertia charact's (Nav Full Ahead→Full Ast'n)
Heading (deg)
Heading
Rate-of-turn
Rate-of-turn (deg/min)
Time (sec)

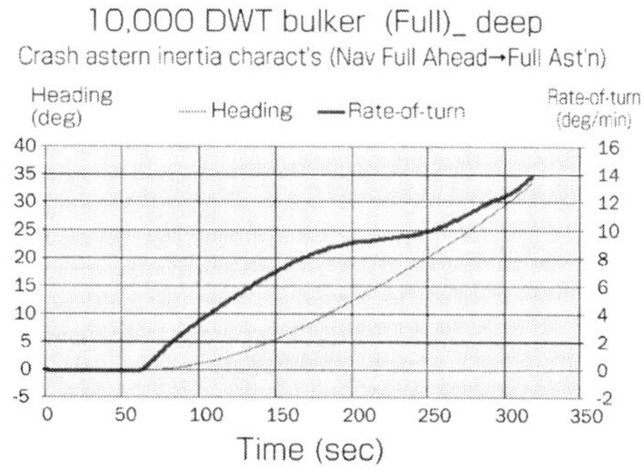
10,000 DWT bulker (Full)_ deep
Crash astern inertia charact's (Nav Full Ahead→Full Ast'n)
Heading (deg)
Heading
Rate-of-turn
Rate-of-turn (deg/min)
Time (sec)

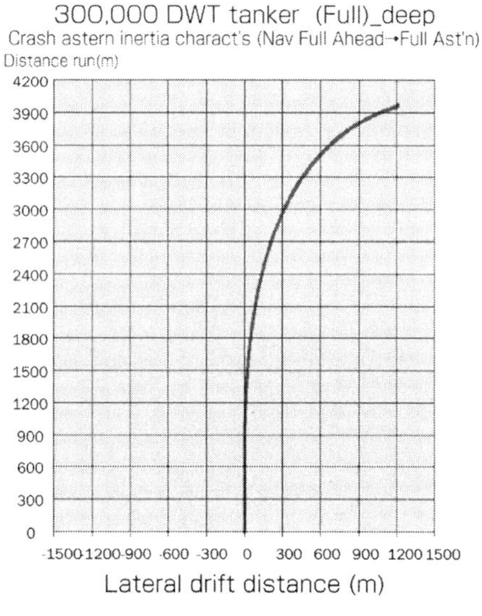
300,000 DWT tanker (Full)_deep
Crash astern inertia charact's (Nav Full Ahead→Full Ast'n)
Distance run(m)
Lateral drift distance (m)

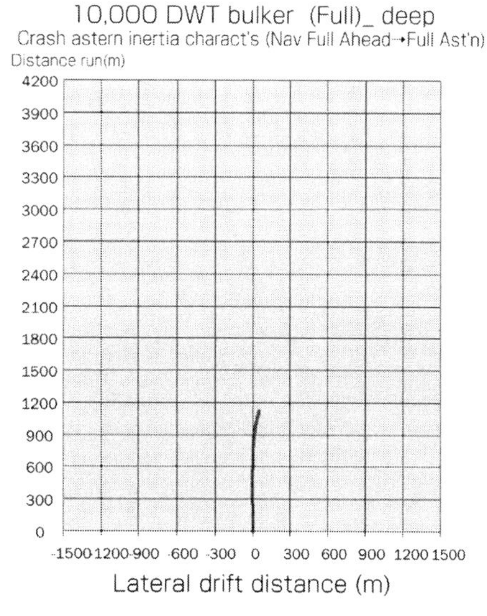
10,000 DWT bulker (Full)_ deep
Crash astern inertia charact's (Nav Full Ahead→Full Ast'n)
Distance run(m)
Lateral drift distance (m)

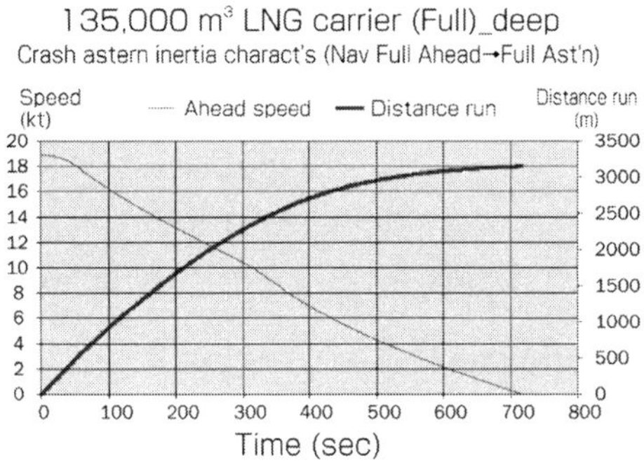
135,000 m3 LNG carrier (Full)_deep
Crash astern inertia charact's (Nav Full Ahead→Full Ast'n)
Speed (kt)
Ahead speed
Distance run
Distance run (m)
Time (sec)

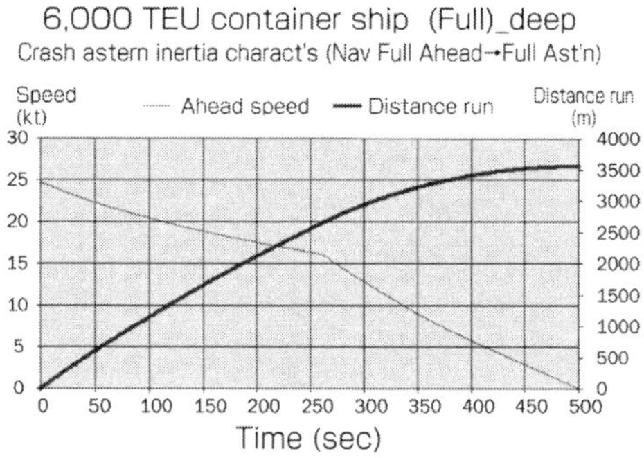
6,000 TEU container ship (Full)_deep
Crash astern inertia charact's (Nav Full Ahead→Full Ast'n)
Speed (kt)
Ahead speed
Distance run
Distance run (m)
Time (sec)

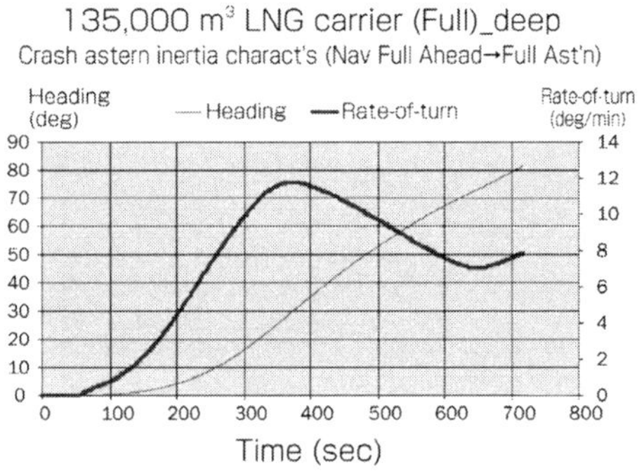
135,000 m3 LNG carrier (Full)_deep
Crash astern inertia charact's (Nav Full Ahead→Full Ast'n)
Heading (deg)
Heading
Rate-of-turn
Rate-of-turn (deg/min)
Time (sec)

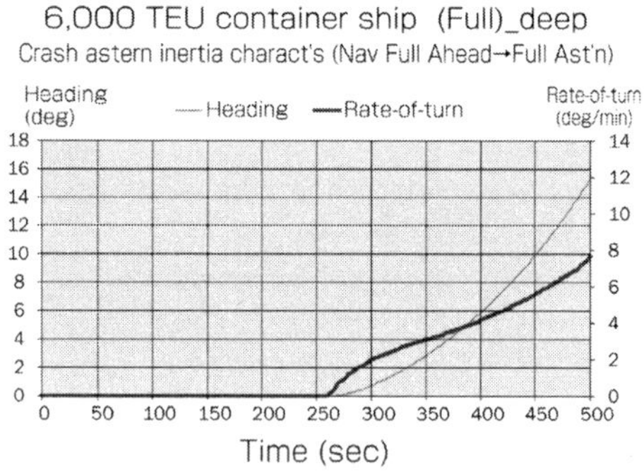
6,000 TEU container ship (Full)_deep
Crash astern inertia charact's (Nav Full Ahead→Full Ast'n)
Heading (deg)
Heading
Rate-of-turn
Rate-of-turn (deg/min)
Time (sec)

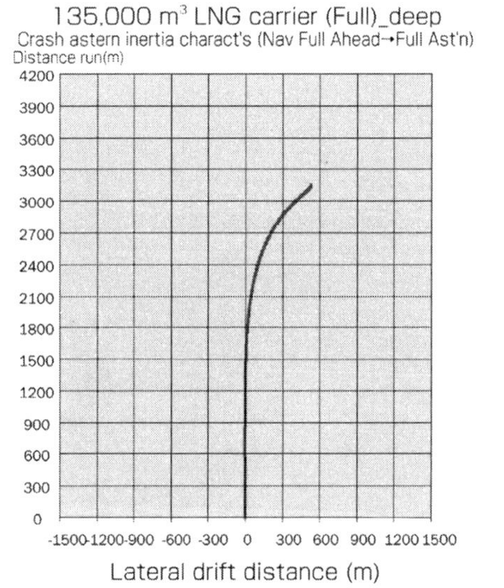
135,000 m3 LNG carrier (Full)_deep
Crash astern inertia charact's (Nav Full Ahead→Full Ast'n)
Distance run(m)
Lateral drift distance (m)

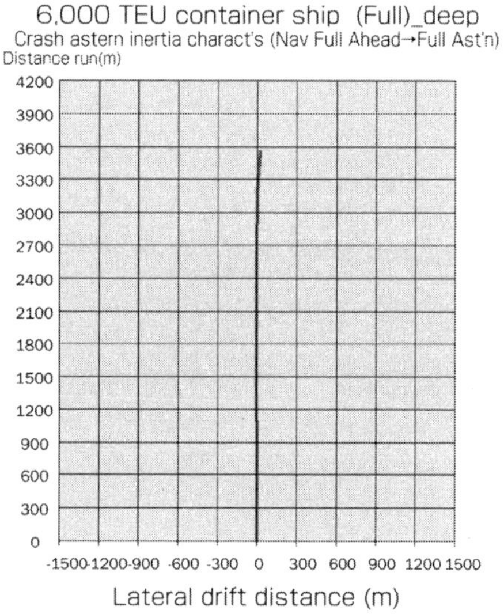
6,000 TEU container ship (Full)_deep
Crash astern inertia charact's (Nav Full Ahead→Full Ast'n)
Distance run(m)
Lateral drift distance (m)

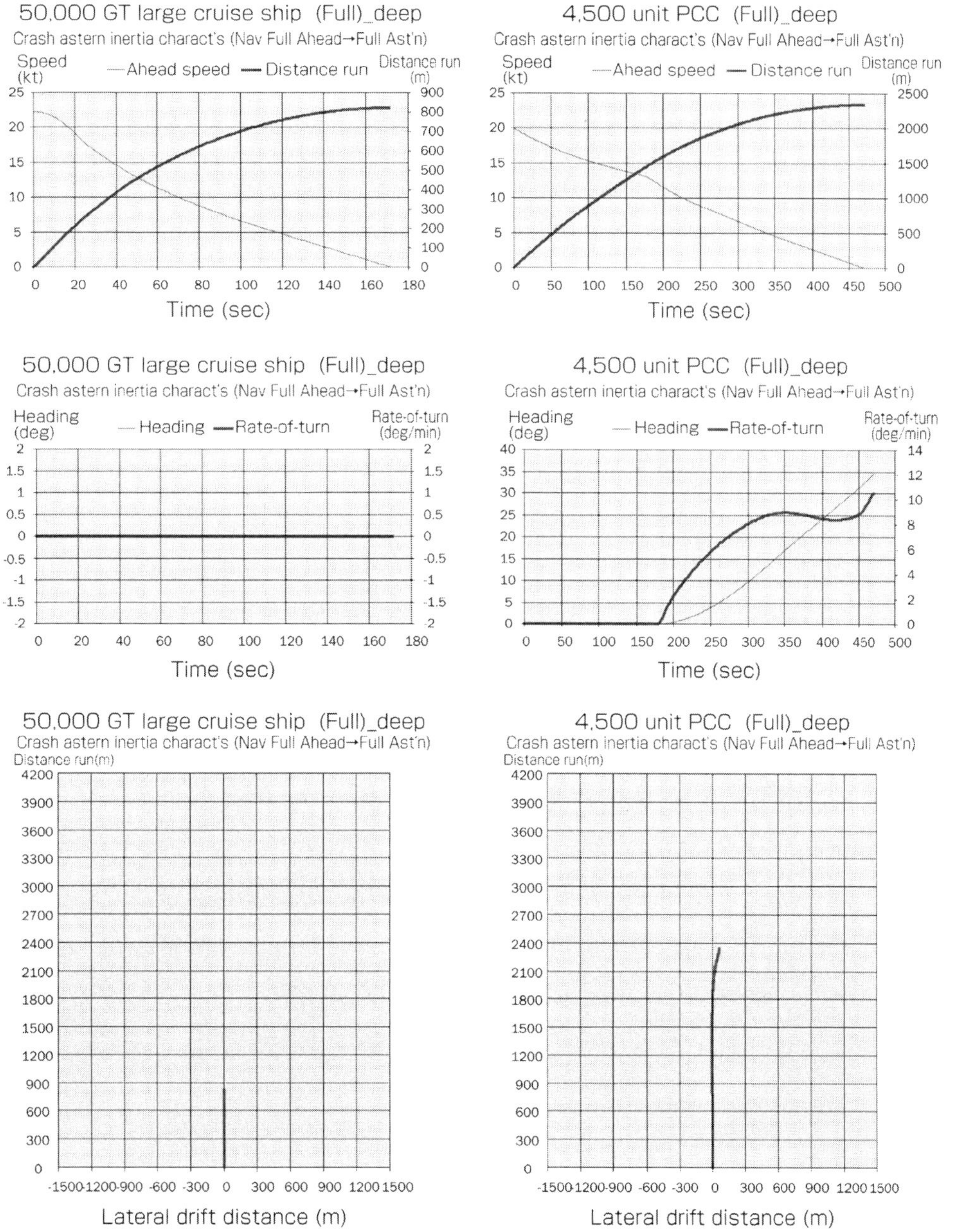

Fig. 4.2.3 Crash astern maneuvering (Initial speed : Nav. Full ahead, h/d = ∞)

Chapter 4 Effect of External Forces on Ship Handling

1. Effects of Wind

When a ship is approaching the berth keeping quite slow headway inertia with its engine stopped and free rotating propeller under the effect of wind, the ship will drift downwind and her heading will tend to turn windward. In such condition on which the ship's own steerageway cannot be relied upon, the most important thing for a person engaged in ship handling is how to estimate the effect of wind and to take necessary countermeasures.

To estimate a ship's potential to drift downwind and its heading deflection, Fig. 4.2.4 has been prepared to provide persons engaged in ship handling with the information derived from the simulation tests on a ship's motions at every minute when their engines are stopped and they are still running ahead at 5 knots with their helms midship. The simulation tests on the motions for 5 typical ship types except small 10,000 DWT bulker listed in Table 4.2.1 were carried out in parametric combination with various wind speed and direction under the condition of the ratio of water depth to ship draft $h/d=1.3$.

In addition to the above information, Fig. 2.4.5 is also prepared to illustrate the angle of heading deflection from the initial heading angle and the angle of course deflection at half ship length and one ship's length running of each ship type on the same approaching conditions as the above.

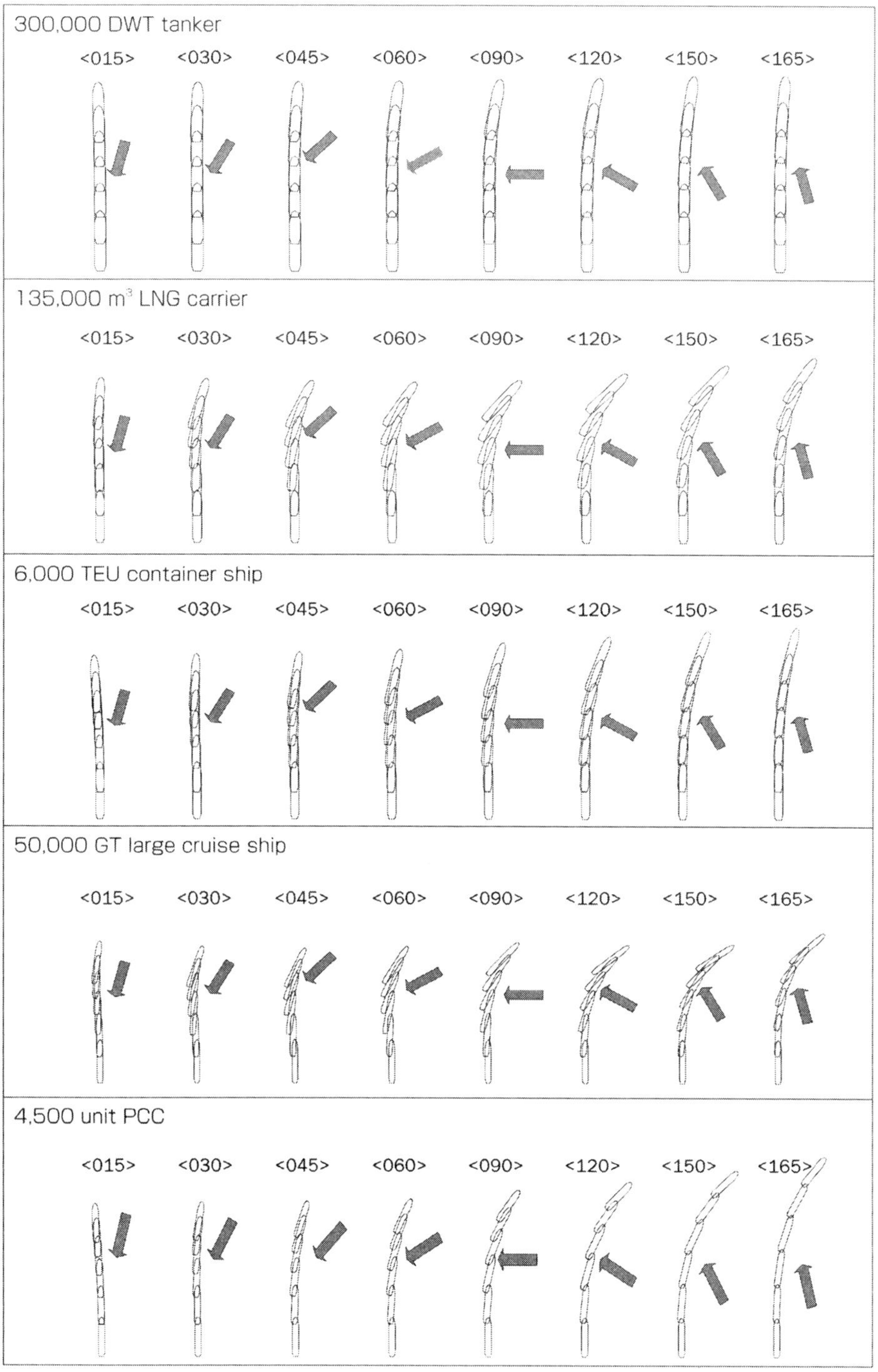

Fig. 4.2.4 Ship's motions under the effect of wind (Engine stop, ahead inertial speed 5 knots, helm midship, wind speed to ship speed ratio 5)

300,000 DWT tanker

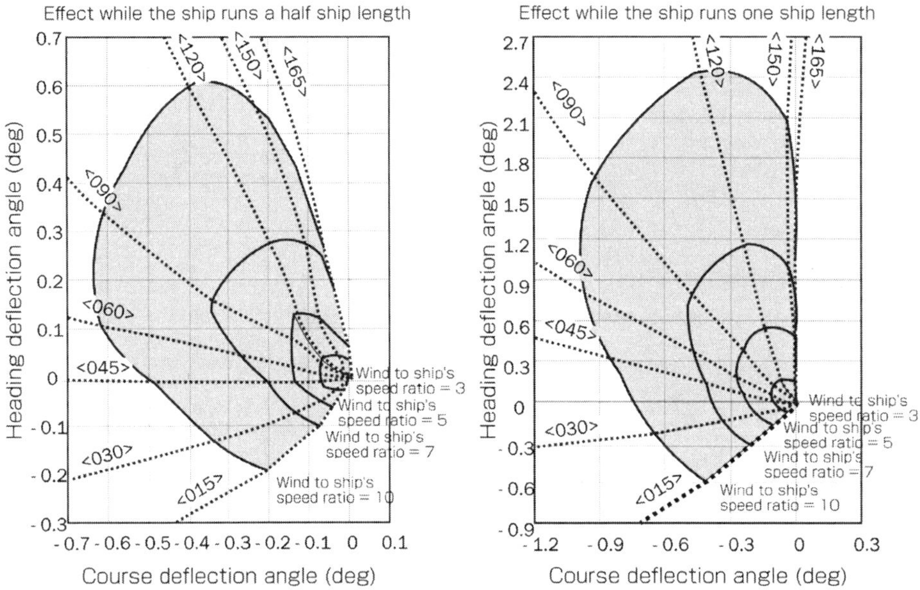

135,000 m^3 LNG carrier

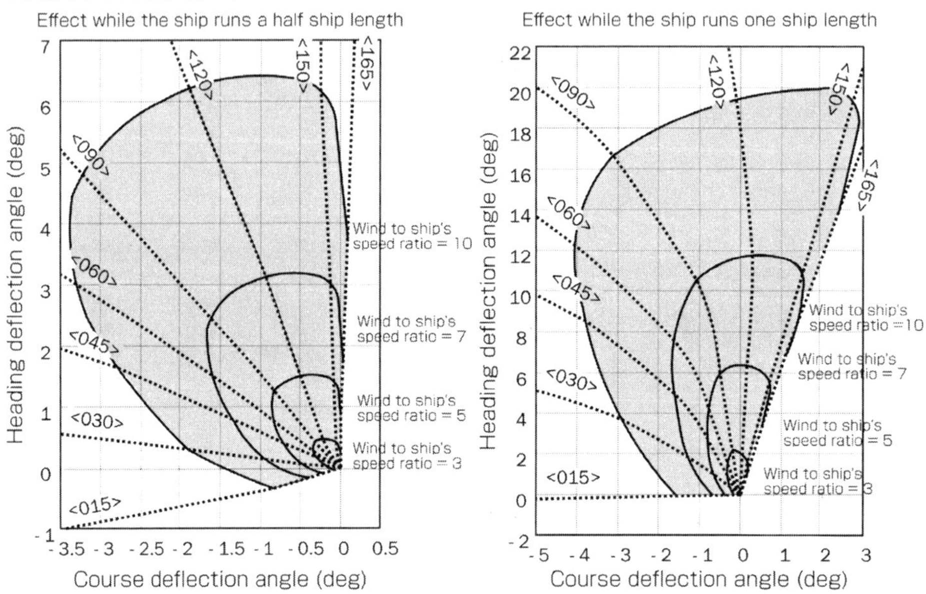

6,000 TEU container ship

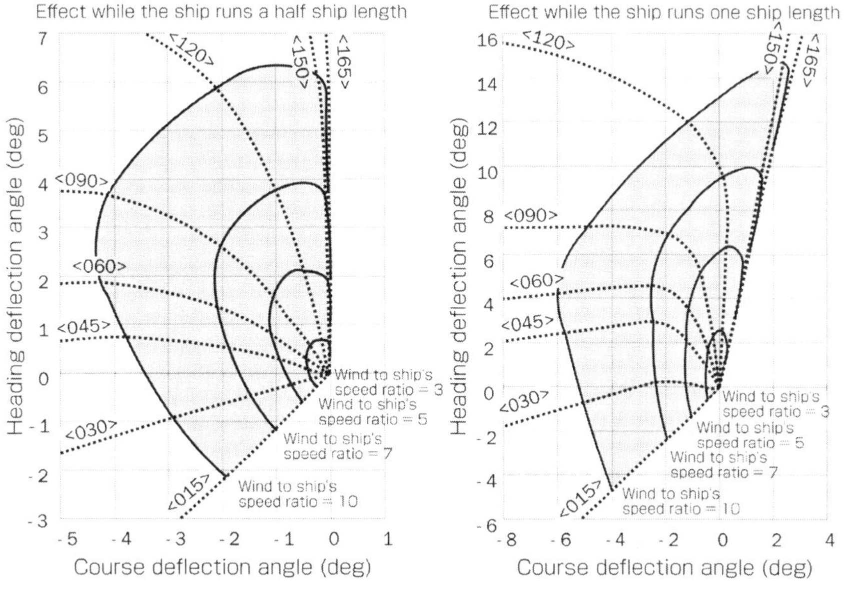

50,000 GT large cruise ship

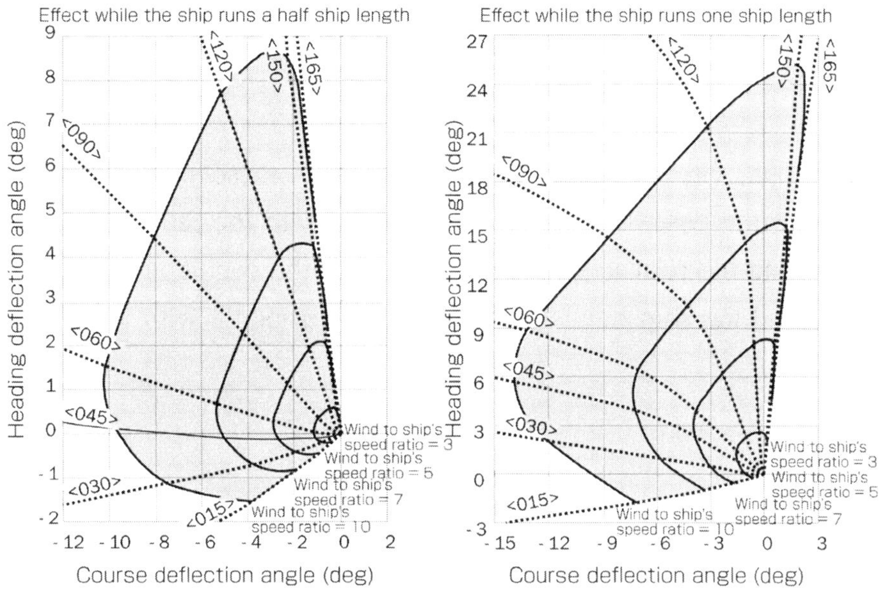

4,500 unit PCC

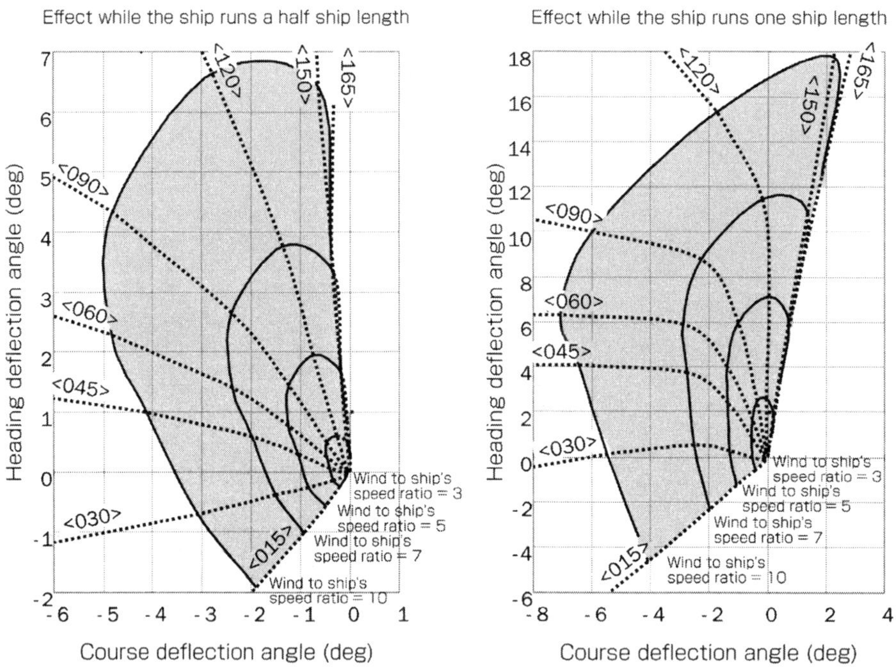

Fig. 4.2.5 Course deflection angles and heading deflection angles created while a ship runs a half ship length and one ship length (Engine stop, ahead inertia, helm midship)

Besides that, Fig. 4.2.6 shows the lateral drift speed of each ship type 60 seconds after the onset of wind in combination with the ratio of wind speed to ship speed and the relative wind direction under the condition where the engine is stopped and the ship runs ahead under its own inertia with its helms midhip.

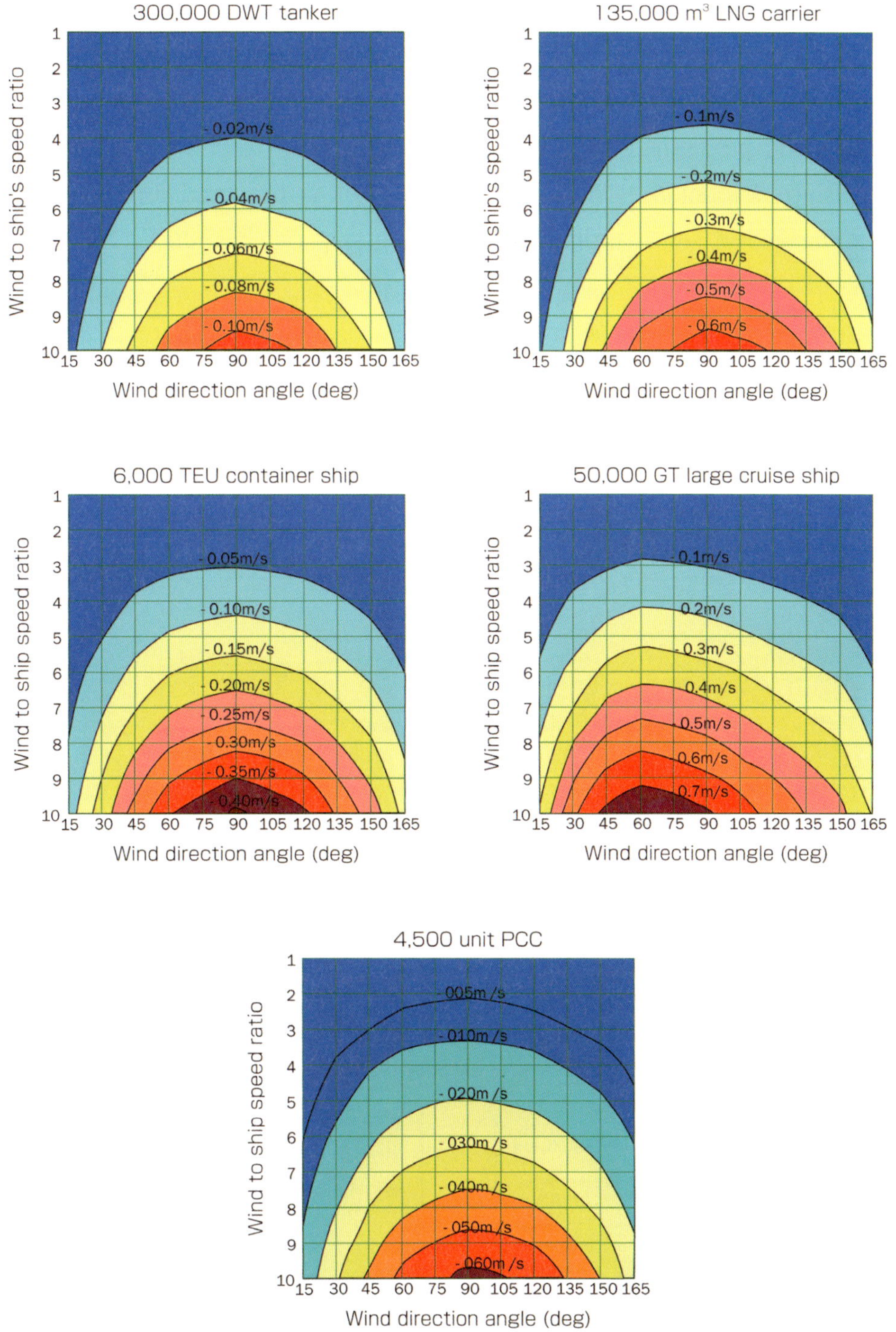

Fig. 4.2.6 Lateral drifting speed of a ship in 60 seconds after receiving the effect of wind (Engine stop, ahead inertia, helm midship)

2. Effects of Current

When a ship is running under the effects of current, in following current the ship will be carried too far down stream and in adverse current the ship's movement will be pushed back. In particular, in beam current it is difficult to turn the ship as intended. The data in this section is prepared for reference for persons engaged in ship handling about how a ship behaves under the effect of current.

Fig. 4.2.7 illustrates the behavior of a 130,000 m^3 LNG carrier derived from simulation tests when it is going to make a U turn under the effects of current with helm angle of 15°and 35°at an initial speed of 12.3 knots in parametric combination with 8 directions to the ship and 2 and 4 knots of current speed under the condition of water depth to ship draft ratio $h/d=1.3$.

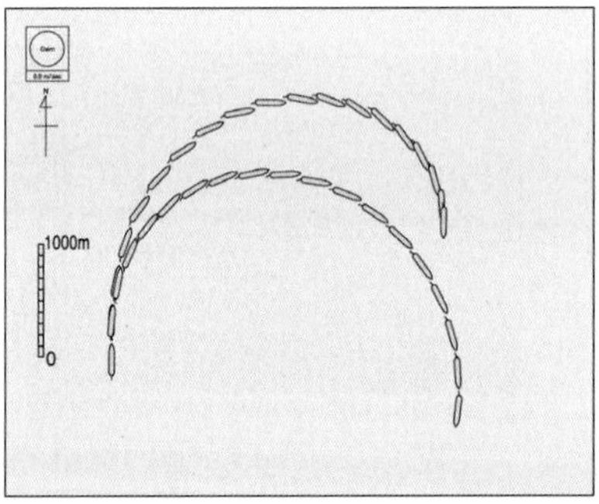

Current speed: 2 kt, Helm angle: 15°,
Current directions: 000°/180°

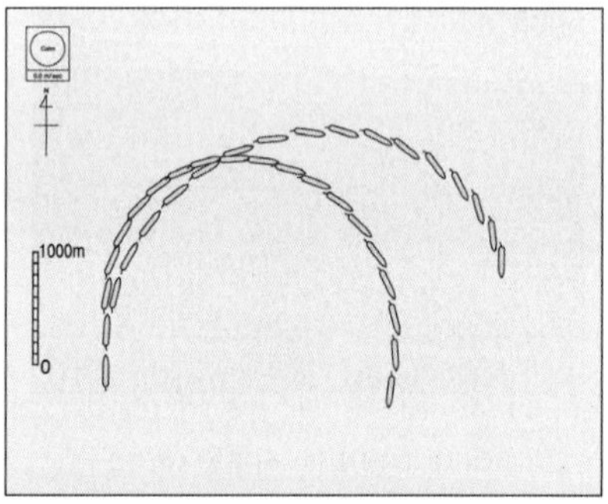

Current speed: 2 kt, Helm angle: 15°,
Current directions: 045°/225°

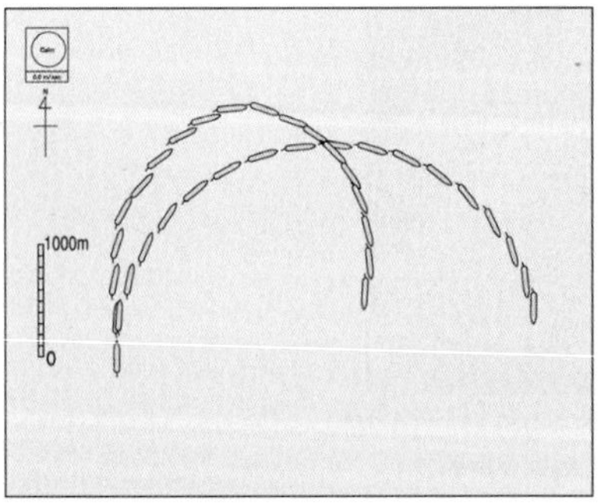

Current speed: 2 kt, Helm angle: 15°,
Current directions: 090°/270°

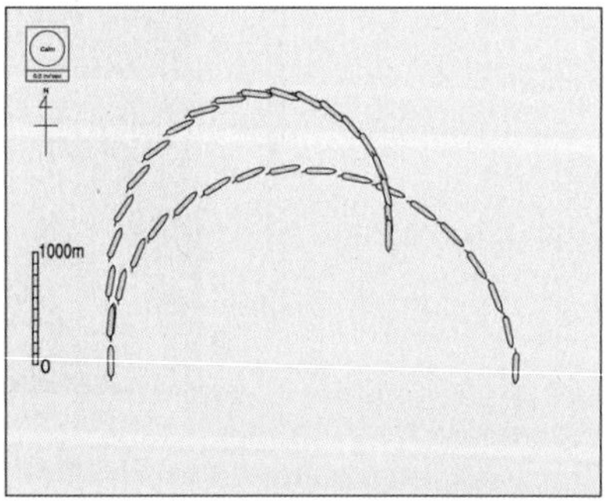

Current speed: 2 kt, Helm angle: 15°,
Current directions: 135°/315°

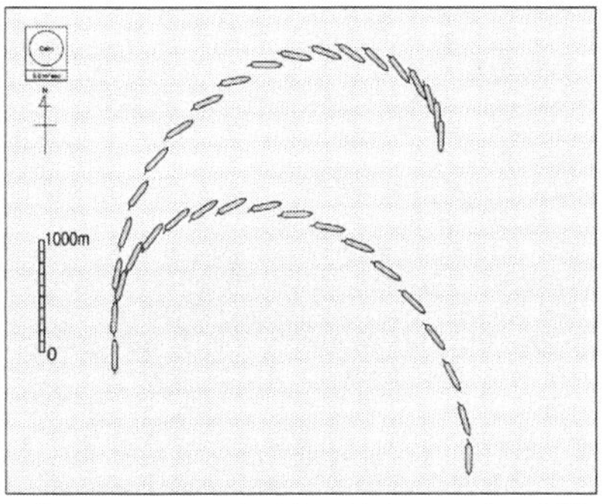

Current speed: 4 kt, Helm angle: 15°,
Current directions: 000°/180°

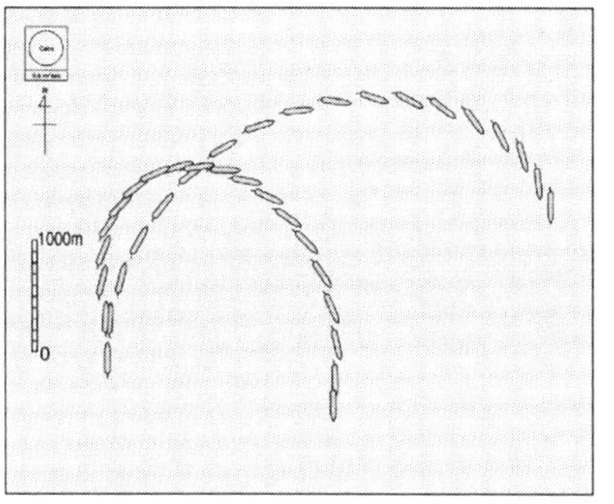

Current speed: 4 kt, Helm angle: 15°,
Current directions: 045°/225°

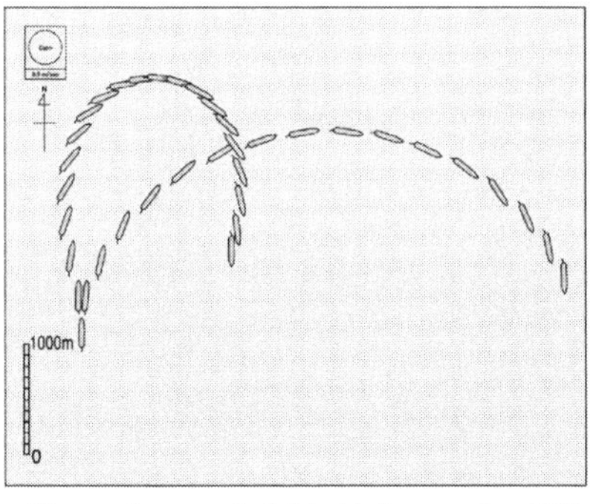

Current speed: 4 kt, Helm angle: 15°,
Current directions: 090°/270°

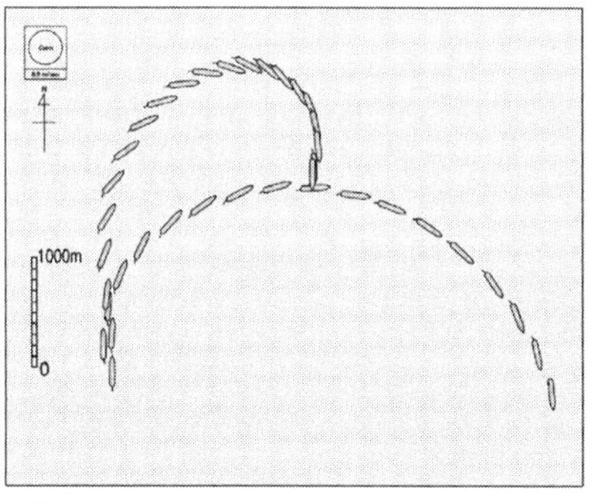

Current speed: 4 kt, Helm angle: 15°,
Current directions: 135°/315°

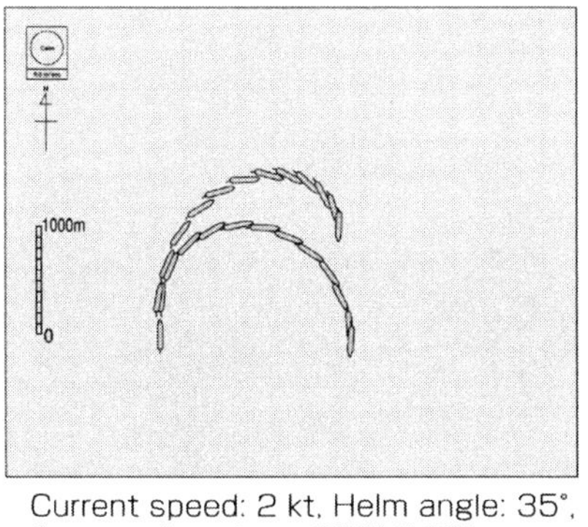

Current speed: 2 kt, Helm angle: 35°,
Current directions: 000°/180°

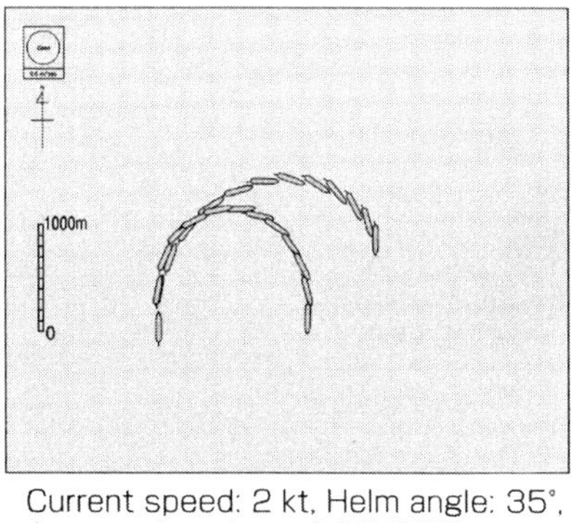

Current speed: 2 kt, Helm angle: 35°,
Current directions: 045°/225°

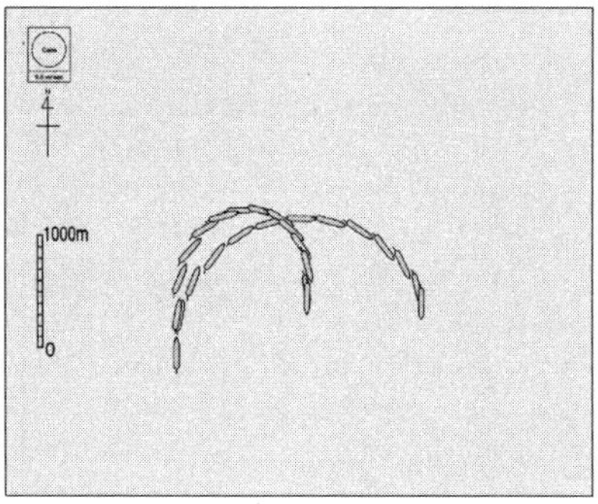

Current speed: 2 kt, Helm angle: 35°,
Current directions: 090°/270°

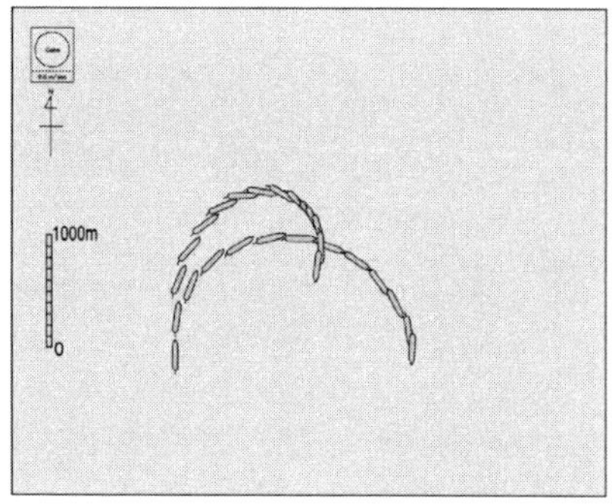

Current speed: 2 kt, Helm angle: 35°,
Current directions: 135°/315°

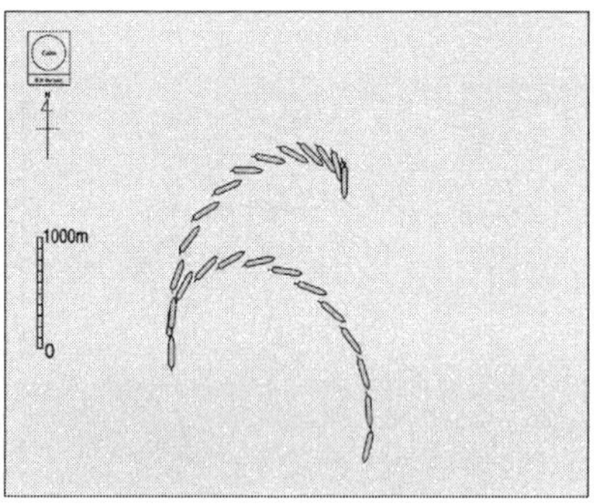

Current speed: 4 kt, Helm angle: 35°,
Current directions: 000°/180°

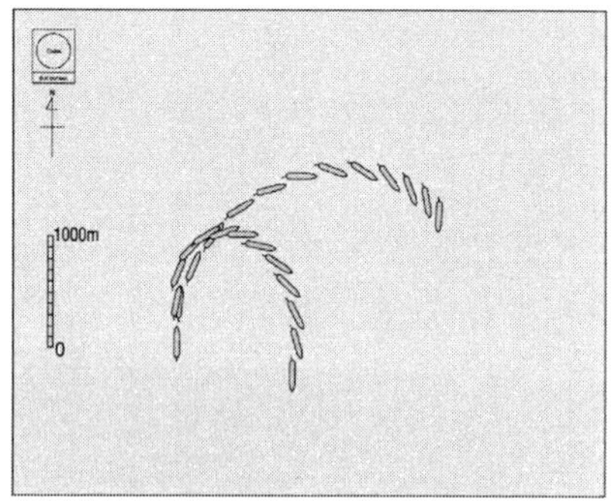

Current speed: 4 kt, Helm angle: 35°,
Current directions: 045°/225°

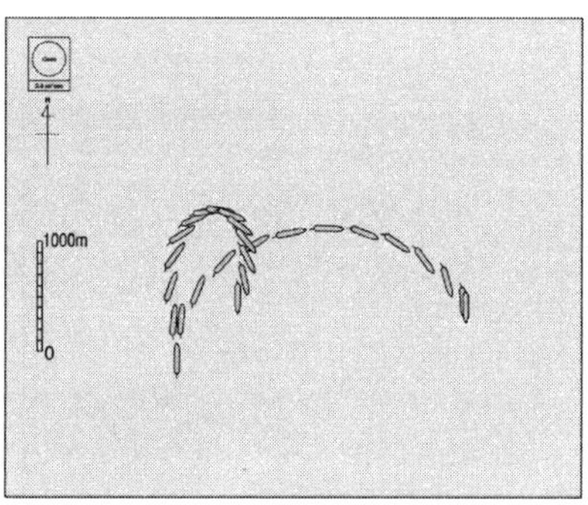

Current speed: 4 kt, Helm angle: 35°,
Current directions: 090°/270°

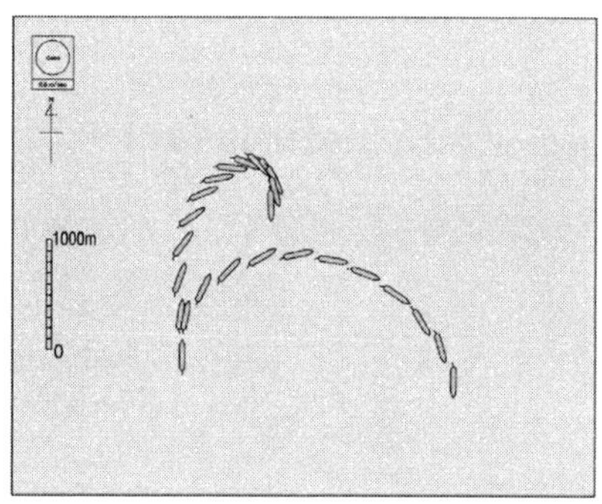

Current speed: 4 kt, Helm angle: 35°,
Current directions: 135°/315°

Fig. 4.2.7 Tracks of a U turn in current
(130,000 m^3 LNG carrier, fully loaded, 12.3 kt, h/d = 1.3, δ = 15° and 35°)

3. Shallow Water Effects

In shallow water, maneuvering characteristics will change as the resistances to the hull and turning moment increase. The shallow water effect under the conditions of the ratio of water depth to the ship draft $h/d=\infty$ and $h/d=1.3$ can be compared on the following maneuvering characteristics as shown below based on the Figs. and Tables in the previous section.

- Scale of turning circle : Ref. Fig. 4.1.4
- Elements of turning circle : Ref. Table 4.1.4
- Reduction of speed : Ref. Table 4.1.5
- Development of rate-of-turn after rudder is taken : Ref. Fig. 4.1.5
- Overshoot angle : Ref. Table 4.1.6
- Driving and stopping inertia : Ref. Fig. 4.2.1
- Emergency stopping inertia : Ref. Fig. 4.2.2

Chapter 5 Characteristics Related to Ship Handling inside Harbors

1. Critical Speed for Steerageway

Maneuverability declines as a ship's headway is reduced. Inside port, it is important for persons engaged in ship handling to know their ship's critical speed beforehand to keep steerageway as ships are required to be handled at quite limited speeds. At slow speeds, chances to take large helm angle instantaneously increase. It would be useful and practical enough to know the rudder effect when hard over rudder is taken.

Fig. 4.2.8 shows the rate-of-turn plotted of 6 typical ship types listed in Table 4.2.1 at one minute and three minutes after 35° rudder was taken under the condition of the ratio of water depth to ship draft $h/d=1.3$. From Fig. 4.2.8, when enough rate-of-turn cannot be gained within a certain amount of time, it can be deemed as the limit of the ship's steerageway. In other words, it can be deemed that the speed is the critical speed for steerageway of the ship. Calculations are made on the assumption that 35° rudder is taken at every 8, 6, 4 and 2 knots of initial speed when each ship is running under headway inertia with its engine stopped.

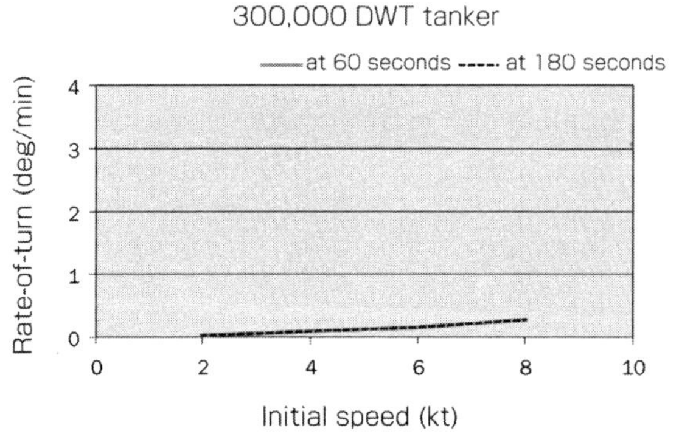

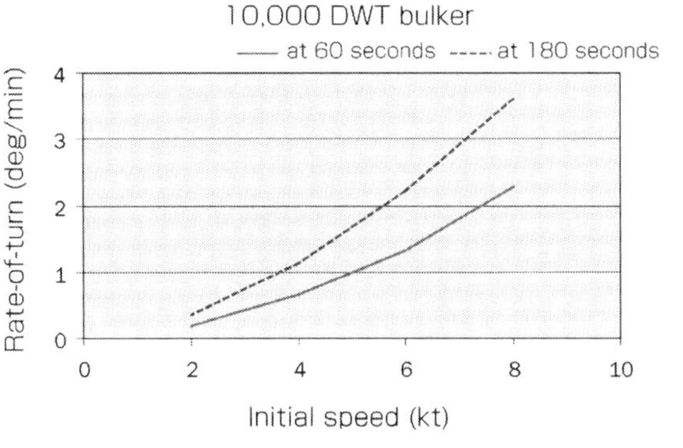
10,000 DWT bulker
at 60 seconds
at 180 seconds
Rate-of-turn (deg/min)
Initial speed (kt)

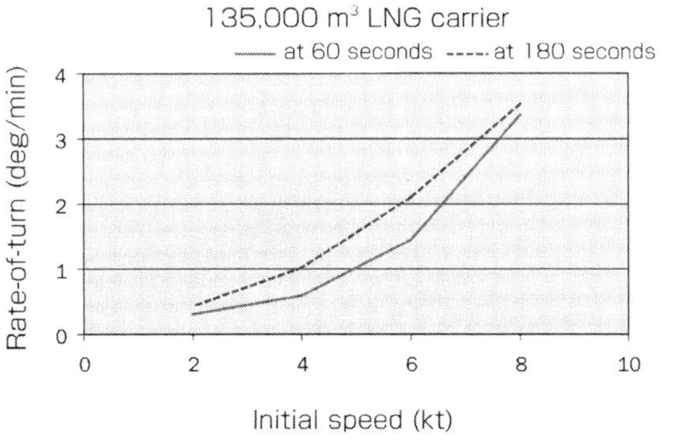
135,000 m³ LNG carrier
at 60 seconds
at 180 seconds
Rate-of-turn (deg/min)
Initial speed (kt)

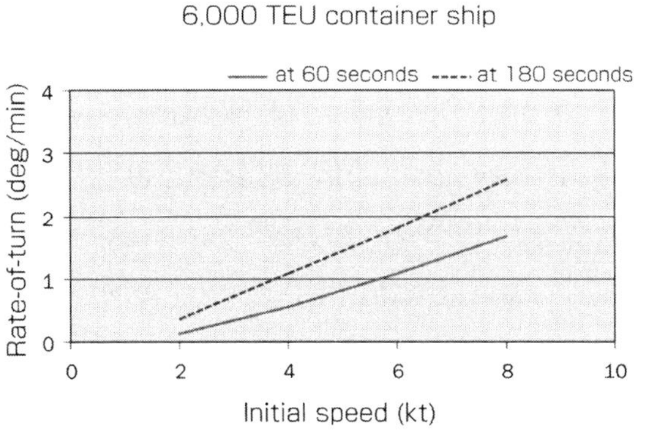
6,000 TEU container ship
at 60 seconds
at 180 seconds
Rate-of-turn (deg/min)
Initial speed (kt)

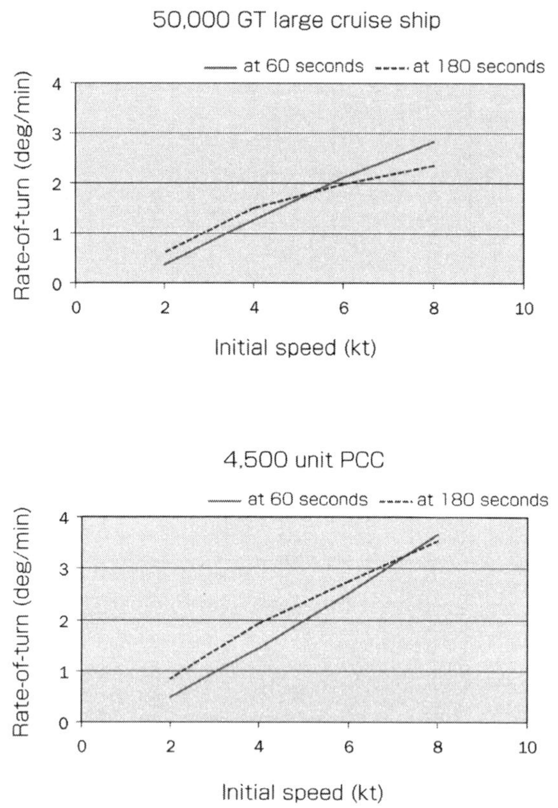

Fig. 4.2.8 Critical speed for steerageway
(Engine stop, ahead inertia, helm angle : 35°)

2. Stopping Distance and Deflection of Heading by Astern Engine

Inside a port a ship is brought before its berth at as low a speed as possible so that it can be stopped in front of the berth by reversing its thrust. The characteristic of this braking performance are one of the most important properties for a person engaged in ship handling.

However, when astern thrust is given while a ship is keeping headway, control of the ship becomes difficult because the ship's heading tends to turn due to the effects of the impingement due to propeller discharge current at the stern and sidewise pressure of the propeller. Although this tendency can be useful in some phases of ship handling, it is unfavorable in keeping posture for berthing operations.

Table 4.2.3 illustrates the stopping distance, lateral drift distance and deflection of heading when an astern engine is used for a ship running straight ahead with the helm midship for 6 typical ship types listed in Table 4.2.1 under the condition of the ratio of water depth to ship draft $h/d=1.3$ until the ship comes to a complete stop in water in combination with various ahead and astern thrusts.

Table 4.2.3 Stopping distance and deflection of heading caused by astern engine

(a) 300,000 DWT tanker

	D. Slow Astern			Slow Astern			Half Astern			S/B Full Astern		
	Advance /L	Transfer /L	Deflection angle of heading	Advance /L	Transfer /L	Deflection angle of heading	Advance /L	Transfer /L	Deflection angle of heading	Advance /L	Transfer /L	Deflection angle of heading
D. Slow Ahead	3.1	0.2	37.7	2.0	0.1	29.4	1.1	0.0	20.7	0.7	0.0	15.2
Slow Ahead	5.5	0.6	48.9	3.8	0.4	41.8	2.4	0.1	32.4	1.6	0.1	25.4
Half Ahead	11.1	1.6	62.4	8.5	1.2	58.9	5.6	0.7	51.0	4.1	0.4	43.9
S/B Full Ahead	16.4	2.6	69.3	12.7	2.0	67.3	9.3	1.4	62.8	6.7	0.9	56.1

(b) 10,000 DWT bulker

	D. Slow Astern			Slow Astern			Half Astern			S/B Full Astern		
	Advance /L	Transfer /L	Deflection angle of heading	Advance /L	Transfer /L	Deflection angle of heading	Advance /L	Transfer /L	Deflection angle of heading	Advance /L	Transfer /L	Deflection angle of heading
D. Slow Ahead	36.3	5.3	32.9	1.8	0.0	29.4	1.1	0.0	20.7	0.9	0.0	15.2
Slow Ahead	43.3	6.0	34.4	3.2	0.1	41.8	2.0	0.0	32.4	1.6	0.0	25.4
Half Ahead	51.4	8.0	41.2	6.0	0.2	58.9	3.7	0.1	51.0	3.0	0.0	43.9
S/B Full Ahead	55.9	9.3	45.4	8.0	0.4	67.3	5.2	0.2	62.8	4.2	0.1	56.1

(c) 135,000 m^3 LNG carrier

	D. Slow Astern			Slow Astern			Half Astern			S/B Full Astern		
	Advance /L	Transfer /L	Deflection angle of heading	Advance /L	Transfer /L	Deflection angle of heading	Advance /L	Transfer /L	Deflection angle of heading	Advance /L	Transfer /L	Deflection angle of heading
D. Slow Ahead	2.9	0.2	25.6	1.9	0.1	20.1	1.1	0.0	13.7	0.9	0.0	11.0
Slow Ahead	4.6	0.4	32.0	3.0	0.2	26.0	1.9	0.1	19.8	1.5	0.0	16.7
Half Ahead	6.8	0.7	38.4	5.2	0.5	33.9	3.5	0.2	27.5	2.9	0.1	24.7
S/B Full Ahead	9.0	1.1	42.8	6.9	0.7	38.6	5.1	0.4	33.5	4.3	0.3	30.1

(d) 6,000 TEU container ship

	D. Slow Astern			Slow Astern			Half Astern			S/B Full Astern		
	Advance /L	Transfer /L	Deflection angle of heading	Advance /L	Transfer /L	Deflection angle of heading	Advance /L	Transfer /L	Deflection angle of heading	Advance /L	Transfer /L	Deflection angle of heading
D. Slow Ahead	2.5	0.0	10.3	1.5	0.0	8.1	0.9	0.0	5.9	0.7	0.0	4.9
Slow Ahead	4.0	0.1	12.3	2.8	0.1	10.6	1.7	0.0	8.3	1.3	0.0	7.3
Half Ahead	6.3	0.2	14.1	4.3	0.1	12.5	2.9	0.1	10.6	2.3	0.0	9.7
S/B Full Ahead	8.3	0.3	14.9	5.6	0.2	13.6	3.8	0.1	11.9	3.2	0.1	11.0

(e) 50,000 GT large cruise ship

	D. Slow Astern			Slow Astern			Half Astern			S/B Full Astern		
	Advance /L	Transfer /L	Deflection angle of heading	Advance /L	Transfer /L	Deflection angle of heading	Advance /L	Transfer /L	Deflection angle of heading	Advance /L	Transfer /L	Deflection angle of heading
D. Slow Ahead	0.5	0.0	0.0	0.3	0.0	0.0	0.1	0.0	0.0	0.1	0.0	0.0
Slow Ahead	2.3	0.0	0.0	1.4	0.0	0.0	0.7	0.0	0.0	0.6	0.0	0.0
Half Ahead	3.7	0.0	0.0	2.4	0.0	0.0	1.4	0.0	0.0	1.1	0.0	0.0
S/B Full Ahead	4.5	0.0	0.0	3.0	0.0	0.0	1.8	0.0	0.0	1.5	0.0	0.0

(f) 4,500 unit PCC

	D. Slow Astern			Slow Astern			Half Astern			S/B Full Astern		
	Advance /L	Transfer /L	Deflection angle of heading	Advance /L	Transfer /L	Deflection angle of heading	Advance /L	Transfer /L	Deflection angle of heading	Advance /L	Transfer /L	Deflection angle of heading
D. Slow Ahead	3.3	0.1	14.4	2.2	0.0	12.6	1.6	0.0	11.0	1.2	0.0	9.6
Slow Ahead	4.7	0.2	16.0	3.6	0.1	14.9	2.8	0.1	13.6	2.1	0.0	12.1
Half Ahead	5.9	0.2	16.9	4.5	0.2	15.8	3.7	0.1	14.9	3.0	0.1	13.7
S/B Full Ahead	7.0	0.3	17.5	5.3	0.2	16.5	4.3	0.2	15.6	3.6	0.1	14.7

3. Turning Ability by Boosting

When headway of a ship is quite slow or a ship is almost dead in the water, to turn the bow without creation of ahead inertia, instantaneous ahead engine and hard over rudder are used simultaneously. This method is called ***boosting***. This is effective in correcting a ship's heading when a ship is almost dead in the water. The effect of boosting is also an important element that persons engaged in ship handling should know beforehand to carry out skillful maneuvering.

Fig. 4.2.9 illustrates rate-of-turn of 6 typical ship types listed in Table 4.2.1, at 30 and 60 seconds after each ship's rudder was turned at 35° to starboard using D. Slow, Slow, Half and S/B Full Ahead engines from the state that each ship is lying dead in the water under the condition of the ratio of water depth to ship draft $h/d = 1.3$.

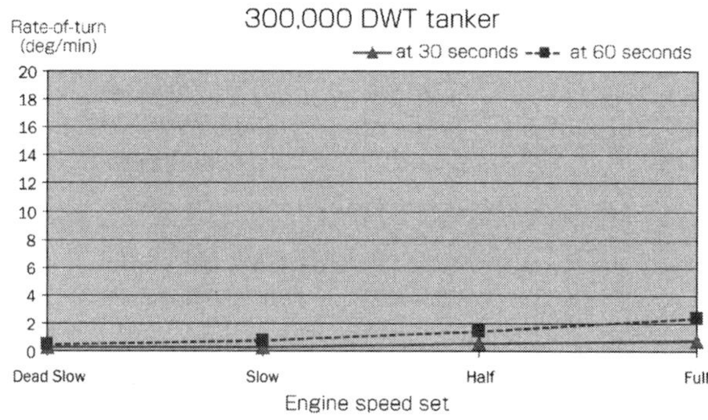
300,000 DWT tanker
Rate-of-turn (deg/min)
at 30 seconds
at 60 seconds
20
18
16
14
12
10
8
6
4
2
0
Dead Slow
Slow
Half
Full
Engine speed set

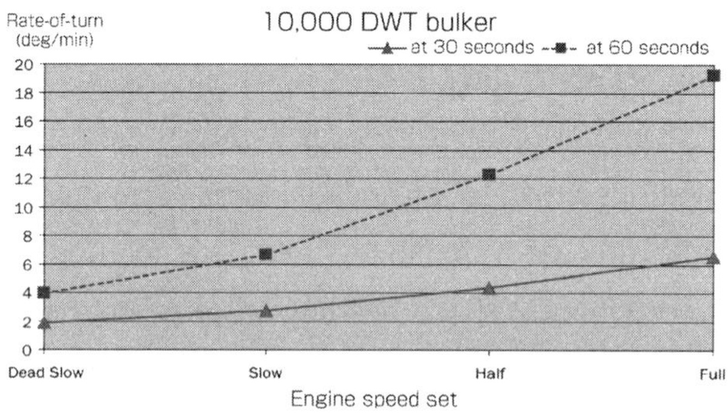
10,000 DWT bulker
Rate-of-turn (deg/min)
at 30 seconds
at 60 seconds
20
18
16
14
12
10
8
6
4
2
0
Dead Slow
Slow
Half
Full
Engine speed set

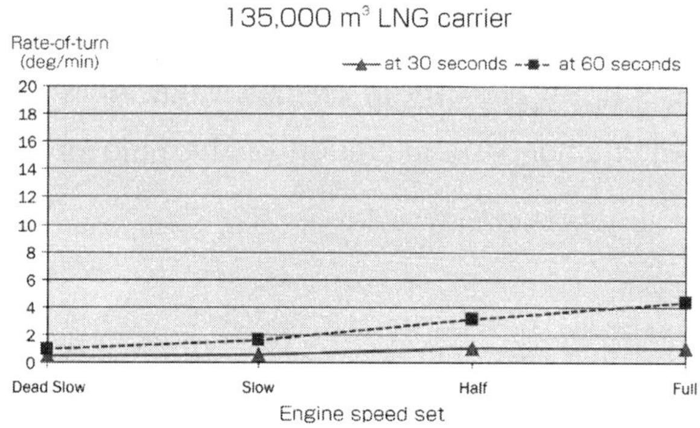
135,000 m³ LNG carrier
Rate-of-turn (deg/min)
at 30 seconds
at 60 seconds
20
18
16
14
12
10
8
6
4
2
0
Dead Slow
Slow
Half
Full
Engine speed set

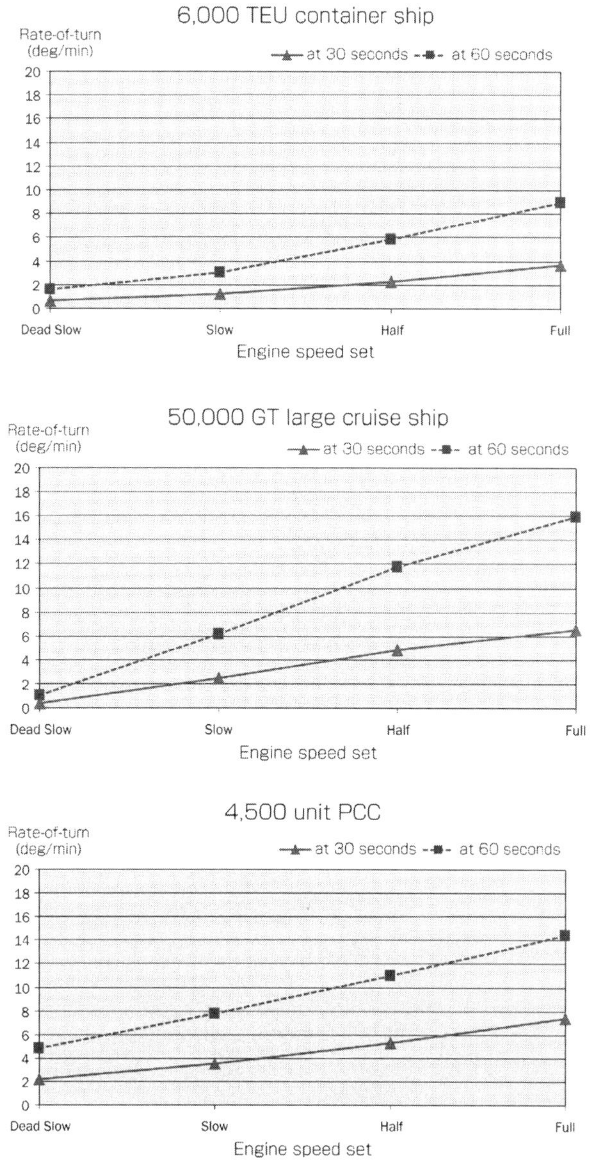

Fig. 4.2.9 Turning ability by boosting

4. Characteristics in Lateral Shifting Maneuvers

When a ship is to be moved slightly to the lateral direction in a restricted waterway without thrusters or tug assistance, its rudder is first turned to the direction of movement then the rudder is turned to the opposite direction to return the ship's heading to its initial heading. During this process the ship moves laterally as intended.

Fig. 4.2.10 illustrates the scale of the lateral shift of 6 typical ship types listed in Table 4.2.1 when they are moved by the above right to left or left to right steering method, so called as ***bang-bang control***, at a helm angle 10°, 20° and 30° at initial speed of Slow and Half Ahead in combination of the heading angle of 10°, 20° and 30° when the rudder is turned to the opposite direction under the condition of the ratio of water depth to ship draft $h/d = 1.3$.

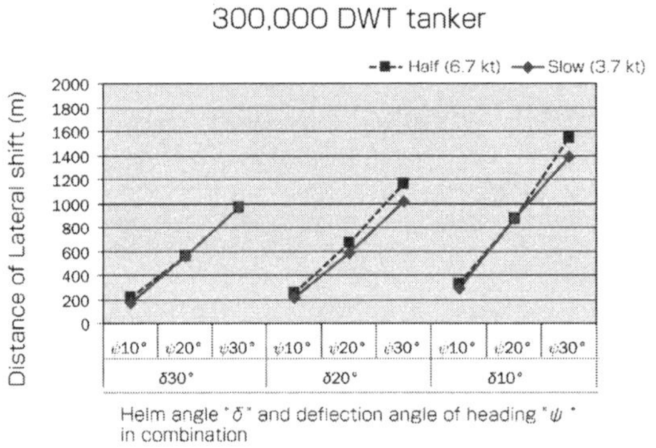

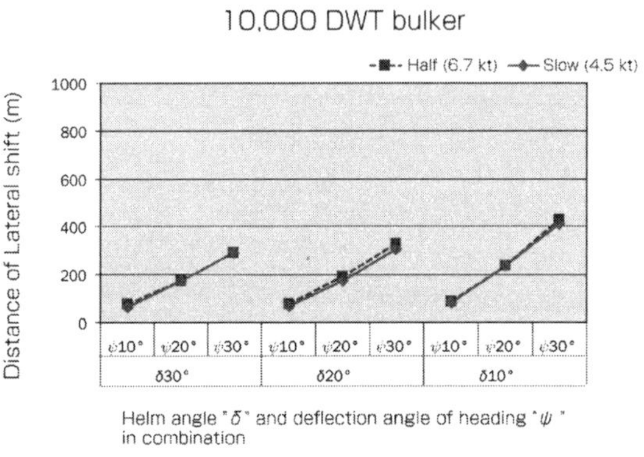

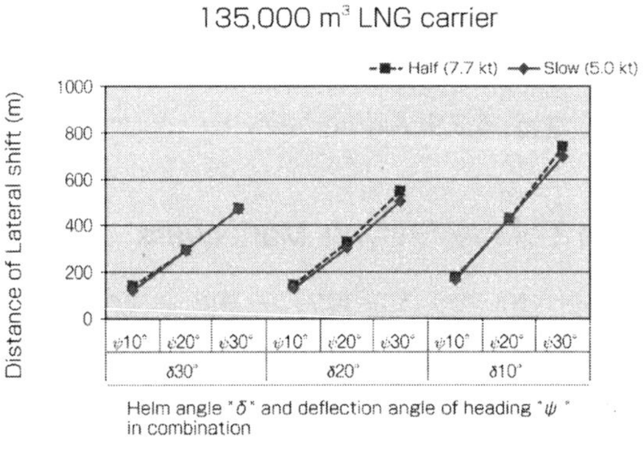

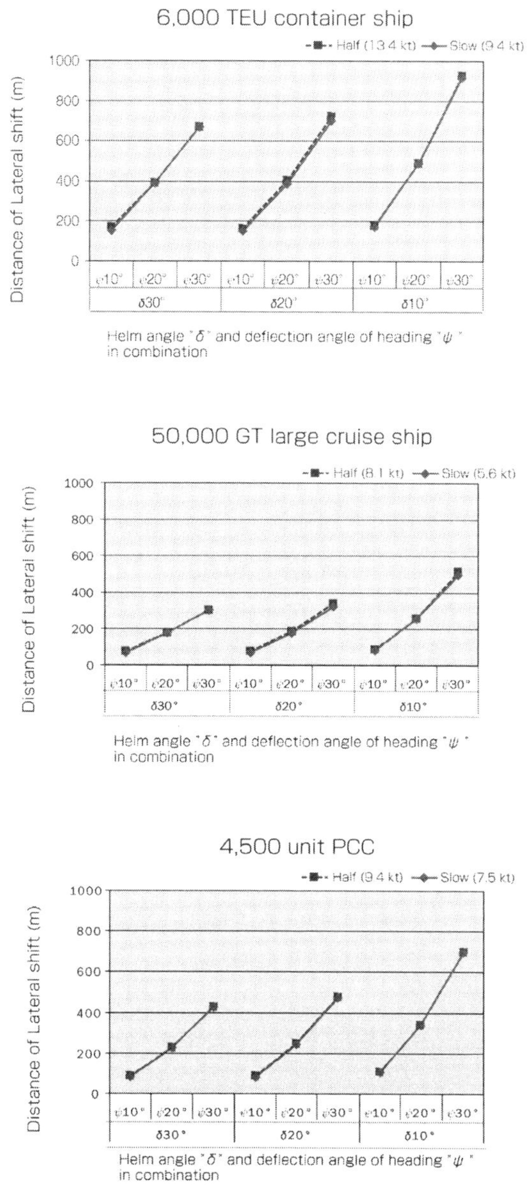

Fig. 4.2.10 Characteristic in lateral shifting maneuvers

5. Swing out of a Ship's Stern by Kick

The scale of kick is an important element to be taken into consideration not only when preventing the ship's stern from coming into contact with any other object or a person overboard, but also when making large turns using hard over rudder while leaving from the berth, although enough clearance is needed for the berth front and forward and the ship needs to be pulled parallel away from the berth face. In this maneuvering technique, persons engaged in ship handling should keep in mind the scale of the stern's swing out accompanied by the kick.

Fig. 4.2.4 shows the scale of kick for 6 typical ship types listed in Table 4.2.1 when their rudders are turned to 35° starboard from the state of running straight ahead with their rudders midship at initial speed of D.Slow, Slow, Half and S/B Full Ahead.

Table 4.2.4 Kick

(h/d = 1.3, helm was taken at 35°, CG : ship's center of gravity)

Ship type	Initial speed (kt)		CG kick (m)	CG kick (L)	Stern kick (m)	Stern kick (L)	at ψ (deg)	at time (sec)
300,000 DWT tanker VLCC (full)	D. Slow	2.4	10.5	0.032	39.7	0.12	7.1	307
	Slow	3.7	11.2	0.035	38.1	0.12	6.7	214
	Half	6.6	6.7	0.021	35.9	0.11	4.4	114
	S/B Full	9.4	6.4	0.020	35.4	0.11	4.1	83
10,000 DWT bulker Bulker (full)	D. Slow	3.1	4.8	0.045	13.9	0.13	9.3	99
	Slow	4.5	4.4	0.041	13.8	0.13	8.7	69
	Half	6.7	4.5	0.042	13.5	0.13	8.6	49
	S/B Full	8.4	3.3	0.032	13.7	0.13	7.5	37
135,000 m^3 LNG carrier LNG (full)	D. Slow	3.6	12.6	0.046	36.4	0.13	10.5	173
	Slow	5.0	12.4	0.045	36.4	0.13	10.4	125
	Half	7.7	12.9	0.047	35.4	0.13	10.2	87
	S/B Full	10.3	10.9	0.039	34.8	0.13	9.1	64
6,000 TEU container ship CNTN (full)	D. Slow	6.3	8.9	0.029	31.4	0.10	6.9	112
	Slow	9.4	9.5	0.031	30.7	0.10	6.9	79
	Half	13.4	7.9	0.026	30.9	0.10	6.2	54
	S/B Full	16.8	9.5	0.031	30.9	0.10	6.9	46
50,000 GT large cruise ship	D. Slow	2.2	12.5	0.061	28.2	0.14	12.6	194
	Slow	5.6	12.9	0.063	26.6	0.13	12.0	82
	Half	8.1	13.3	0.065	26.2	0.13	11.9	60
	S/B Full	9.8	13.9	0.065	25.6	0.12	11.9	52
4,500 unit PCC PCC (full)	D. Slow	5.3	3.5	0.021	19.9	0.12	4.7	69
	Slow	7.5	3.6	0.016	19.8	0.12	4.6	51
	Half	9.4	2.8	0.016	19.8	0.12	4.1	39
	S/B Full	11.0	3.2	0.019	19.9	0.12	4.4	35

Postscript

Two textbooks of ship handling, "Theory of Ship Handling" written by Prof. Iwai and "General Theory of Ship Handling" written by Prof. Honda, have long been the two greatest in this field in Japan. Nearly a half century has passed since those first editions were published. But it is certain that ship handling techniques have changed following technological innovations in the maritime industry.

I feel that these textbooks should not be left without any improvements for over three to five decades in this swiftly changing world, and that an updated textbook for ship handling conforming to the advancements of maritime technology for coming generations has long been anticipated.

Principles of ship handling are an indispensable part of any curriculum for sea pilots, masters, officers and students who wish to get a seafarer's license. In addition, I believe it is also useful for naval architects and designers of ports and harbors to study how ships are handled and resultantly build ships with good performance or construct safe havens for ships.

I'd been thinking for long time that in the future, if I would take on the task of writing a textbook on ship handling, it would have to include "how ships behave in transitional and steady phases of ship handling under the effects of various external disturbances" from a dynamic point of view in addition to a conventional static point of view based on other ship handling text books, not only to meet seafarers' needs but also to draw the interest of naval architects and designers of ports and harbors in the construction of efficient ships in terms of performance or constructing facilities such as ports and harbors.

In this book, I tried to realize my long-standing concept by compiling data of a ship's motion from the transitional phase to the steady phase through fast time ship maneuvering simulations and by means of a real time ship handling simulator. I would be happy if this textbook is of some use for the next generation of seafarers, naval architects and designers of ports and harbors to study the theoretical and practical aspects of ship handling.

Truthfully, for a long time I'd been urged by Prof. Keinosuke Honda and Mr.

Minoru Ogawa, Chairman, Seizando Shoten Co. Ltd. to write a new textbook on ship handling. And just recently I, because of their interest and support, managed to complete this textbook titled "Theory and Practice of Ship Handling". Therefore, I would like to express my heartfelt thanks to them, and all those who offered valuable assistance in writing this textbook.

February, 2011

Dr. Kinzo Inoue

Author

References

(1) General Theory of Ship Handling : Keinosuke Honda 1986 Seizando Shoten

(2) Theory of Ship Handling : Satoshi Iwai 1967 Kaibundo Shuppan

(3) 10 Chapters on VLCC : VLCC Workshop 1977 Seizando Shoten

(4) Practical Naval Architecture : Kiyoshi Takashiro 1975 Kaibundo Shuppan

(5) Handbook for Ship Design and Building : Kansai Shipbuilding Association 1976 Kaibunndo Shuppan

(6) Ship Design : Masahiko Mori 1997 Japan Ship Technology Research Association

(7) Basics of Merchant Ship Design : Shipbuilding Textbook Workshop

(8) Theory of Naval Architecture : Masanobu Okushi 1958 Kaibundo Shuppan

(9) Rudder and Turning : Hisamitsu Shiba 1971 Seizando Shoten

(10) Study of Ship Turning : Shigeru Akasaki 1975 Kaibundo Shuppan

(11) Tug and Utility : Kyoichi Yamagata 1976 Seizando Shoten

(12) Ship Handling by Tugs−Quintessential Technique - : Toshio Nakajima 1979 Kaibundo Shuppan

(13) Reference Paper for Ship Handling No.1 : Japan Captains' Association 1994

(14) Reference Paper for Ship Handling No.2 : Japan Captains' Association 1995

(15) Reference Paper for Sheltering from Typhoon in Osaka Bay : Kobe Marine Casualty Prevention Institute 1992

(16) Study for Safety Management at Sea : Kinzo Inoue 2006 commemorating publication to celebrate author's 60th birthday

(17) Scientific Marine Safety Management : Kinzo Inoue 2008 Seizando Shoten

(18) Maneuverability of Ship : Kensaku Nomoto 1964 Text for Ship Maneuverability Symposium No.1 The Society of Naval Architects of Japan

(19) Estimating the Position of the Pivot Point in Ship Handling and Analyzing its Characteristics : Inoue, Jiang, Ishihata 1993 Japan Institute of Navigation Paper No. 89

(20) Unstable Yawing Motion of a Ship Equipped with CPP - I : Tachi, Inoue, Kondo 1992 Japan Institute of Navigation Paper No. 86

(21) Unstable Yawing Motion of a Ship Equipped with CPP - II : Inoue, Tachi, Yasuda 1992 Japan Institute of Navigation Paper No. 87

(22) A Study on the Supporting Information in Approaching Maneuver under Wind Effect : Hara, Inoue, Park, Sakurai 2006 Japan Institute of Navigation Paper No. 115

(23) A Study on the Supporting Information in Approaching Maneuver under Wind Effect - II : Hara, Inoue 2007 Japan Institute of Navigation Paper No. 116

(24) A Study on the Supporting Information in Altering Maneuver under Wind Effect : Hara, Inoue 2007 Japan Institute of Navigation Paper No. 117

(25) Modeling of Mariners' Senses on Minimum Passing Distance between Ships in Harbor : Inoue, Usami, Shibata 1994 Japan Institute of Navigation Paper No. 90

(26) Assessment of Safety Margin of Approach Maneuver : Inoue, Hanabusa 1991 Japan Institute of Navigation Paper No. 84

(27) Guidelines for Speed Reduction in Berthing Maneuver : Inoue, Seta, Masuda 2002 Japan Institute of Navigation Paper No. 107

(28) Safety Assessment of Berthing Maneuvering Assisted By Tugs : Kinzo Inoue 1987 Japan Institute of Navigation Paper No. 76

(29) Guidelines for Desirable Berthing Operation : Kinzo Inoue 1990 Japan Institute of Navigation Paper No. 82

(30) An Investigation of Reducing Cable Tension Caused by Swing Motion of a Ship Moored at Single Anchor in Wind - I - On the Factor Affecting the Magnitude of Cable Tension : Kinzo Inoue 1981 Japan Institute of Navigation Paper No. 65

(31) An Investigation of Reducing Cable Tension Caused by Swing Motion of a Ship Moored at Single Anchor in Wind - II - On Reducing Cable Tension by Application of Bow Thruster : Kinzo Inoue 1982 Japan Institute of Navigation Paper No. 66

(32) Countermeasures to Assure the Safety of the Outside - Harbor - Refuge : Kinzo Inoue 1988 Japan Institute of Navigation Paper No. 78

(33) Safe Assessment of Mooring at Anchor under Stormy Weather : Nagao, Kuroda,

Inoue, Honda 1986 Japan Institute of Navigation Paper No. 74

(34) Guideline for Safe Anchoring Based on a Risk - Index of Dragging Anchor : Inoue, Usui 1994 Japan Institute of Navigation Paper No. 90

(35) Experimental Study on the Behavior of Dragging Anchor of AC14 Type Anchor : Inoue, Usami, Tachibana 1995 Japan Institute of Navigation Paper No. 92

Index

Author affiliation

Kinzo Inoue

Born in Kyoto, 1946. Currently professor emeritus at Kobe University, Faculty of Maritime Sciences. Earned a Bachelor's degree in 1968 and Master's degree at Kobe University of Mercantile Marine. Also earned Ph.D. in engineering at Kyoto University. Worked for Nippon Yusen Kaisha (NYK Line Shipping Company) and later for Toyoma National College of Maritime Technology as an associate professor before working for Kobe University of Mercantile Marine in 1981 as an associate professor, and promoted to a professor in 1990. Held the captain's position for the school's training ship "Fukae-maru" in 1992 and in 2000 served as a vice president of the university. In 2004, when the university was integrated with Kobe University, became a professor at Kobe University and in 2005 held the dean position of the Faculty of Maritime Sciences of Kobe University. Currently an associate fellow of The Royal Institute of Navigation in England and a honorary member of Japan Institute of Navigation after serving as a chairman in 2002. Has received best paper awards for eight publications from domestic and international academic organizations. A major contributor and establishing member for International Association of Maritime Universities (IAMU) providing the basic concept for the association. Presided on multiple international study projects while serving as an affiliate professor at Dalian Maritime University and Shanghai Maritime University.

Theory and Practice of Ship Handling

2023년 3월 20일 1판 3쇄 발행

저 자 Kinzo Inoue
발행인 남 승 우

발행처 도서출판 상尙 학學 당堂
서울시 동작구 사당로9가길 6(사당동)
TEL : 02) 595-1692~4
FAX : 02) 595-1394
E-mail : shdbooks@naver.com
신고 : 2-155호 (1968. 11. 29)

정가 32,000원

ISBN 978-89-6587-084-5 93500

〈디자인〉 편집 엄해숙 / 표지 토틀컴